The Facts On File
DICTIONARY of FORENSIC SCIENCE

Suzanne Bell, Ph.D.

☑®
Checkmark Books®
An imprint of Facts On File, Inc.

The Facts On File Dictionary of Forensic Science

Checkmark Books
An imprint of Facts On File
132 West 31st Street
New York NY 10001

Library of Congress Cataloging-in-Publication Data
Bell, Suzanne.
The Facts on File dictionary of forensic science / Suzanne Bell.
    p. cm.
    ISBN 0-8160-5131-3(hc)—ISBN 0-8160-5153-4(pbk.)
1. Forensic sciences—Dictionaries. I. Title: Dictionary of forensic science. II. Title.

HV8073.B426 2004
363.25'03—dc222003015735

Checkmark Books are available at special discounts when purchased in bulk quantities for businesses, associations, institutions, or sales promotions. Please call our Special Sales Department in New York at (212) 967-8800 or (800) 322-8755.

You can find Facts On File on the World Wide Web at http://www.factsonfile.com

Text and cover design by Cathy Rincon
Illustrations by Sholto Ainslie

Printed in the United States of America

MP Hermitage 10 9 8 7 6 5 4 3 2 1

This book is printed on acid-free paper.

*This work is dedicated to a fellow scientist
whose devotion and sacrifices
made my life and career possible:
Thanks for everything, Dad.*

CONTENTS

PREFACE

Forensic science is the ultimate inter-disciplinary science; accordingly, the reader will find within terms from biology, chemistry, geology, physics, anthropology, and archaeology, to name a few.

With more than 1,800 entries, *The Facts On File Dictionary of Forensic Science* is designed not only for use in school and public libraries but also can serve as a handy pocket reference for anyone interested in forensic science or working in a related field. Illustrations accompany many of the entries, and cross-referencing will assist the reader in obtaining a quick but complete definition of any term. The Appendices provide useful supplemental information including drawings of the human skeleton and skull as well as an extensive bibliography and list of websites.

The *The Facts On File Dictionary of Forensic Science* is best thought of as a distillation and compilation of the rich forensic literature such as is listed in Appendix I. This volume is not intended as a primary reference in the field, but rather as a pocket guide where readers can find information about forensic terms or concepts they encounter. The entries are, of necessity, brief, and interested readers are encouraged to seek more information, starting with the cited materials in the first appendix. All reasonable efforts have been extended to use current, common, and correct terminology as generally used in the field. Where terms have several potential meanings, definitions most relevant in the forensic arena are emphasized.

ACKNOWLEDGMENTS

Although he did not review this text directly, I would like to thank Mr. Max Houck of West Virginia University for comments and ideas in the general context of forensic science. Also thanks to my husband, Mike, for tireless hours reading and rereading rough pages. His help and suggestions were invaluable.

A

ABI PRISM 310 An automated sequencing system widely used in forensic laboratories for DNA TYPING. The instrument is manufactured by Perkin Elmer, Applied Biosystems Division.

ABO blood group system A blood group system proposed by Karl Landsteiner in 1900. The ABO system consists of ANTIGENS found on the surfaces of red blood cells (also called erythrocytes and commonly abbreviated RBCs) and corresponding ANTIBODIES in the serum. In the U.S. population, the approximate frequencies of the types are as follows:

- Type A 42 percent
- Type O 43 percent
- Type B 12 percent
- Type AB 3 percent

In addition, a large percentage of people (~80 percent) are SECRETORS, meaning that the antigens present in their blood are also found in other body fluids such as saliva. Until the introduction of DNA TYPING in the late 1980s, forensic serology made extensive use of this system for typing blood, bloodstains, and body fluids.

absorbance In spectroscopy, a measure of the amount of electromagnetic energy that is absorbed by a given sample. The amount of energy absorbed depends on the concentration of the sample, the amount of sample through which the energy travels, and a constant called the molar absorptivity coefficient. This coefficient depends on the structure of the sample molecule and the wavelength of the energy. This relationship is summarized as Beer's law: $A = \varepsilon b c$ where ε is the molar absorptivity, b is the path length, and c is concentration. Many spectrophotometric techniques used in forensic science take advantage of this relationship to determine the concentration of a sample. In other cases, such as infrared SPECTROPHOTOMETRY (IR), the pattern of absorbance across many wavelengths is used to help identify the compounds present in a sample.

absorption The taking in of material or energy by some substance, compound, or molecule. For example, cotton swabs can be used to absorb blood, a process that does not result in any chemical change to the blood. Similarly, matter can absorb energy, a phenomenon called absorptivity. This contrasts with ADSORPTION, in which something attaches to a surface but is not taken into the substrate.

absorption-elution and absorption-inhibition tests Two tests that are used to type blood and BODY FLUIDS for ABO and other BLOOD GROUP SYSTEMS. Absorption-inhibition was developed in 1923 in Italy by Vittorio Siracusa, and absorption-elution followed in the 1930s. Many modifications and variants have appeared, and the general procedures have been applied to other blood group systems. Absorption-inhibition works by reducing the strength of an antiserum on the basis of the type and amount of antigens present in the stain. Conversely, absorption-elution is based on the elution of antibodies that bind to antigens present in the stain.

absorption spectrum A graph that plots the absorbance of ELECTROMAGNETIC RADIATION (EMR) by a selected material as a function of the wavelength of radiation. Although commonly associated with the visible portion of the electromagnetic spectrum, in which colors correlate with wavelengths, an absorption spectrum can be generated in any range. In forensic analysis, the most common types of absorption spectra used are those in the visible (VIS) range, ultraviolet (UV), and infrared (IR).

accelerant In ARSON cases, an accelerant is the flammable material that is used to start the fire. Accelerants can be solids, liquids, or gases; gasoline is most common. Solid accelerants include paper, fireworks, highway flares, and black powder. Butane (cigarette lighter fuel), propane, and natural gas are examples of gaseous accelerants, which do not leave any chemical residue at a fire scene.

accidental characteristics Characteristics in a material that are acquired by wear or by some accidental or other nonrepeatable circumstance during manufacture. Tire treads, bullets, shoe soles, plastic bags, glass, and a host of other materials of forensic interest can acquire accidental characteristics when they are manufactured. For example, glass that is made by pouring molten material into a mold may develop bubbles, which would be considered an accidental characteristic. Because they are often unique, accidental characteristics are often valuable in forensic examinations.

accidental pattern In FINGERPRINTS, a ridge pattern that either includes two or more patterns (such as LOOPS and WHORLS) or is not one of the standard ridge patterns.

accident reconstruction A type of forensic engineering involving the study of automobile accidents and related accidents involving pedestrians, motorcycles, trucks, bicycles, boats, buses, trains, and other vehicles. Reconstructions can be used in civil or criminal cases and can become crucial when an accident has no witnesses. For example, if a car crashes into a light pole in the middle of the night, reconstruction could be the only method of assessing what may have happened. Points of investigation in traffic accidents commonly include speed of the car(s), positions, directions of travel, braking, and points of impact.

accounting, forensic The application of accounting techniques to criminal and civil matters. Forensic accountants study financial records and other financial evidence, prepare analyses and reports, assist in investigation, and as other forensic professionals are, can be called on to relate findings to a court of law. Most often, forensic accountants are certified public accountants (CPAs) who specialize in fraud or other investigative accounting. With the near-universal adoption of electronic and computer-based accounting tools, aspects of forensic accounting and forensic COMPUTING often overlap. Financial institutions, insurance companies, and governmental agencies (notably the Internal Revenue Service [IRS], Federal Bureau of Investigation [FBI], General Accounting Office [GAO], and Securities and Exchange Commission [SEC]) employ forensic accountants, as do some law enforcement agencies.

accreditation The process that confers approval of laboratory practices and procedures. For forensic laboratories, the accreditation body is the AMERICAN SOCIETY OF CRIME LABORATORY DIRECTORS/LABORATORY ACCREDITATION BOARD (ASCLD/LAB). A lab that has been accredited has completed a lengthy review of its practices, procedures, and personnel and has been deemed to be in compliance with the standards set by that organization.

accuracy The closeness of the result of a given analysis to the correct or true value. This term is often confused with PRECISION, which refers to the reproducibility of results. Although the goal of any analysis, forensic or other, is the most accurate result possible, accuracy can be difficult to define. In drug analysis, for example, an accurate result might be "cocaine, 52.3 percent," when the true percentage is 52.2 percent. However, unless the sample has a reliable known composition, defining the "true value" is difficult or impossible. Similarly, in forensic analyses in which visual comparison is used, as in the microscopic examination of hair, the concept of accuracy becomes less quantitative and more difficult to apply.

acetone-chlor-hemin test (Wagenar test)
A confirmatory test for blood that is based on the presence or absence of HEMOGLO-BIN. Procedures for the test which are fairly simple, were published in 1935. A few drops of acetone (a common ingredient in nail polish removers) are added to a suspected bloodstain, followed by a drop of diluted hydrochloric acid (HCl). If hemoglobin is present, characteristic crystals form and are then observed under a microscope.

acid   Most commonly defined as a substance that can donate a proton (H⁺) species in water. Example acids include hydrochloric (HCl, also called muriatic acid), nitric (HNO_3), sulfuric (H_2SO_4), and carbonic acid (H_2CO_3). The relative acidity of any aqueous solution is defined as the opposite of the log of the concentration of H⁺ present as measured in MOLAR-ITY. In a forensic context, "acid" is also an older slang term that refers to the drug LSD (LYSERGIC ACID DIETHYLAMIDE).

acid phosphatase (AP, ACP, EAP)
See ERYTHROCYTE ACID PHOSPHATASE.

acute   An effect, such as induced by a POISON, that occurs quickly and is usually the result of a large single dose. The health consequences are usually immediate and serious. A person given a large dose of thallium, for example, would quickly (within hours) become very ill, and if the dosage were large enough, would die; all are acute effects.

ACVE   In the evaluation of fingerprint evidence, an abbreviation for a four-step process that can be used: *a*nalysis, *c*omparison, *e*valuation, and *v*erification. It is pronounced "ace vee."

adaptive elastic string matching   *See* ELASTIC MATCHING.

adenine (A)   One of four NUCLEOTIDE bases that compose DNA and ribonucleic

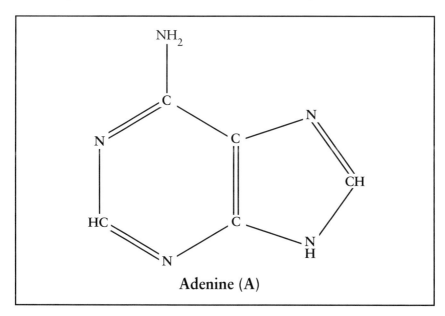

Adenine (A)

The structure of adenine (A), one of the four bases found in deoxyribonucleic acid (DNA).

acid (RNA). Because of its molecular structure, adenine associates with thymine (T), and the two are referred to as complements of each other.

adenosine deaminase (ADA) *See* ISOENZYMES.

adenylate kinase (AK) An ISOENZYME system with three common types, 1, 2-1, and 2.

adiabatic flame temperature The theoretical maximum temperature at which combustion of a fuel occurs. The condition of the fuel and presence of oxygen (or other oxidant) determine actual flame temperature. The concept is applicable in fire investigation and ARSON cases.

adipocere A grayish waxlike substance that forms as a result of a slow chemical reaction between body fat and water (hydrolysis) that occurs after death. The word comes from a combination of the words for fat (adipose tissue) and wax. The consistency of adipocere is very much like that of soap. Adipocere formation can occur in bodies that are left in damp environments such as mud, wet soil, swamps, or water.

adjudicated An adjudicated matter has been settled by or through a court of law, either civil or criminal.

admissibility and admissibility hearing The process of determining which evidence and expert testimony will be heard by a court. The standards that courts use to determine admissibility of evidence vary among the jurisdictions. Those following the *Frye* standard (*Frye v. United States*) require that new methods be generally acceptable to a significant proportion of the scientific discipline to which they belong. Jurisdictions that follow the *Federal Rules of Evidence* and the *Daubert v. Merrell Dow Pharmaceuticals* decision use more flexible guidelines in which the judge plays the role of "gatekeeper." Essentially, under *Daubert,* the trial judge is responsible for determining whether the scientific evidence is useful and relevant and that the expert presenting it is qualified to discuss the results and offer an opinion.

adsorption The process of adhering to a surface but not penetrating it. This contrasts with ABSORPTION, in which energy or material is consumed or taken into the interior of a structure. The adherence of paint to a surface is an adsorption; staining in which penetration occurs would be absorption.

aerobic A process occurring in or requiring the presence of air, specifically oxygen.

affidavit Written testimony taken from an individual who is under oath before an authorized representative of a court. Occasionally, forensic scientists and other expert witnesses offer testimony by way of an affidavit.

AFIS (Automated Fingerprint Identification System) Computerized system for searching FINGERPRINT databases and identifying suspects. The program locates and identifies major characteristics of the print and searches the database for the closest matches. A fingerprint examiner makes the final decision and identification. In 1999, the Integrated Automatic Fingerprint Identification System (IAFIS) became operational at the Federal Bureau of Investigation (FBI) Laboratory, allowing automatic searching of the world's largest collection of fingerprints, which includes prints of more than 35 million people.

AFTE (Association of Firearms and Toolmark Examiners) A professional organization devoted to impression evidence. It was formed in 1969 and publishes the *AFTE* journal as well as overseeing a certification program. The association maintains an extensive website at www.afte.org.

agarose and starch-agarose Gel media that are used for immunodiffusion tests and for ELECTROPHORESIS. Agarose, also called agar, is obtained from seaweed and is classified as a polysaccharide, or long-chained sugar polymer.

age-at-death estimation Determine of the approximate age of a deceased person. The three common methods of determining the age are based on skeletal development and measurements, dental development and condition, and AMINO ACID RACEMIZATION (AAR), usually in teeth. Given that skeletal and dental formation follows a known and consistent pattern of development, estimates based on these techniques are reliable to within a year for younger people. However, once growth and development are complete, estimates become more difficult, and in general, the older the person is, the larger the uncertainty in the age estimates. *See also* ORDER OF ERUPTION.

agglutination A clumping of red blood cells that occurs when cells with one type of antigen on their surface are placed into a solution containing antibodies to that antigen. When red blood cells from a person with type A blood are placed into a solution containing anti-A antibodies, the cells clump together. Thus, agglutination is an antibody–antigen reaction, also called an immunological reaction.

agglutinin In an agglutination reaction, the antibody. In the case of clumping of red blood cells, the antibody in the blood plasma is the agglutinin.

agglutinogen In an agglutination reaction, the antigenic substance. In the case of clumping of red blood cells, the antigen on the surface of the cells is the agglutinogen.

aging of bloodstains Application of analytical chemistry and biochemistry to analysis of bloodstains to determine how much time has passed since they were deposited. Analysis may target stain color, composition, or breakdown products of proteins, fats, and other components in the sample. To date, no reliable model has been developed.

aging of latent prints Application of analytical chemistry and biochemistry to analysis of latent fingerprints. *Aging* can refer either to the age of the print (how long it has been on a surface) or to the age of the person who deposited it (child or adult).

alcohol (ethanol) A term used in forensic toxicology to refer to ethanol or ethyl alcohol. Ethanol is a central nervous system (CNS) depressant that is a factor in approximately 40 percent of fatal traffic accidents. Approximately 20 percent of ingested ethanol is absorbed through the stomach wall and the rest through the walls of the small intestine. Ethanol can be removed from the body by metabolic processes (~90 percent) or by exhalation or in urine, perspiration, or saliva (~10 percent). Alcohol intoxication can be detected by using the BLOOD ALCOHOL CONCENTRATION (BAC) or breath BREATH ALCOHOL.

alcohols A class of organic compounds defined by the presence of the OH functional group. The use of the term *alcohol,* particularly in forensic contexts, usually refers to ethanol. Other common alcohols such as isopropyl (rubbing alcohol) and methyl alcohol (methanol or wood alcohol) are more toxic than ethanol; however, large doses of ethanol can be fatal. Methanol is occasionally encountered as a poison found in homemade or bootleg liquors.

aldehydes A class of organic compounds defined by the presence of the CHO functional group. Formaldehyde, once widely used as a preservative, is the most familiar aldehyde. These compounds often have strong odors and are responsible for many common fragrances such as those in flowers or perfumes.

algae A microscopic plant found in fresh and salt water that can be useful in associating a body or evidence with a place such as pond, lake, or even moist soil. DIATOMS, a form of algae, have been used as part of the investigation of drowning.

algor mortis The rate of cooling of a body after death occurs. Many variables affect the cooling rate including the ambient temperature, the amount of fat of the victim, and the amount of exposed surface area. The temperature of a body can be useful in determination of the POST-MORTEM INTERVAL (PMI) but cannot be used alone as an infallible estimator.

alkaloids A class of chemical compounds that are extracted or obtained primarily from seed plants. The pure compounds, which are usually colorless and bitter tasting, are encountered in forensic work as drugs or POISONS.

allele frequency The percentage of the total collection of a version or alternate form of a gene (an allele) present in a given population. This percentage is often referred to simply as frequency.

alleles Alternative forms of a gene or base pair sequence that occur on a chromosome.

allelic markers *See* SIZE MARKERS.

alligatoring A pattern seen in burned wood that resembles the rough skin of the reptile for which it was named.

alloenzyme The allelic form of an enzyme, which is determined by a given gene.

allometry The growth or development of one part of the body in relation to the whole body. The term is applied in forensic anthropology in estimations of stature from partial skeletal remains. If, for example, a long bone is recovered and measured, that information can be used, along with a database of previous measurements, to estimate the height of the person.

alpha radiation A form of radioactive decay emitted from the nucleus of an atom in the form of a small particle. An alpha (α) particle consists of two protons and two neutrons; it is the equivalent of a helium nucleus. Alternative notations include $^4_2$He and $^4_2\alpha$.

alternate light sources (ALS) Lighting provided by something other than a typical room light (white light) or sunlight. An ALS is used to help make visible objects or impressions that cannot be otherwise seen. An ALS unit typically provides several different selectable wavelengths of light that are chosen on the basis of the application. An ultraviolet (UV) light ("black light") induces FLUORES-CENCE in materials such as semen or treatments applied to bloodstains, allowing them to be seen.

alu **repeat** "Jumping genes" or sequences of DNA or ribonucleic acid (RNA) that are widely distributed and may be found in many areas of the genome. *See also* REPET-ITIVE DNA.

alveoli sacs Structures deep in the lung where exchange of oxygen and carbon dioxide takes place. This is also where blood alcohol diffuses into the gas phase, becoming breath alcohol.

amalgam The material used as fillings for cavities or as part of restorative dentistry. In metal fillings, the amount of mercury present determines the "softness" of the filling and its capacity to flow at elevated temperatures.

amelogenin gene A genetic locus that can be used in conjunction with DNA TYP-ING techniques to identify a person's sex. The gene codes for tooth pulp; often abbreviated as AMEL.

American Academy of Forensic Sciences (AAFS) A professional society for forensic science established in 1949 with a current membership of approximately 5,000. It is headquartered in Colorado Springs, Colorado, and publishes the *Journal of Forensic Sciences* through the AMERICAN SOCIETY FOR TESTING AND MATERIALS. The academy maintains an extensive website at www.aafs.org.

American Board of Criminalistics (ABC) Board formed in 1989 as a means to develop a national certification program for criminalists. The ABC administers a General Knowledge Examination

(GKE) as well as specialty examinations in fields such as fire debris, forensic biology, and drug identification.

American Board of Forensic Anthropology An organization formed in 1977 to certify practitioners in forensic anthropology. Certification is based on academic credentials, casework, and testing. Currently there are fewer than 70 anthropologists certified by the board.

American Society for Testing and Materials (ASTM) An organization founded in 1898 devoted to the development of voluntary standards and specifications for numerous materials, systems, services, and procedures. The Committee on Forensic Sciences, formed in 1970, has subcommittees dealing with criminalistics, questioned documents, pathology and biology, toxicology, engineering, odontology, jurisprudence, physical anthropology, psychiatry and behavioral science, interdisciplinary forensic science standards, long-range planning, terminology, awards, and liaisons. The ASTM is the publisher of the *Journal of Forensic Sciences.*

American Society of Crime Laboratory Directors/Laboratory Accreditation Board (ASCLD/LAB) An organization representing crime laboratory directors formed in 1974 to improve crime laboratory operations and procedures. ASCLD coordinates a voluntary accreditation program for forensic laboratories that addresses facilities, management, personnel, procedures, and security, among other aspects of the field. Membership is open to current and former laboratory managers and forensic science educators.

amido black A reagent used to help visualize latent prints. It is a protein dye that stains proteins that are present in the fingerprint residue.

amino acid racemization (AAR) A technique used in archaeology, geology, anthropology, and forensic science to date materials and to determine age at death. AAR has been applied to tissues including the disks between vertebrae, the lens of the eye, and parts of the brain; forensic applications focus on the analysis of the aspartic acid in teeth. It is based on a known rate of conversion of amino acids from the *l*-form isomer to the *d*-form.

amino acids The molecular building blocks of proteins including DNA. As the name indicates, all of these molecules have at least one acidic site (functional group) as well as an NH_3 (*amino,* as in ammonia) group. Proteins are polymers of amino acids, meaning they are built by linking many (*poly*) amino acids in a long chain.

ammonia (NH_3) A gaseous compound with a distinctive odor that is a common chemical reagent and by-product of chemical reactions. One PRESUMPTIVE TEST for urine is based on heating the sample and sniffing for this odor.

ammonium nitrate (NH_4NO_3) A salt used as a fertilizer that can also be used in drug synthesis and in the manufacture of explosives such as ammonium nitrate-fuel oil (ANFO). Such an explosive was used in the 1995 bombing of the Murrah Federal Building in Oklahoma City.

ammunition For modern firearms, ammunition consists of a projectile (BULLET or pellets) and a CARTRIDGE CASE containing PROPELLANT and the PRIMER that ignites it. The function of ammunition is to exploit the chemical energy stored in the propellant (gunpowder) by igniting it. The burning releases heat and rapidly expanding gases that are trapped behind the projectile in the breach and barrel of the weapon. When sufficient pressure is built up, the pressure accelerates the projectile forward.

amphetamines Illegal drugs (all synthetic) that stimulate the sympathetic nervous system, which controls heart rate, blood pressure, and respiration. Excessive use can lead to severe effects such as hallucinations, convulsions, prickling of the skin, unpredictable emotional swings, extreme aggression, and death. Methamphetamine, which is currently the most widely abused, is produced in CLANDES-TINE LABORATORIES.

amplicon In DNA TYPING, the copies of an original DNA segment that are produced by the amplification step in the procedure.

amplification The process of copying a target segment of DNA for DNA TYPING. This is accomplished by using THERMAL CYCLING and a polymerase chain reaction (PCR).

amylase An enzyme that catalyzes the breakdown of starches. Alpha amylase is found in saliva and is the basis of a common PRESUMPTIVE TEST for its presence.

Amy model A statistical model for fingerprints that described probabilities associated with the types and position of MINUTIAE.

anabolic steroids A class of synthetic steroids related to the male sex hormone testosterone that were declared controlled substances in 1991. Anabolic steroids are synthetic steroids related to testosterone, a male sex hormone that promotes the development of secondary male characteristics called androgen effects such as deepening of the voice. Dangers of anabolic steroid misuse include kidney and liver damage, liver cancer, masculinization and infertility in women; impotence in men; and unpredictable emotional effects, including mood swings and extreme aggression. Some of these effects are irreversible.

anaerobic A process that occurs in or requires a surroundings devoid of oxygen. Anaerobic processes often involve bacteria that thrive in low-oxygen or zero oxygen environments.

anagen phase or stage The active growth phase in the life cycle of a hair, which can last up to six years. The phase of a hair may be important because hair in the TELOGEN phase is shed naturally, whereas hair in the other two growth stages (anagen and CATAGEN) may have been forcefully removed.

analgesics A class of drugs that relieve or reduce pain by depressing the central nervous system (CNS). Aspirin and acetaminophen (Tylenol) are common over-the-counter (OTC) analgesics. Many narcotic drugs, including opium alkaloids such as MORPHINE and CODEINE, are powerful analgesics; their abuse can lead to physiological and psychological dependence.

analyte The substance, compound, or element that is the target of a specific test or analysis.

analytical balance A scale used in the laboratory to obtain accurate weights, typically to one-tenth of a milligram (0.0001 g).

ANFO An explosive consisting of ammonium nitrate (AN, 95 percent) and heavy fuel oil (FO, 5 percent). The ammonium nitrate is the oxidant and the oil is the fuel. Variations of ANFO explosives were used in the first bombing of the World Trade Center in 1993 and the bombing of the Murrah Federal Building in Oklahoma City in 1995.

angle of extinction *See* EXTINCTION ANGLE.

angle of impact A term applied to wound biomechanics and in bloodstain patterns. In blood spatter, the angle of impact is formed by the trajectory of blood when it strikes a surface. As shown in the figure, a drop of blood striking at a 90° angle is essentially circular. As the angle becomes more oblique, the resulting spot become more elongated; the amount of elongation can be used to estimate impact angle.

anion A negatively charged ion such as Cl⁻ (chloride, or less commonly, the chlorine anion). In an electrochemical cell or electrical field such as those employed in electrophoresis, anions migrate toward the anode.

anionic surfactants Substances with negative chemical functional groups (anions) that, when added to water,

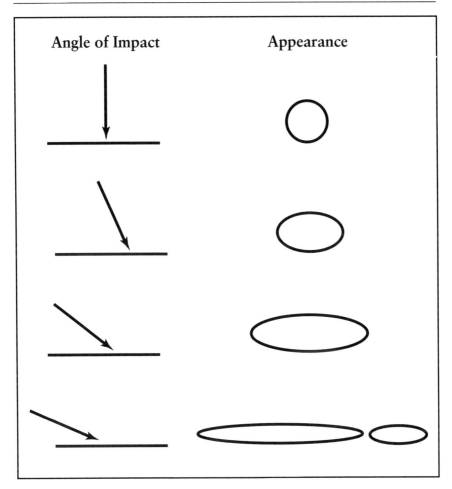

The effect of angle of impact on the appearance of a bloodstain pattern.

reduce the surface tension. Soaps and detergents are surfactants; surfactants are used in analytical techniques such as capillary zone electrophoresis.

anisotropy An optical property of some crystals and fibers useful in the forensic analysis of evidence such as dust, soil, and fibers. A material that is isotropic for a given optical characteristic has the same value of that characteristic regardless of the direction of the source of light. In contrast, anisotropic materials have a nonuniform distribution of such characteristics. Solid materials that are made up of molecules that are randomly placed or molecules that are not symmetric are isotropic. Many types of glass are also isotropic. Other kinds of crystals and many polymers (which consist of ordered subunits bonded together) are anisotropic. The term BIREFRINGENCE is also used to describe anisotropy.

annealing In DNA TYPING procedures (polymerase chain reaction [PCR] techniques), the step in which PRIMERS are added to the DNA sample in which the double-helix structure has been broken

(DENATURATION). The DNA primers, which are specific for certain DNA regions, bind to the unzipped DNA and prepare it for addition of bases that will complete the copying operation. Annealing is sometimes also referred to as HYBRIDIZATION.

anode An electrode with a positive charge or potential that attracts negatively charged species. As an example, in forensic science, anodes are used in electrophoresis and its variants.

ANOVA Analysis of variance, a statistical technique that is used to separate different individual contributions of variance to the total variance. It is used to compare within-sample variances and between-sample variances to determine whether the difference is significant. This procedure can best be illustrated with an example shown in the figure. Assume that a foren-

sic chemist receives three plastic bags of diluted cocaine, all from different cases. A question important to the investigation might be, Are these three bags from the same source? One crucial piece of evidence would be the concentration of cocaine in each, so the chemist would take each sample, mix it thoroughly, and draw three replicates, for a total of nine samples, three per bag. In this case, the analyst would start with the NULL HYPOTHESIS that the concentrations of the three bags do not differ and would use the results of the analyses to determine whether the null hypothesis were valid.

A set of hypothetical results are shown; although the mean values for all bags are similar, variations in the results for the three bags are evident; the three percentages for bag 1 are closer together than those of bag 3, for example. An ANOVA analysis could be used to determine

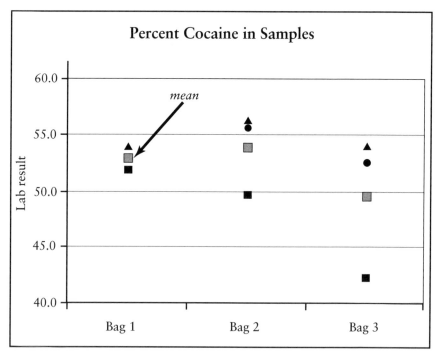

Hypothetical data described in the text. The gray boxes show the mean values of each of three sets of measurements; the circle, diamond, and squares show the three actual measurements for each sample.

whether the percentages could be considered the same at some confidence level, which is usually 95 percent. A single-factor ANOVA on these data shows that the means are indeed the same and that the null hypothesis was valid. In other words, there is a 95 percent certainty that the three bags all have the same percentage of cocaine. The ANOVA test is not without limitations, but it does provide a relatively simple way to explore variation and is implemented in many common software packages.

antemortem "Before death." For example, identifications based on dental work are accomplished by comparing postmortem records and X rays with antemortem records obtained from the victim's dentist.

anthracene An organic compound with the chemical formula $C_{14}H_{10}$ and a formula weight of 178.22 g/mole. It is found in abundance in coal tar and is considered to be carcinogenic. It is used as a starting point for the manufacture of dyes. The compound is also fluorescent and has been investigated for visualizing latent fingerprints.

anthrax A deadly bacterium that was sent through the U.S. mail during October and November 2001. The bacterium *Bacillus anthracis,* which is found mostly in domesticated animals such as sheep and cattle, causes the anthrax disease. The spores of the bacteria can lie dormant in soil for years and infect humans; the spores can also be manipulated to form a potent biological weapon.

Anthropological Research Facility (ARF) Also known informally as the "Body Farm," this facility was founded in 1972 at the University of Tennessee at Knoxville. Major objectives of research are to improve estimates of the POST-MORTEM INTERVAL (time since death) and to provide a working laboratory for forensic anthropologists.

anthropology A diverse field that studies many aspects of human culture and existence from their earliest roots. The discipline can be divided into cultural anthropology and physical anthropology, the branch that examines, among other areas osteology. Osteology is the study of the variability, development, growth, and evolution of the human skeleton; it is from osteology that forensic anthropology has emerged. *See also* APPENDIX V.

anthropology, forensic The analysis and study of skeletal remains that are or become involved in legal procedures. In 1972, the AMERICAN ACADEMY OF FORENSIC SCIENCES (AAFS) added a forensic anthropology section, and in 1977, the American Board of Forensic Anthropology (ABFA), which regulates practices and provides certification for practitioners, was formed. The common techniques used in identification include the use of dental records (ODONTOLOGY), FACIAL RECONSTRUCTION, and analysis of mitochondrial DNA. *See also* APPENDIX V.

anthropometry The use of body measurements to identify individuals. Most often the term is associated with a system of body measurements developed by ALPHONSE BERTILLON and used for identification purposes until it was replaced by fingerprinting in the early 1900s. Anthropometric measurements are still used occasionally in forensic ANTHROPOLOGY, in which measurements of bones or bone fragments can be used to determine height and stature of deceased people.

antigen A substance that provokes production of an antibody; material recognized as "foreign" by an organism. A person's ABO blood type is determined by the type of antigen found on the surface of the red blood cells.

antimony (Sb) A chemical element that can be found forensically as a component of GUNSHOT RESIDUE, as an ingredient in copier toners, and as a poison.

AOAC (Association of Analytical Communities, International) An organization devoted to the practice of analytical chemistry. It was founded in 1884, when it was named the Association of Official Agricultural Chemists. It operated as part of the U.S. Department of Agriculture (USDA). Today the organization oversees a number of publications, assists in accreditation, and is particularly active in the areas of food and drug analysis.

aperture An opening. In forensic science, the term is most often associated with forensic photography and forensic microscopy, in which the term *aperture* applies to a lens opening or size.

apocrine gland A type of sweat gland that secretes sweat with a high fat (oil) content. Apocrine glands are found on the side of the nose or in other "oily" areas of the face and scalp. *See also* FINGERPRINTS.

apothecary units Units of measurement such as the DRAM or GRAIN that are still used occasionally for measurement of drugs and medication. Other apothecary units include the scruple, pound, and ounce. *See also* APPENDIX IV.

appeal The process of removing a case from a lower court and taking it to a higher court for review and potentially for reversal of the decision rendered.

aqueous A solution in which the solvent (the component present in the largest amount) is water. Body fluids such as blood and urine are aqueous solutions, as are beer and soda pop.

archaeology, forensic Although often considered interchangeable with forensic ANTHROPOLOGY, forensic archaeology is emerging as a separate related discipline. In general, forensic anthropologists concentrate on the analysis of skeletal remains, whereas forensic archaeologists focus on the location and excavation of these remains. Archaeological procedures are ideally suited for processing CLANDESTINE GRAVES and for CRIME SCENE analysis and reconstruction, particularly for scenes that are undiscovered for long periods.

arch/arch patterns One of the main FINGERPRINT patterns in which the ridge pattern enters from the side and travels in an arching pattern but lacks a central core or delta pattern. *See also* PLAIN ARCH; TENTED ARCH.

aromatic hydrocarbon An organic molecule containing a benzene ring (C_6H_6) or some derivative of such a stable ringed structure.

arrest warrant A warrant based on probable cause that is issued by a court to direct a law enforcement officer to arrest a person suspected of a criminal offense and to take him or her before the court.

arsenic (As) A heavy metal that was widely used as a poison until advancements in forensic toxicology in the mid-1800s allowed toxicologists to detect it in body tissues. Arsenic is a metal, which is found in the same chemical family as ANTIMONY, another poison. It exists in many forms, all of which are toxic. The first reliable chemical test for arsenic was the MARSH TEST. Because arsenic persists in hair, nails, and to a small extent bone, cases of arsenic poisoning can be detected even in skeletonized remains.

arson The act of purposely setting a fire with criminal intent. According to the National Fire Protection Agency (NFPA, www.nfpa.org), intentionally set fires in 2002 accounted for 8.6 percent of all structure fires and 12.4 percent of all vehicle fires. Fire investigators determine whether a fire can be assigned to natural causes, accidents, arson (incendiary), or indeterminate causes. In the case of incendiary fires, the usual motive is profit through insurance fraud. The role of the forensic chemist in arson investigation focuses on detection of ACCELERANTS such as gasoline, EXPLOSIVES, or INCENDIARY DEVICES that might have been used to start and sustain a suspicious fire.

arterial spurting When an artery is punctured by a knife, bullet, or other

Plain Arch Pattern

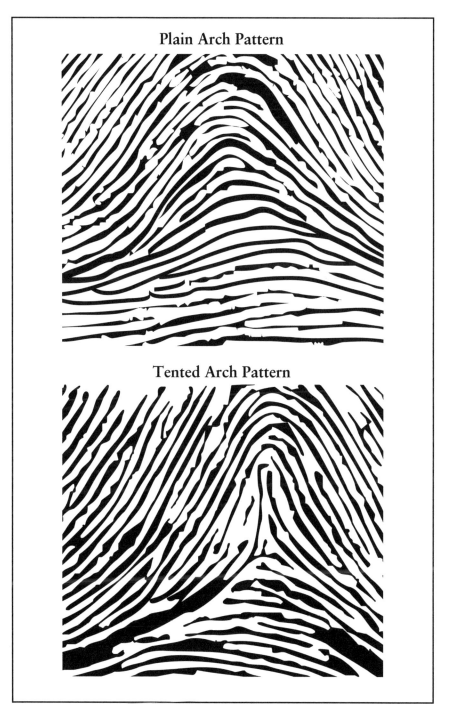

Tented Arch Pattern

The two types of arch fingerprint patterns. The tented arch shows a more distinctive peak shape than the plain arch.

method, the arterial spurt produces a distinctive wavelike pattern, assuming that the heart is still beating. Arterial spurting patterns are useful in crime scene analysis and reconstruction.

art, forensic Application of drawing, sculpture, and other visual techniques to forensic casework. Areas within forensic art include composite imagery, in which interviews and witness statements are used to generate a sketch of a missing person or suspect; image enhancement; aging progressions, in which images are generated to indicate what a child will look like as he or she grows, or how an adult's appearance will change as he or she ages; postmortem drawings; superimposition, in which computers are used to superimpose photographs of a person's face over the computerized representation of a skull, also for identification purposes; FACIAL RECONSTRUCTION; and preparation of graphical or visual information for courtroom presentation.

asbestos A mineral fiber that when inhaled can cause lung cancer. It was at one time widely used as an insulating material, and can be identified by microscopic examination.

aspermia A condition in which a man's seminal fluid does not contain any sperm. Before the use of DNA TYPING and the p30 test for semen, analysis of sexual assault evidence produced from a man with aspermia was complicated by the inability to find sperm and thus to identify a stain conclusively as semen.

asphyxia Death caused by lack of oxygen to the brain. Asphyxia results from suffocation, strangulation, drowning, crushing of the airway, or swelling of the airway in response to injury. Suffocation can occur when the airway is blocked by an object (choking or smothering with a pillow) or in confined spaces where oxygen is depleted or displaced by another gas such as nitrogen or carbon dioxide.

association The process of linking one person, place, or object to another person,

place, or object to establish a relationship between them. Association is a key part of forensic analysis.

Ativan (lorazepam) An antianxiety drug in the same chemical family (the BENZODIAZEPINES) as Valium.

atomic absorption (AA) An instrumental technique used for ELEMENTAL ANALYSIS. In forensic science, target elements include lead (Pb), barium (Ba), antimony (Sb), and copper (Cu) in suspected GUNSHOT RESIDUE (GSR). Heavy metal poisons such as ARSENIC can also be detected and quantitated by using AA. Other terms used to describe this technique include *flame absorption spectrophotometry* (FAS) and *atomic absorption spectrophotometry* (AAS). In place of a flame, a graphite furnace can be used for heating and atomization.

atomic emission Instrumental techniques for ELEMENTAL ANALYSIS that detect ELECTROMAGNETIC RADIATION emitted by metal atoms when they are heated to extreme temperatures. The instrument most commonly used in forensic applications relies on an inductively coupled plasma torch to induce emission. The technique, inductively coupled plasma–atomic emission SPECTROSCOPY, is referred to as ICP-AES.

atomic weight The weight of an atom as expressed as either the atomic mass unit (amu or daltons) or as grams per mole. This information is easily obtained from the Periodic Table of Elements. *See* APPENDIX III.

Atrocine *See* SCOPOLAMINE.

attenuated total reflectance spectroscopy (ATR) A variation of INFRARED SPECTROPHOTOMETRY (including Fourier transform infrared [FTIR]) used in drug analysis and the analysis of trace evidence such as paint and FIBERS. ATR spectroscopy differs from traditional IR spectrophotometry in that the ELECTROMAGNETIC RADIATION penetrates only a small distance into

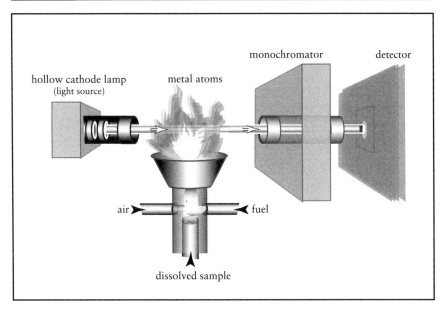

Principles of atomic absorption spectrophotometry (AAS). A hollow cathode lamp containing the element of interest is the source of monochromatic light directed into the flame. The flame breaks chemical bonds, leaving free atoms of the element, which absorb the monochromatic light in proportion to the number of atoms present. The monochromator is a light filter that prevents stray light from entering the detector.

the sample, making it an ideal technique for surface analysis.

auger spectrophotometry A technique related to SCANNING ELECTRON MICROSCOPY (SEM) and X-RAY PHOTOELECTON SPECTROSCOPY in which an X ray interacts with an atom, causing an inner shell electron to be ejected. The ejection causes an outer shell electron to "fall" into the inner shell to take the place of the ejected electron. This falling frees energy, which in turn ejects another electron from an outer level. Auger (pronounced "oh-jay") spectrometry is useful for surface characterization.

autoerotic death Accidental death resulting from ASPHYXIA that occurs as a victim uses a ligature to reduce blood flow to the brain to heighten sexual pleasure.

autolysis The postmortem breakdown of cells caused by enzymes in the cells themselves. This process is sometimes referred to informally as self-digestion and is part of the decomposition process.

automotive finishes Primers, PAINT, and coatings used in layers to cover the bare metal of a vehicle. These coatings may be applied by an electrochemical process or by dipping, brushing, or spraying; the layer pattern can be useful in analysis and PHYSICAL MATCHING.

autopsy A postmortem medical examination, including dissection, that is performed to determine the cause of death and, to the extent possible, the circumstances surrounding the death. The word *autopsy* is derived from Greek and is roughly translated as "to see for one's self" or "to see with one's own eyes," but the term has evolved to the current usage, which refers to a postmortem dissection. Autopsies are normally performed when a death is suspicious, unattended, or otherwise unexplained. The cause of any death can be broadly classified as natural, accidental, SUICIDE, HOMICIDE (NASH), or indeterminate.

autoradiograph (autorad) An X-ray image that reveals labeled (DNA) fragments produced during certain types of DNA TYPING procedures. A membrane that contains an image of the band pattern of a DNA TYPING analysis using RESTRICTION FRAGMENT LENGTH POLYMORPHISM (RFLP) techniques. DNA fragments called probes and labeled with a radioactive material (or sometimes a luminescent material) are allowed to react with DNA that has been separated by using electrophoresis and transferred to a membrane. After the probes bind to DNA on the membrane, the pattern can be visualized by placing the membrane beside X-ray film. The bound probes irradiate the film, producing the autorad. Although effective, RFLP techniques and autorads have become outdated and are rarely used in DNA TYPING.

autoradiogram *See* AUTORADIOGRAPH.

autosome Any chromosome that is not one of the sex chromosomes. Humans have 23 pairs of chromosomes, of which one pair determines sex and the remaining 22 are autosomal.

axial illumination In forensic microscopy, illumination of a specimen by light that is traveling parallel to the optical axis of the microscope.

Azostix A prepacked test strip used to detect blood in urine. It works by detecting a pH change when urea is catalytically converted to CO_2 and ammonia. The stick can also be used to detect the presence of urine.

B

background color The color of the backing used for depositing lifts of latent FINGERPRINTS. When a print is discovered on a piece of evidence, one approach to its recovery is to visualize it and then lift it by using tape. The tape is then placed on a firm backing of contrasting color that will make studying the print easier.

back spatter A type of bloodstain pattern that can occur when a person suffers a gunshot wound inflicted from a distance of a few inches or less. Back spatter can also result from forceful blows. The back spatter is the blood that moves backward toward the weapon and not in the line of the blow. The amount of back spatter is less than the amount of forward spatter.

bacteria A broad class of microorganisms that are active in decomposition processes. Bacteria can cause degradation of blood and body fluid evidence and can be broadly and informally categorized as aerobic (living in or requiring an oxygen environment) and anaerobic (needing little or no oxygen).

ballistics Strictly defined, ballistics is the study of projectiles in motion. However, the term is often used to describe the forensic study of FIREARMS and the use of striation and other toolmarks to analyze firearm evidence.

ball powder A type of SMOKELESS POWDER. To make ball powder, the ingredients are mixed in a solvent base and allowed to form small spherical particles that are then allowed to dry. The spheres can be further processed by pressing into disks.

Balthazard model A model for fingerprint description and classification proposed in 1911 by VICTOR BALTHAZARD. This model assigned probabilities to minutia patterns and was historically important as the foundation for modern techniques of fingerprint comparison.

Balthazard, Victor (1852–1950) French forensic scientist who served as the medical examiner for the city of Paris and helped advance fingerprint, firearm, and hair analysis at a time when forensic science was emerging as a distinct scientific discipline. In 1910 he, along with Marcelle Lambert, wrote the first comprehensive book on hair analysis, *The Hair of Man and Animals*. In it, they advocated MICROSCOPY and the careful evaluation of microscopic structures, which in a broad sense are still the standard today. Balthazard also developed an advanced photographic method of comparing markings on bullets and in 1912 testified in a case using photos and point comparison techniques to identify bullets involved in a fatal shooting. He was also among the first to note other distinctive markings in firearms including firing pin impressions and fabric impressions that result when a soft lead bullet passes through woven fabrics. Further demonstrating the breadth of his knowledge, in 1939 he made a presentation in Paris discussing the value of BLOODSTAIN PATTERNS as physical evidence.

band shifting In DNA TYPING performed by using gel electrophoresis, movement of one of the separated components from the location where it should be. Control samples help identify band shifts.

Barberio test A presumptive micro-crystal test that was once used to detect spermine in seminal fluid. The reagent used to create the crystals was picric acid.

barbiturates A class of drugs based on barbituric acid that act to depress the central nervous system (CNS) and are therefore classified as CNS DEPRESSANTS. Administered primarily by ingestion of pills, barbiturates produce a general feeling of well-being and promote sleep. Adolph Von Bayer, the German chemist who first synthesized barbituric acid in 1863, reportedly named the compound after a woman. It was not until 1903 that the first derivative (veronal) was marketed as a sedative, and several others followed. Barbiturates are classified in terms of the duration of their effect; pentobarbital and secobarbital are short-acting, amobarbital is intermediate-acting, and barbiturates such as phenobarbital are long-acting. Abuse of barbiturates can lead to dependence, and an overdose can kill by altering the pH of the blood and disturbing the system that regulates breathing. Since barbiturates are acidic, overdoses can cause inflammation of the stomach lining and small intestine, where absorption takes place. Barbiturates are listed on Schedules II, III, and IV of the CONTROLLED SUBSTANCES ACT.

barium (Ba) A metallic element found in GUNSHOT RESIDUE. It originates from the PRIMER.

Barr bodies Small structures found in the nucleus of female cells that have been used in SEX DETERMINATION. The female sex chromosomes differ from those of males (XX vs. XY respectively), and in many female cells, the inactive X chromosome shrivels. This structure is called a Barr body, a sex chromatin that male cells lack. Barr bodies absorb fluorescent dye strongly and under a microscope resemble baseball bats or drumsticks. If an abundance of these structures are seen in a sample, determining the sex of the donor is possible. However, given the difficulty of the test and unreliability of results, forensic use of Barr bodies has been limited and replaced with DNA TYPING techniques.

barrel A cylindrical piece of metal in which a bullet or other projectile travels after a gun is fired. Modern firearm barrels are rifled, meaning they have lands and grooves that twist down the length of the barrel. As a result, spin is imparted to the bullet, increasing accuracy.

basal cells (basal layer) In the skin, the lowest cell layer of the epidermis. It is also referred to as the stratum germinativim. This layer along with the next layer closer to the skin surface (the prickle cell layer or stratum spinosum) are collectively referred to as the Malpighian layer. *See also* FINGERPRINTS.

base A chemical base is a substance that can donate an OH⁻group (such as lye, NaOH); a compound that can accept an acidic proton (H⁺; $NH_3 \rightarrow NH_4^+$); or, in the case of a Lewis base, a compound that can donate a pair of unshared electrons. In forensic biology, the term *base* also is used to describe one of the four NUCLEOTIDES (adenine, thymine, cytosine, and guanine) found in DNA.

baseline (base line) In forensic chemistry, many instruments such as gas chromatographs produce readouts that have a baseline. Although the specific meaning and interpretation of the baseline vary with the instrument, in general, the baseline indicates no response from the instrument. The term *baseline* is also used to refer to a set reference line used at a crime scene. The position of objects is reported in relation to this baseline.

base pair (BP) A pair of complementary nucleotides that bond together through hydrogen bonds and impart the characteristic double-helix shape to DNA. Because of their respective structures, adenine (A) pairs with thymine (T), cytosine (C) pairs with guanine (G), and the corresponding base pairs can be referred to as AT and CG. *See also* DNA; DNA TYPING; NUCLEOTIDES.

base sequence (base pair sequence)
The sequence of nucleotide bases in a segment of DNA that are analyzed in DNA TYPING. POLYMERASE CHAIN REACTION (PCR) techniques target SHORT TANDEM REPEATS (STRS), or short base sequences that are repeated. The term *base pair sequences* is often used to describe a base sequence since DNA strands are found as pairs. *See also* DNA; DNA TYPING; NUCLEOTIDES.

Bayesian statistics A method of comparing hypotheses (theories) that takes into account prior knowledge and modifies it by using information gathered from evidence. Bayes's theorem can be stated informally as follows:

Posterior odds = prior odds *
LIKELIHOOD RATIO

One of the advantages of a Bayesian approach is that it requires the comparison of two scenarios. For example, consider a hypothetical case in which blood is found at a scene and is typed as AB blood in the ABO BLOOD GROUP SYSTEM. A suspect who also has blood type AB is identified. The traditional statistical approach to interpreting these results would involve citing POPULATION FREQUENCIES, which show that about 3 percent of the population is type AB. Alone, this information, which is based on prior knowledge of population frequencies, supports the hypothesis that the suspect deposited the blood. However, in a Bayesian approach, this information could be modified to take into account new information gathered through investigation or analysis. Perhaps the suspect has no wounds or other scars that would support the idea that he or she lost blood at the scene. This information would decrease the importance of this suspect's having the relatively rare AB blood type.

bear claw An informal term for a morphological feature of marijuana called CYSTOLITHIC HAIRS.

bearing surfaces Surfaces of a structure or a mechanism that bear weight, such as load-bearing beams in a building. This term is used in forensic engineering.

Becke line (Becke line method) A method used in microscopic analysis to determine the relative differences in refractive index of two adjacent media, such as a particle and the surrounding mounting media. The Becke line appears as a bright halo of light surrounding a specimen that is immersed in a liquid. When the REFRACTIVE INDEX (RI) of a specimen is the same as the refractive index of the liquid, the Becke line vanishes. This phenomenon can be further exploited by moving the specimen relative to the objective of the microscope. The Becke line method is used in the analysis of particulates such as GLASS, minerals, and FIBERS.

Beer's law (Beer Lambert law) An equation that describes the relation of the concentration of an analyte (dissolved in a solvent) to the amount of electromagnetic energy that sample will absorb. For example, if a tiny drop of red food coloring is placed into a test tube of water and that tube is held up to the light, most of the light will pass through. If several drops of food coloring are added, the solution becomes darker (more concentrated), and, as a result, less light is able to pass through it, meaning that more light is absorbed. Beer's law is stated as $A = \varepsilon bc$ where ε is the MOLAR ABSORPTIVITY, b is the path length, and c is concentration. Many spectrophotometric techniques used in forensic science take advantage of this relationship to determine the concentration of samples.

beginning stroke The initial stroke of a pen or other writing instrument, a characteristic that can be studied as part of QUESTIONED DOCUMENT examination.

behavioral evidence A broad category of forensic disciplines and investigative tools that includes forensic psychiatry, psychology, neurology, neuropsychiatry, neuropsychology, deception analysis, and polygraph testing. Some of the disciplines arc routinely accepted and used by law enforcement and the courts; others are considered less reliable or even questionable in validity and reliability.

Benedict's reagent A chemical reagent that is used to test for the presence of reducing sugars. A reducing sugar is capable of causing a REDUCTION of other species in an OXIDATION-REDUCTION reaction. Most common sugars such as fructose are reducing sugars; table sugar (sucrose) is not. This reagent which is used as part of the analysis of cutting agents that may be found in illegal drugs, contains Cu^{2+} ions in a basic buffer solution. When a reducing sugar is present, the distinctive rusty red solid Cu_2O is produced. The sugar has caused the reduction of Cu^{2+} to Cu^{+1}.

benzidine (benzidine test) A PRESUMPTIVE TEST for blood that works by detecting the presence of HEMOGLOBIN. The heme group in hemoglobin has the ability to catalyze certain oxidation reactions; this is called peroxidase activity. The peroxidaselike activity of heme is the basis of the benzidine test, as well as several other presumptive blood tests. In this case, benzidine, which is colorless, is oxidized in the presence of hemoglobin and changes to a bluish color. However, the test is not specific, and many other substances can give a positive result (or FALSE POSITIVE result). Since benzidine has been shown to be a potent carcinogen, this test is rarely used.

benzodiazepines Among the most widely prescribed drugs in the world, benzodiazepines are used as mild tranquilizers and as anticonvulsants. Benzodiazepines produce a sense of well-being and reduce anxiety and generally cause less sleepiness than BARBITURATES do. The most famous member of this family is probably Valium (diazepam); other examples include lorazepam (Ativan), Xanax, Halcion, and Klonopin, which are used to control seizures. Benzodiazepines can induce physical and psychological addiction and are listed on Schedule IV of the CONTROLLED SUBSTANCES ACT.

benzoylecgonine (BZ) A primary METABOLITE of COCAINE that was first isolated from cocoa leaves (the natural source of cocaine) in 1923. Once cocaine is metabolized to BZ, the metabolite has a half-life (time required to excrete half the remaining compound) of 7.5 hours, meaning that it can be detected in the urine up to 48 hours after cocaine in administered.

Bertillon, Alphonse (1853–1914) French forensic scientist who developed the first systematic method for the identification of suspects and criminals, setting the stage for fingerprinting, which ultimately replaced it. The system, called ANTHROPOMETRY or Bertillonage, used 11 body measurements along with descriptive information and photographs stored on a card, similar to modern fingerprint cards.

Bertillonage See ANTHROPOMETRY.

beta radiation A form of radioactive decay that consists of an electron ejected from the nucleus of a radioactive element or isotope. The electron originates from a neutron, which is converted to a proton as a result of the loss of the electron.

beyond a reasonable doubt The legal standard for reaching a decision in a criminal case as to the guilt or innocence of the accused. This standard contrasts with the standard in a civil case, which is the PREPONDERANCE OF EVIDENCE.

bias In scientific analysis, a tendency to obtain results that are offset from the correct or true result. Bias can be psychological in the sense that an analyst may, consciously or unconsciously, expect or desire a certain result for a test. As a result, the probability of obtaining the desired outcome may increase. Bias can also be introduced, purposely or not, when test results are interpreted in a report or courtroom testimony. To combat bias, laboratory methods and protocols include the use of CONTROLS and other QUALITY ASSURANCE/QUALITY CONTROL (QA/QC) procedures designed to detect, minimize, and correct biases.

bifurcation A division or fork in a ridge pattern of a fingerprint. A bifurcation occurs whenever a single ridgeline reaches a point from which it divides into two separate paths.

binary Something that consists of a two components, such as a binary mixture of cocaine and sugar. In computers and forensic computing, *binary* refers to a numbering system based on 1s and 0s or on-off notation.

binning The process of dividing a collection of measurements or values into regions or ranges called bins. For example, a collection of M&M candies can be categorized by color, in a form of binning. In forensic science, binning is usually associated with DNA TYPING using the restriction fragment length polymorphism (RFLP) technique. Here, interpretation of results is based on a band pattern in gel (AUTORADIOGRAPH) that is created by ELECTROPHORESIS. The bins correspond to locations in the gel that in turn reflect the size of the DNA fragments. Thus, the DNA is binned on the basis of size ranges.

biological microscope *See* COMPOUND MICROSCOPE.

biological substances Any type of evidence of biological origin. Examples include blood and body fluids; insects; plant material such as seeds, leaves, or wood; algae; feathers; starches; diatoms; and vegetable fibers such as cotton.

biology, forensic The analysis of evidence for biological composition or characteristics using biological and biochemical techniques. Forensic SEROLOGY (analysis of blood and body fluids) and forensic ENTOMOLOGY are examples of subdisciplines in forensic biology. The term has become more commonplace since about 1990, with the use of DNA TYPING. In common usage, the term *forensic biology* usually refers to DNA typing, although technically the field includes many more disciplines.

biomechanics, forensic The application of biomechanical principles and prac-

tices to legal matters and proceedings. Biomechanics is the study of the mechanics of motion in biological organisms, primarily muscle-driven motion. These considerations can become important in many cases in both civil and criminal law. In assaults, suicides, mass disasters, and homicides, biomechanical investigations can provided detailed information on how injuries might have been inflicted, how an injured person moved or was able to move, or whether a proposed motion was feasible.

biometrics The use of unique physical characteristics for individual identification by an automated system. An example of a biometric device is a door lock that requires a person to place the thumb on a reader pad for verification of identity. Other physical features that have been studied for use in biometric devices include the pattern of blood vessels in the eye (a retinal scan), the pattern of the iris in the eye, facial features, speech recognition, veins in the palm, thermal face image, body odor, and the geometric characteristics of the hand. FINGERPRINTS and the AFIS system represent biometric identification systems.

birefringence In a substance with two refractive indices, the difference between the two. Birefringence is determined by using a POLARIZING LIGHT MICROSCOPE. For example, the refractive index of the fiber can differ when observed parallel to the long axis of the fiber and when observed perpendicular to it. The calculated difference between these two values is the birefringence of the fiber. Forensically, birefringence is useful in the analysis of minerals such as are encountered in soil or dust, glass, and fiber evidence.

bite marks (bitemarks, bite-marks) A type of IMPRESSION EVIDENCE that can be left in the skin of a victim, but also in food, chewing gum, and even pencils and pens. Given the variability in the dental structure and such factors as distance and angles between teeth, missing teeth, fillings and other dental work, and unique wear patterns, bite marks are often considered to be individually unique. Bite marks in victims are common in sexual assaults, homicides, domestic assaults, and child abuse cases, and courts have accepted bite mark evidence since the 1950s.

black powder (blasting powder) Also called gunpowder, black powder was used as a PROPELLANT for early FIREARMS. It consists of a mixture of 75 percent potassium nitrate (KNO_3 or saltpeter), 15 percent carbon (charcoal), and 10 percent sulfur. Black powder is a low explosive that is generally prepared by a wet mixing stage, followed by pressing into a cake, drying, and breaking up of the residue into granules. When ignited, black powder produces copious smoke and accordingly was replaced in the late 1800s with SMOKELESS POWDER, which is the propellant used in modern AMMUNITION.

Black's Law Dictionary A widely used reference guide containing concise definitions of legal terms and concepts. It is published in hardcover and paperback by West Publishing Group (Saint Paul, Minnesota).

blasting agent An explosive that requires some other charge or EXPLOSIVE to detonate it.

blasting cap Device used to initiate detonation of high EXPLOSIVES. The detonator contains a low explosive that is ignited either by a burning fuse or by an electrical charge.

blast pressure effect The SHOCK WAVE produced by the detonation of an explosive. When an explosive is detonated, a rapid decomposition reaction that produces copious quantities of hot, expanding gases occurs. These gases move out from the center of the explosion, compressing and heating air molecules that it forces out of its way. The results are the blast effect and the extremely loud sound associated with explosions. This shock wave is responsible for most of the damage caused by an explosion.

blind sample A QUALITY ASSURANCE/QUALITY CONTROL sample submitted to an analyst as if it were a real case sample. Blind samples are used to verify analyst performance as part of ACCREDITATION and/or CERTIFICATION.

blood An extracellular fluid (a fluid found outside the cells) that is a complex mixture of organic and inorganic materials including electrolytes such as sodium, proteins, and several different kinds of cells. The characteristic color of blood is produced by the complex formed between HEMOGLOBIN in red blood cells (RBCs) and oxygen. Spinning a blood sample in a centrifuge separates it into a cellular component (approximately 45 percent of the total volume) and a noncellular component called plasma, which makes up the remaining 55 percent. The composition of blood is illustrated in the figure. Plasma can be further subdivided into serum and fibrinogen, the material that forms clots. Serum, a clear straw yellow in color, carries electrolytes; the sodium ion (Na^+) and the chloride ion (Cl^-) are the most concentrated (sodium chloride, NaCl, is table salt). Proteins (albumins and globulins) are also carried in the serum. The word SEROLOGY is derived from the word *serum*.

The cellular portion of blood can be divided into three types of cells: red blood cells (RBCs, also called erythrocytes); white blood cells (WBCs, leukocytes); and platelets (thrombocytes). RBCs, which transport oxygen and bicarbonate, are the most numerous and are unique in that they lose their nucleus before entering the circulatory system. WBCs (several types exist) are the next most numerous and are active in fighting diseases. Platelets are needed for clot formation.

All portions of blood contain GENETIC MARKER SYSTEMS that have been used in forensic serology and biology. Serum, a yellowish liquid, contains serum blood group systems such as HAPTOGLOBIN (Hp) and GROUP-SPECIFIC COMPONENT (Gc) that are POLYMORPHIC (occur in many forms). Within the cellular component, white blood cells (leukocytes) contain the

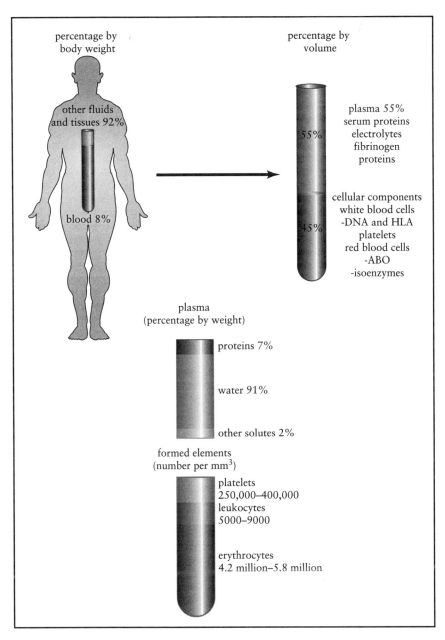

percentage by
body weight

percentage by
volume

other fluids
and tissues 92%

blood 8%

plasma 55%
serum proteins
electrolytes
fibrinogen
proteins

55%

cellular components
white blood cells
-DNA and HLA
platelets
red blood cells
-ABO
-isoenzymes

45%

plasma
(percentage by weight)

proteins 7%

water 91%

other solutes 2%

formed elements
(number per mm^3)

platelets
250,000–400,000
leukocytes
5000–9000

erythrocytes
4.2 million–5.8 million

The composition of blood separated by centrifuging.

HUMAN LEUKOCYTE ANTIGEN (HLA) system, which includes many different factors and types. Both the serum blood group systems and the HLA system were difficult to type in stains and were not routinely used in forensic casework.

Unlike red blood cells, the white blood cells have a nucleus, which is the source of

DNA used in most DNA TYPING. The 13 loci that are usually typed in current practice can also be classified as genetic markers since they are inherited and polymorphic. Red blood cells are the richest source of non-DNA genetic marker systems that were once widely used in FORENSIC SEROLOGY. These cells (erythrocytes) have on their surface the antigens that make up blood group systems such as the ABO and Rh-groups. Within the cell are found the ISOENZYME SYSTEMS such as phosphoglucomutase (PGM) and esterase D (ESD), as well as variations of the hemoglobin molecule. The ABO blood group and isoenzymes were used in casework before DNA TYPING.

blood alcohol concentration (BAC)

The concentration of ethanol detected in a blood sample and reported as grams per deciliter (g/DL), or as a percentage. Most states have adopted a legal limit of 0.08 percent, meaning that 0.08 percent of a person's blood by volume is ethanol. Anyone with a higher BAC is considered legally intoxicated. The blood alcohol test is considered an evidentiary test, meaning the results can be used as evidence in a prosecution; in contrast, field tests produce approximate results only and are used only to determine whether a BAC test should be performed.

blood-borne pathogens

Harmful microorganisms such as bacteria or viruses that are transmitted by blood. Forensic scientists in many disciplines must be conscious of the risks these pathogens present and take precautions to prevent contact with potentially infected blood. The human immunodeficiency virus (HIV), which causes acquired immunodeficiency syndrome (AIDS), is an example of a blood-borne pathogen.

blood group systems

Blood group systems are based on ANTIGENS that are POLYMORPHIC, meaning that more than one variant exists, and have known frequencies in the population. The ABO system is the best known, but many more exist, including the MNSs, Rh, Kidd, Duffy, P, Kell, and Lewis systems, all of which have been used in forensic serology. However, none is easy to type in stains and none is as persistent as the A and B antigens of the ABO system. More than 40 secondary blood group systems have also been discovered but none has been used forensically. Research into typing techniques for forensic work faltered once the ISOENZYMES were discovered and simple typing techniques using ELECTROPHORESIS were developed for the isoenzyme GENETIC MARKERS. In turn, isoenzyme systems have given way to DNA TYPING, which is much more successful in individualizing blood than isoenzymes or blood group systems were.

bloodstain patterns/bloodspatter patterns

Patterns of blood deposits found at crime scenes that can be useful in analysis and reconstruction of events that produced them. Although laws and principles of physics, mathematics, and biology underlie the interpretation of spatter patterns, it still involves an element of subjectivity and requires training and experience. Spatter patterns can be classified by the force, measured in feet per second, required to produce the drop or droplets that strike the wall, ceiling, floor, or other stained objects. Types of spatter include CAST-OFF, BACK SPATTER, FORWARD SPATTER, and ARTERIAL SPURTING.

bloodstains

Stains found on objects, on bodies, or at crime scenes. Often the stains are clearly visible, but occasionally stains must be visualized by using reagents such as LUMINOL. This may be necessary when a perpetrator has attempted to clean up a scene. Collection of BLOOD, as well as other BODY FLUID evidence, must be done carefully to prevent putrefaction, which is degradation caused by microorganisms. Drying is critical to preclude this degradation. Analysis of stains progresses from PRESUMPTIVE TESTS through to DNA TYPING.

blood sugar   *See* DEXTROSE.

blood types   *See* BLOOD GROUP SYSTEMS.

blotting The process of transferring a liquid or other sample from one medium to another. Blotting techniques can be used to absorb liquid blood onto a cotton swatch or to transfer small amounts of dried stain materials such as semen or urine to a moistened cloth or swab. The material transferred by blotting can then be tested without destroying the original stain. *Blotting* can also refer to the practice of Southern blotting, which was used in some older DNA TYPING procedures.

blowback A term that can refer to back spatter created by close-range gunshot wounds or to a principle exploited in modern semiautomatic or automatic firearms. When such a gun is fired, the forward expansion of gases also forces the CARTRIDGE CASE back into the BREECHBLOCK. Some of the expanding gas from the burning propellant is directed into a piston chamber, which operates an EXTRACTOR mechanism that grabs the empty cartridge and ejects it from the breech area.

blunt trauma Injury inflicted by an object or surface that is not sharp. Blows to the head can produce brain hemorrhage, an example of a blunt trauma injury. Blunt trauma can also create rips or tears in the skin, which can be distinguished from cuts or stab wounds (INCISED WOUNDS) on the basis of the appearance of the LACERATION.

body fluids A type of evidence that can be analyzed by using serological techniques and, in some cases, DNA TYPING. Body fluids and body fluid stains encountered as evidence include saliva, semen, sweat, urine, feces, vomit, vaginal fluid, and human milk. In 1932, an inherited characteristic that determines whether a person secretes substances such as the A and B ANTIGENS of the ABO BLOOD GROUP SYSTEM into body fluids was discovered. Approximately 80 percent of the Caucasian population are secretors; that means that their body fluids (saliva, semen, and vaginal fluids) can be typed by the same techniques used to type blood.

With the use of DNA TYPING, secretor status has become less critical.

body temperature (algor mortis) The rate of cooling of a body, which can be used to estimate the time of death or POSTMORTEM INTERVAL and provides the best estimate if a body is discovered soon after death. In general, a body reaches ambient temperature in 18 to 20 hours, but the rate of cooling is not necessarily fixed or constant. It depends on many factors, including temperature of the environment, humidity, submergence (and in those cases, water temperature), temperature at time of death (which may not have been 98.6°F), clothing or cover on the body, body fat, and the ratio of surface area to weight. Body temperature can be measured by rectal thermometer or by insertion of a thermometer into the liver.

bolt *See* BREECHBLOCK.

bone Along with teeth, the components of the body that endure longest after death. Forensic ANTHROPOLOGISTS study bone (OSTEOLOGY) and from their observations attempt to determine the race, sex, stature, and age of the deceased. The analysis starts by determining whether the bone is human or not and making a rough estimate of how long a person has been dead. For very old bones, carbon dating techniques can be used, but for more recent deaths, age estimation can be attempted by a microscopic examination of structures in the bone called OSTEONS. The figure on page 26 shows some of the important morphological features of a typical long bone such as one in the leg or arm.

booster *See* PRIMER.

bore In firearms, the open cylindrical portion of the BARREL. The nominal bore diameter is equivalent to the CALIBER of handguns and rifles.

borosilicate glass A type of glass used in industrial and some chemical applications as well as in some automobile headlamps.

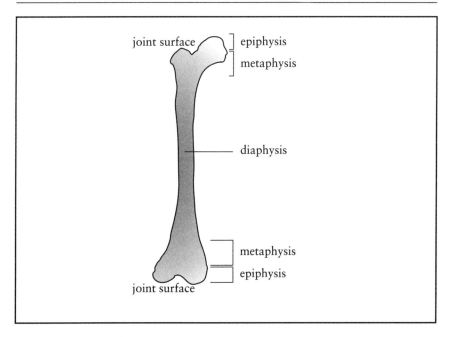

joint surface — epiphysis

metaphysis

diaphysis

metaphysis

epiphysis

joint surface

The structure of a long bone illustrating features of bone structure.

Borosilicate glass is a type of SODA LIME GLASS with boron substituted for some of the sodium.

botanical evidence Any evidence that is of plant origin such as pollen, fungus, molds, plants, stems, and leaves.

botany, forensic The analysis of evidence obtained from or related to plants as applied to legal matters. Plant matter has been used in civil and criminal cases and has been applied to tasks such as estimation of a time of death and the POST-MORTEM INTERVAL, identification of plant matter in stomach contents to characterize a last meal, identification of plant poisons, linking of a suspect to an outdoor scene, determination of whether a body has been moved, determination of whether a person was alive when placed into water, and detection of CLANDESTINE GRAVES.

bovine serum albumin (BSA) A serum derived from the blood of cattle

that is sometimes used in serological techniques and DNA TYPING.

brass An informal term referring to empty CARTRIDGE CASES used in modern ammunition.

breath alcohol and breath analysis Content of ethanol detected in exhaled breath. ALCOHOL (ethanol) in blood can evaporate from the blood into exhaled air deep in the lungs. As such, exhaled air contains a concentration of alcohol that is proportional to the concentration of alcohol in the blood. The concentration is governed by HENRY'S LAW, which states that when a fluid such as blood is in equilibrium with a gas such as air, the concentration of a volatile substance (ethanol) in the gas is proportional to the concentration in the fluid, as long as the temperature remains constant (as it does in the body). For ethanol in blood in contact with air at body temperature (98.6°F or 37°C), that ratio is 2,100:1, meaning that the blood contains 2,100 times the concentration of alcohol present in the air. Since this ratio is known, it is possible mathematically to relate breath alcohol concentration to blood alcohol concentration. Thus, a field breath test can be used to determine whether a person is probably intoxicated and typically this result is followed up by a BLOOD ALCOHOL test.

Breathalyzer A commercial device, produced by Draeger Industries, that is used to measure BREATH ALCOHOL content.

breechblock (breechblock, breech face) The breechblock is the part of a gun (pistol, rifle, shotgun) that cradles and supports a CARTRIDGE when it is inserted into the chamber before firing.

breech face markings (breechblock markings) Markings produced on the base of a CARTRIDGE CASE as a result of firing of a gun. When the trigger is pulled, the firing pin strikes the PRIMER, igniting it and the PROPELLANT. The rapid expansion of gas accelerates the bullet down the barrel, but it also drives the cartridge case backward into the breechblock. Since the breechblock is a machined or filed surface, it possesses a pattern of markings that can be transferred to the cartridge case (IMPRESSION EVIDENCE) if it collides with sufficient velocity. These markings can be examined by using a COMPARISON MICROSCOPE in much the same procedure used for bullets to determine whether a cartridge was fired from a specific gun. Complications can arise if the cartridge has been reloaded and fired more than once, since each firing produces a separate set of breechblock impressions.

broach A type of machined tool that can be used to cut the rifling into the barrel of a firearm.

brucine A highly toxic ALKALOID that has been used as a POISON. Also known as dimethoxystrychnine, it is similar to strychnine in action and has a bitter taste.

bubble ring (vacuole) A pattern that can develop in a bloodstain or bloodstain pattern when the blood had bubbles in it when deposited. After drying, the stain retains the ring-shaped pattern associated with the bubbles.

buccal swab (mouth swab) A swabbing collected from the inside surface of the cheek. Because buccal cells (cheek cells) are recovered, this is a noninvasive method of collecting DNA samples.

buffers Chemical solutions that contain acids, bases, and other ions and that are designed to resist changes in pH. Buffers are used in ELECTROPHORESIS and are generally prepared by combining a weak acid with its salt. For example, a phosphate buffer can be prepared by combining phosphoric acid (H_3PO_4) and sodium phosphate (Na_3PO_4).

building materials Materials used in construction can serve as transfer evidence (TRACE EVIDENCE) and are often found in burglary cases. A partial list of construction materials that may be encountered includes glass, minerals such as gypsum found in plasterboard (Sheetrock) and plaster, mineral fibers such as asbestos, wood, cement and mortars, stucco, brick, insulation materials, and metals. Much of the forensic work on such evidence is done microscopically and may involve PHYSICAL MATCHING.

bullets The projectiles fired from rifles and pistols. The primary component of bullets is lead, but there are many types and configurations of bullets available, varying by shape and degree of jacketing, among other features. The lead that is used also varies with the metals alloyed with it. Bullets made of softer leads tend to break up on impact, whereas harder lead alloys resist fragmentation. Fully jacketed bullets (those that have a "full metal jacket") consist of a harder metal shell (copper alloy or steel) that encases the lead core. Semiautomatic pistols and rifles use jacketed ammunition to prevent lead fouling of the chambering mechanisms and to increase the ease of bullet feed. Semijacketed bullets have the front portion of lead exposed and are much more prone to fragmentation. Hollow point bullets have the center portion of the nose removed, promoting a mushrooming effect on impact. Other bullet variations include soft point, bronze point, and synthetic resin (Teflon) bullets.

bullet trap A device designed to capture a bullet after it is fired from a gun without marking or damaging it. The most common type of trap consists of a large steel tank filled with water.

bullet wipe *See* BULLET WOUNDS.

bullet wound Injury produced by gunshot. When a bullet strikes flesh, the skin is stretched and then broken as the projectile penetrates. As the bullet enters, material on its surface such as dirt and dust, lubricants, powder and primer residue, and lead is wiped onto the skin in a pattern called bullet wipe or smudge ring. The bullet also scrapes off skin cells, creating an injury called a contusion ring. These features may be obscured or altered by the presence of clothing, and in some cases the bullet wipe pattern may obscure the contusion ring. The shape of the bullet wipe and contusion ring can provide clues about angles and relative positions: in the case of straight-on shooting, these features are roughly circular; they can be more oval shaped if the shot is fired from an angle or is offset from center. Beyond the bullet wipe and contusion ring there is a dispersed deposit of material (GUNSHOT RESIDUE) that contains flakes of unburned powder and other residues. This is called stippling. The con-

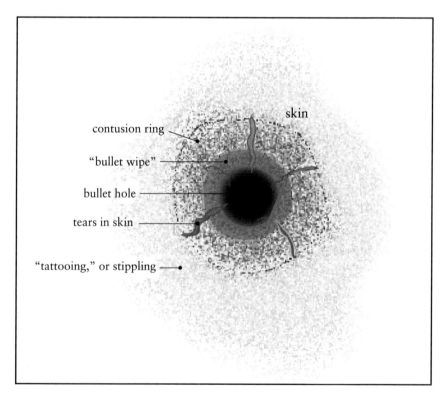

Characteristics of the exterior features of a bullet wound.

centration and spread of these residues depend primarily on the distance between the shooter and the victim. Once the bullet passes through the skin, its path is not predictable. *See also* STELLATE PATTERN.

burden of proof   The responsibility for presenting evidence and testimony to support a position. In the American legal system, the burden of proof is on the prosecution, meaning that the prosecution must prove the charges are true. If the burden of proof were on the defense, it would mean that the defendant would be responsible for disproving the charges.

Bureau of Alcohol, Tobacco, Firearms, and Explosives (ATF)   A federal agency housed in the Justice Department, having been moved there in 2003 as a result of the Homeland Security Bill. Previously, ATF was housed in the Treasury Department. In 2003, the agency split: Firearms, arson, and explosives programs were now overseen by the ATF in the Justice Department, while the newly created Tax and Trade Bureau (TTB) in the Treasury Department was to oversee taxation and regulation for tobacco and alcohol. Both the ATF and the TTB have forensic laboratories, and forensic services in their predecessors date back to 1886, when chemists performed simple analyses of butter to determine whether or not it had been adulterated with margarine.

burn rate (burning rate)   The speed at which a PROPELLANT burns. Burn rate is an important characteristic of propellants used in ammunition, which can be manipulated through changes in the size and shape of powder granules and by chemical additives.

byte   In computers the basic unit of memory used to store one character. A byte consists of eight binary bits, each of which holds either a 0 or a 1. Units of computer storage are often given in bytes, for example, a 120-gigabyte hard drive.

C4 *See* PLASTIQUE.

cadaver A dead body. The term is generally used to describe a corpse that is going to be autopsied or dissected.

cadaver dogs Specially trained dogs used to locate buried, concealed, or scattered human remains hidden in CLANDESTINE GRAVES and to find bodies or portions of bodies in mass disasters such as airplane crashes.

cadaveric spasm *See* RIGOR MORTIS.

cadmium selenide (CdSe) A QUANTUM DOT (tiny photoluminescent particle) being studied for use in visualizing latent FINGERPRINTS. These nanoparticles are chemically modified to bind with the amino acid content of fingerprints and are then exposed to a laser light source that excites the quantum dot, producing an intense emission of light.

cadmium sulfide (CdS) A compound that has been used to visualize latent FINGERPRINTS. As most commonly used, the compound is incorporated in a polymer (DENDRIMER) that reacts with the amino acid components in fingerprints. The print is then exposed to a laser light source that excites the quantum dot, producing an intense emission of light. CdS has also been studied for use as a QUANTUM DOT for the same purpose.

caffeine An ALKALOID and stimulant found in coffee, tea, and cola beverages. It is a white powder. Caffeine is sometimes encountered in forensic drug analysis when it is used as a CUTTING AGENT (dilu-

ent) for drugs such as COCAINE, AMPHETAMINE, and methamphetamine.

CA fuming (cyanoacrylate fuming) A technique used to visualize latent fingerprints on a variety of surfaces ranging from nonporous materials such as glass to paper and skin. Fuming is usually initiated by heat or by a strong base such as sodium hydroxide (NaOH) and is conducted in some type of enclosure such as a FUMING CABINET.

caliber Originally, a term that referred to the diameter of the barrel of a rifled pistol or rifle; however, it can also refer to the size of CARTRIDGES used in FIREARMS. Caliber is measured from the top of the LANDS and is given in hundredths or thousandths of an inch or in millimeters. Common calibers include .22, .38, .40, .45, and 9 mm for pistols and .22 and .30-06 for rifles. The caliber of a gun is considered to be a nominal measurement, meaning that the actual barrel diameter may vary slightly from the caliber measure used to describe it.

calibration (calibration curve) The process of relating an instrument reading or measurement to a reliable standard or set of standards. For example, an analytical balance should be frequently calibrated by using standardized weights provided by the NATIONAL INSTITUTE OF STANDARDS AND TECHNOLOGY. Similarly, many analytical instruments and devices such as microscopes must be calibrated so that the instrument response can be properly and reliably interpreted. A calibration curve for an instrument is obtained by determining the instrument response to a series of calibration standards of known concentrations and then plotting concen-

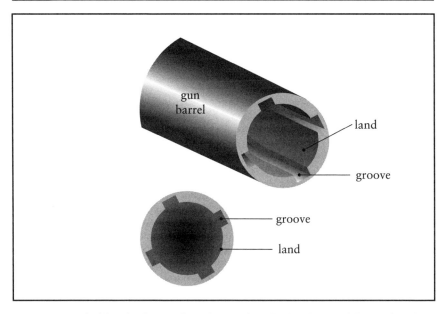

Measurement of caliber, the distance from the top of one land to the top of the one directly across from it in the gun barrel.

tration on the *x*-axis of the curve versus the response of the instrument at each of those concentrations on the *y*-axis.

California Association of Criminalists (CAC) The organization, founded in 1954, that is the oldest regional forensic science association in the United States. It was instrumental in developing and implementing certification examinations for CRIMINALISTS.

cannabis *See* MARIJUANA.

cannelures Small grooves imprinted by rolling on the base of a BULLET or near the top of a CARTRIDGE CASE. For bullets, cannelures can hold lubricant or can be used as a seat around which the throat of the cartridge case can be crimped closed. In cartridge casings, the cannelures prevent the bullet from being forced backward into the cartridge case.

capillary electrophoresis (CE) Instrumental techniques used to separate and identify a variety of substances of interest in forensic science. They evolved from ELECTROPHORESIS carried out on horizontal slabs of gel or other media and are based on the same basic principles. Many types of capillary electrophoresis exist; the ones used most in forensic science are CAPILLARY ZONE ELECTROPHORESIS, CAPILLARY GEL ELECTROPHORESIS, and MICELLULAR ELECTROKINETIC CAPILLARY CHROMATOGRAPHY (MEKC or MECC). The common element in all of these techniques is the use of a narrow capillary tube filled with a conductive water-based solution or medium. As in traditional gel slab electrophoresis, separation of the individual compounds in the sample occurs in the gel as these molecules move under the influence of an applied electrical field. Molecules with a net negative charge move toward the positive (+) end of the capillary; molecules with a net positive charge move toward the negative (–) end of the tube. Neutral molecules (those without charge) are immobile unless additional modifications are made. The larger the molecule, the slower its rate of migration

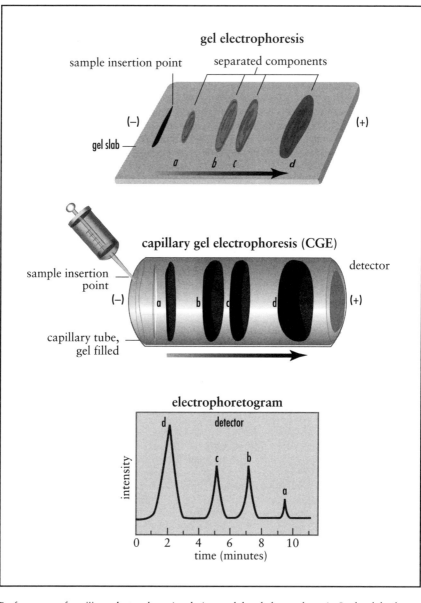

gel electrophoresis

sample insertion point

separated components

(−)

gel slab

a   b   c   d

(+)

capillary gel electrophoresis (CGE)

sample insertion point

detector

(−)   a   b   c   d   (+)

capillary tube, gel filled

electrophoretogram

d

detector

c   b

a

intensity

0   2   4   6   8   10

time (minutes)

Performance of capillary electrophoresis relative to slab gel electrophoresis. In the slab, the proteins migrate under the influence of the electrical field with their speed dependent on their size. The process occurs analogously inside the capillary tube, and the detector signal is used to create the electropherogram.

through the gel; thus, CE separates on the basis of both size and charge. *See also* ELECTROOSMOTIC FLOW.

capillary gel electrophoresis (CGE) A form of capillary electrophoresis in which sample is introduced into the capillary tube via an injection system (usually a syringe). Separation of individual compounds occurs by "sieving"—the larger the molecule, the more slowly it moves through the gel medium.

capillary zone electrophoresis (CZE) A capillary electrophoresis technique used in applications such as ink, drug, and gunshot residue analysis. One of the distinguishing characteristics of CZE is ELECTROOSMOTIC FLOW. When a sample is introduced into the capillary, any components that are positively charged move quickly toward the detector and separate on the basis of factors such as relative sizes and charges. Not only are the positively charged species attracted to the negatively charged cathode, but in addition these species are caught in the electroosmotic flow already moving in that direction. Neutral species move at the same speed as the electroosmotic flow, and all neutrals arrive at the detector at the same time. Although the negatively charged species are actually attracted to the opposite end of the capillary (the anode), the electroosmotic flow is sufficient to overcome this attraction, eventually delivering the negatively charged species to the detector. Thus, negatively charged species move more slowly than the electroosmotic flow, but in the same direction.

carbonic anhydrase II (CA II) An ISOENZYME system that is polymorphic in black populations and has three types that can be separated by using ELECTROPHORESIS. It was used occasionally in forensic serology before the introduction of DNA TYPING.

carbon ink (India ink) A black ink that consists of charcoal in a binder or glue medium. Despite the name, it was first used in China. It is still used today.

carbon monoxide (CO) A colorless, odorless, and tasteless gas generated as a by-product of COMBUSTION reactions. Carbon monoxide is also highly flammable and toxic and is the leading cause of poisoning (accidental and intentional) in the United States. The mechanism of poisoning arises from carbon monoxide's ability to bind with HEMOGLOBIN 200–300 times as strongly as oxygen.

carbon paper A blackened paper that was at one time widely used to create duplicate copies of typed or handwritten documents. The sheet of paper on which the original writing is done is placed on top, followed by a carbon sheet, and a second sheet of paper. Pressure applied on the top (original) sheet by a pen or typewriter pushes down on the carbon paper, which in turn creates the same pattern on the sheet beneath. Many forms still use a type of carbon paper technology, and carbon papers can be useful in QUESTIONED DOCUMENT examinations.

carrier gas In gas chromatography, the inert gas, typically helium (He), hydrogen (H$_2$), or nitrogen (N$_2$), that is used to propel a sample down and through a chromatographic column.

cartridge cases Also called shells or casings. The cartridge is the part of a round of AMMUNITION that encloses the PROPELLANT. The BULLET is seated at the forward end and the PRIMER at the base of the cartridge case. Casings are usually made of brass (which can be reloaded), nickel-coated brass, or aluminum (which is not designed to be reloaded). Given the composition, cartridge casings are often referred to generically as "brass." In SHOTGUN ammunition, the cartridge case is made of plastic or cardboard and crimp-sealed at the top.

casting A technique to preserve and replicate IMPRESSION EVIDENCE made in soft material such as soil or snow. Casts can be made of TIRE PRINTS, SHOE PRINTS, and occasionally TOOLMARKS and BITE MARKS. When done properly, casting techniques produce excellent positive replicates of the impression, but not exact duplicates. Dental stone is frequently used for impressions in soil, epoxylike materials for toolmarks, and Snow Print Wax for impressions in snow.

cast-off pattern (castoff pattern) A BLOODSTAIN PATTERN created when a bloody weapon or a limb such as a fist is swung or otherwise moved fast enough to eject blood. As shown in the figure, this cast-off blood impacts surfaces in a characteristic pattern that can be used in crime scene analysis and reconstruction.

catagen stage The intermediate stage of a hair growth cycle. It is the transition state between active growth (ANAGEN STAGE) and dormancy (TELOGEN STAGE).

catalyst A material that serves to speed up a chemical (or biochemical) reaction without itself being consumed.

cathode In an electrochemical cell, the pole or area to which positively charged species or cations migrate. In electrophoresis, the zone that has a negative charge is the cathode since positively charged cations are attracted to it.

cation An ion that has a positive charge created by the loss of electrons. For example, the species Ca^{2+}, the calcium cation, is created when elemental calcium loses two electrons.

cationic surfactants Substances with positively charged chemical functional groups (CATIONS) that, when added to water, reduce the surface tension. Surfactants are used in analytical techniques such as capillary zone electrophoresis. Soaps and detergents are examples of surfactants.

cause of death The immediate reason for a death; the action or injury that most directly caused a person to die. In criminal matters, a medical cause of death is not necessarily the legal cause. For example, if a victim is stabbed and dies as a result of complications such as infection, the legal cause of death remains the stabbing and the case remains a homicide even though the immediate cause of death is infection.

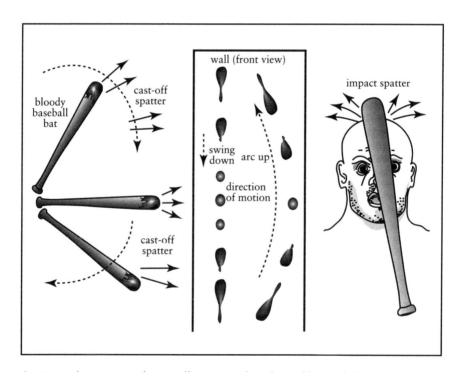

Creation and appearance of a cast-off pattern resulting from a blow to the head.

Cause of death is also distinct from the circumstances of death, which comprise the situation and conditions that led up to the fatal encounter.

CCD (charge-coupled device) A semiconductor device used in cameras and in detectors for some types of spectrophotometers. When a PHOTON strikes a CCD, it is converted into an electron and the charge is stored in the semiconductor. The amount of charge that is stored is proportional to the number of photons striking the detector and can be related to the intensity of the light signal.

celioscopy The study of the patterns formed by wrinkles, skinfolds, and other structures on the fleshy part of the lips. "Lip prints" have not been accepted by courts as evidence.

cell potential (cell voltage) In an electrochemical cell made of an ANODE and a CATHODE, the total potential of the cell. Cell potential can be measured by a voltmeter or, in some cases, calculated.

cementum The material coating the root structure of a TOOTH that is in contact with the gum.

centerfire cartridge A type of firearm AMMUNITION in which the primer is centered in the base of the cartridge. This contrasts with the RIMFIRE design used in the smaller CALIBER of ammunition.

Centers for Disease Control and Prevention (CDC) A federal agency headquartered in Atlanta, Georgia, that is tasked with protecting the health and safety of the American public, broadly defined. Recently, it has become a key element in preparing for and combating bioterrorism.

Central Identification Lab (CIL, CILHI) A United States Army laboratory located at Hickam Air Force Base in Honolulu, Hawaii. The lab uses a variety of techniques including forensic ANTHROPOLOGY and DNA TYPING to locate, identify,

and repatriate the remains of American soldiers and other armed services personnel.

central pocket loop A type of ridge pattern found in fingerprints, one of eight classes of ridge patterns used in AFIS fingerprint classification. The other seven are the arch, tented arch, whorl, radial loop, ulnar loop, double loop, and accidental.

certification A process employing written and laboratory testing that ensures the proficiency of forensic analysts and the reliability of the data and results they produce. Certification can be divided into two areas: the certification of individual analysts through written and lab testing, and laboratory certification, which is referred to as laboratory ACCREDITATION. The goals of both accreditation and certification are the same, to ensure that the analyses being conducted in forensic laboratories are being done properly and that the results produced are accurate and trustworthy. In general, certification is obtained through professional associations.

CFNE (clinical forensic nurse examiner) A nurse active in FORENSIC NURSING in settings such as emergency rooms, prison clinics, and forensic (psychiatric) hospitals.

chain of custody Procedures and documentation used to ensure the integrity of evidence from collection to courtroom presentation and through to final disposition or destruction. The paramount goal of the process is to prevent any breaks in the chain, which would put in question the reliability of the evidence and the link between the evidence and the person or scene from which it was obtained.

chamber The portion of a firearm at the rear of the barrel where the cartridge is placed before firing.

charcoal strips A method employed in arson analysis. In this approach, a strip of charcoal is suspended over the surface of fire debris suspected of containing an ACCELERANT. This occurs in a closed container that is gently heated, driving off residual accelerant, which then collects and concentrates on the charcoal. The charcoal strip is then removed and the concentrated vapors extracted into a small amount of solvent. A portion of this solution is injected into a gas chromatograph for further analysis.

char depth The depth of burning on or in a piece of material that has been exposed to fire. The depth of charring can be related to the duration and intensity of the fire.

chemical analysis Techniques used to identify the chemical components present in a material (QUALITATIVE ANALYSIS) and often the amounts or concentrations of some or all of those components (QUANTITATIVE ANALYSIS).

chemical fuming A group of techniques used to develop and enhance LATENT FINGERPRINTS. One of the earliest such procedures, still used, is fuming the evidence with iodine (I_2). Currently, the most common fuming method uses CYANOACRYLATE, also known as Super Glue. In a fuming process, the evidence is placed into an enclosed container and reagents that quickly evaporate and produce fumes are added. Components of the fumes bind with substances in the fingerprint residues, making visualization easier. Additional treatments may follow to stabilize or further enhance the print.

chemical properties Properties of matter that cannot be determined without carrying out a chemical reaction. This is in contrast to PHYSICAL PROPERTIES, which can be observed or measured without changing the original composition of the material. Examples of physical properties include density, color, size, mass, and boiling point; flammability is a chemical property.

chemical shift A term used in NUCLEAR MAGNETIC RESONANCE. The chemical shift of a given atomic nucleus is the frequency at which it absorbs electromagnetic energy while being exposed to a magnetic field. The chemical shift can be related to the structure of organic molecules.

chemiluminescence The release of light as a result of a chemical reaction. This property is exploited in some development methods for fingerprints. It is also the basis of visualization of blood using LUMINOL.

chemistry, forensic The application of the principles and techniques of chemistry, particularly analytical chemistry, to situations in which the legal system is or may be involved. Drug analysis, arson analysis, and toxicology are the main areas in which forensic chemistry is practiced. Other areas of forensic chemistry include analysis of ARSON evidence and fire debris, EXPLOSIVES, PAINT, FIBERS, GUNSHOT RESIDUE, and other kinds of trace evidence.

chirality The "handedness" (as in left-handed or right-handed) of a molecule. Molecules of the same substance that exhibit chirality are those that exist in two forms that cannot be superimposed on each other. Such a molecule has two forms, or two ENANTIOMERS. Amino acids are examples of chiral molecules and most exist in the d-form or the l-form: d meaning "dextrorotatory" and l meaning "levorotatory." This naming convention arises from the way the two enantiomers rotate PLANE POLARIZED LIGHT. For example, d-alanine and l-alanine are the two forms of the amino acid alanine.

chi-squared test A statistical test used to compare the frequency of occurrences of some action, or most commonly in the forensic context, genetic frequencies. The value of χ^2 is calculated by using the following formula where O is the observed frequency, E is the expected frequency, and i is the number of observations.

$$\chi^2 = \sum_i \frac{\left(O_i - E_i\right)^2}{E_i}$$

The calculated value chi-squared is compared to table values at different probability levels, and from this comparison it is possible to say whether the observed frequency differs from the expected and a specified level of significance, typically 95 percent.

chloroform (CHCl$_3$) An organic solvent that was at one time widely used in laboratories and in medicine as an anesthetic. Use has decreased dramatically since the compound was identified as a carcinogen. In movies, books, and television, chloroform-soaked rags are sometimes used to knock out people; however, in reality doing so is extremely difficult and can lead to convulsions and death in the victim.

choke The constriction built into the barrel of a shotgun. The greater the degree of choke, the smaller the diameter of the dispersal of the shot pellets.

choline/choline test A component of semen that can be detected by using a microcrystal test called the choline test or Florence test. For this procedure, the sample or stain extract is treated with iodine and if choline is present, characteristic choline iodide crystals can be seen under the microscope.

Christmas tree stain A histological staining procedure used principally for visualizing sperm cells. The staining proceeds in two parts, using a dye (nuclear fast red) and picroindigocarmine made from picric acid and indocarmine dye. Different parts of the sperm cell take up different colors, simplifying microscopic identification. The tail shows as a yellow-green and the head shades of red and pink, leading to the name.

chromatogram The printed or digital output of a chromatographic instrument. A chromatogram is a plot of the response of the detector as a function of time.

chromatography Literally, "color writing," a class of separation techniques used extensively in forensic science. The name *color writing* originated from early development work in which plant pigments were separated into their components, producing bands of color in the separation column, a process called column chromatography (shown in the figure on page 38). Similar color bands can be seen in the modern version of THIN LAYER CHROMATOGRAPHY when applied to inks. All chromatographic separations are based on SELECTIVE PARTITIONING; their mechanisms and details vary with the specific application. GAS CHROMATOGRAPHY (GC) for example, depends on differential partitioning of sample molecules moving between a liquid phase and a gas phase.

chromophore A chemical structure or species capable of absorbing electromagnetic radiation that is used in spectrophotometric analysis. For example, in INFRARED SPECTROSCOPY, the chromophores are chemical functional groups such as C=O. Similarly, a fluorophore is a chemical group or species that fluoresces.

chronic A disease, condition, or response that occurs over time, typically in response to continual small doses of an irritant or poison. Chronic effects contrast with acute effects, which occur immediately and in response to larger doses. For example, chronic exposure to low levels of carbon monoxide can cause breathing problems and headaches, whereas exposure to a large amount in a short period can cause acute effects including death.

circumstances of death The situation and conditions that led up to a fatal encounter or incident. For example, if a man's body is found at the base of a large building, the cause of death is the fall; however, the circumstances relate to whether he jumped, fell, or was pushed.

circumstances of the event An expression used to describe the known details and circumstances of a crime that might be relevant to a forensic analysis. For example, if a fiber is found on clothing, the story

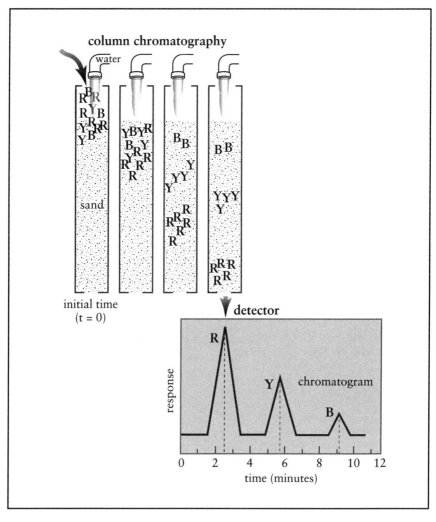

Column chromatography, illustrating the basic principles of chromatographic separation. This example is of a colored solution containing red (R), yellow (Y), and blue (B) components that are separated.

of how it arrived there can be considered to be the circumstances of the event.

circumstantial evidence Evidence that does not reflect directly on the question at hand, but rather can be used to draw conclusions. Much of the evidence produced during forensic analyses is circumstantial evidence.

Civil Aeronautics Board (CAB) An agency created by Congress in 1940 charged with investigating aviation accidents, determining probable causes, and issuing findings and recommendations to prevent recurrences. The CAB was merged into the newly formed NATIONAL TRANSPORTATION SAFETY BOARD (NTSB) in 1967.

civil law   *See* CRIMINAL LAW VS. CIVIL LAW.

clandestine graves   Grave sites associated with crimes, which often are located in remote areas and that may not be found for weeks, months, or years after the burial occurred. Searches for clandestine graves are made by personnel from many disciplines including botanists, entomologists, and geologists; increasingly, archaeologists are involved in the excavation of these sites. Many of the techniques used to locate clandestine graves rely not on detecting the body directly, but on finding soil and ecological disturbances above it.

clandestine lab   Illicit laboratory established to manufacture or process illegal drugs. Although the types of drugs made vary by region and by current availabilities, the most common types of clandestine labs are those making METHAMPHETAMINE.

class characteristics and evidence   A class can be considered a discrete group or subgroup of items (or individual items) that are similar as a result of reproducible processes used to make them. Classification of evidence is a process of assigning it to these groups or categories. Members of a given class share the same class characteristics, whereas evidence that can be individualized possesses characteristics that make it unique. For example, handwriting has distinctive class characteristics; in writing a capital *A*, two slanted lines meet at a point and are linked by a horizontal line about halfway up. Thus, the class of letters designated as *A* are easily distinguished from letters that belong to class *B*. However, all people write the letter *A* slightly differently, imparting individual characteristics. Almost every type of physical evidence can be classified in some way, but not all evidence can be individualized.

classification   The process of dividing objects or images into distinct groups based on characteristic features, morphological characteristics, or other criteria. For example, bullets can be classified by characteristics such as caliber, composi-

tion, and jacketing, and fibers can be classified as natural or synthetic, and so on.

clinical forensic nurse   *See* CFNE.

CMC (critical MICELLE concentration)   When surfactants such as soap are added to a water-based solution, structures called micelles form, when the concentration of the soap exceeds the CMC. When micelles form, the surface tension of the water is broken, and water will no longer form beads on a smooth surface. Micelles are exploited in a form of electrophoresis called MICELLULAR CAPILLARY ELECTROPHORESIS.

CNS   An abbreviation for *central nervous system*, which consists of the brain and spinal cord.

coatings   *See* PAINT.

cobalt thiocyanate   A chemical reagent used as a PRESUMPTIVE TEST for COCAINE. There are three common variations of the test. In the first, the reagent is prepared by dissolving cobalt thiocyanate (CoSCN) in distilled water, sometimes with additional ingredients such as ammonium thiocyanate or glycerin. When this solution is added to cocaine powder, a blue solid (blue precipitate) is formed. To increase the specificity of the test, a variation called the Scott or Ruybal test has been used. In the test three reagents are employed: a cobalt thiocyanate solution containing glycerin, a hydrochloric acid solution, and chloroform, an organic solvent that does not mix with water (similarly to vinegar and oil). If cocaine is present, addition of the first reagent causes the blue precipitate to form; addition of the second causes the blue to turn to a clear pink. Adding the final reagent, chloroform, causes two layers to form in the test tube, the aqueous layer (the water-based solution) and the chloroform layer. If cocaine is present, a blue color reappears in the chloroform layer.

cocaine   a powerful central nervous system stimulant derived from the leaves of the coca plant (*Erythroxylon coca*). Cocaine can also be synthesized, but the process is difficult and expensive and so far has not replaced the coca plant as the primary source in the illegal drug market. The coca plant grows in the Andes Mountains and in some parts of Asia; the largest source of raw coca is South America, principally Colombia.

codeine An opiate alkaloid that is found naturally in OPIUM at concentrations of approximately 0.7–2.5 percent. Codeine is taken orally as an ANALGESIC (pain reliever) and as a cough suppressant, and it is usually synthesized from morphine rather than extracted from opium. Many codeine mixtures and preparations are listed on Schedule III of the CON-TROLLED SUBSTANCES ACT, although in many countries, codeine preparations are available over the counter.

code of ethics A list of expectations, responsibilities, and rules of conduct typically issued by a professional association or organization. Any and all members of the issuing body are expected to abide by the provisions of such a code.

CODIS (Combined DNA Indexing System) A national program coordinated by the FBI that assists state and local labs in establishing DNA databases from unsolved crimes, missing persons, and convicted offenders. The system electronically stores the results of a specific set of DNA TYPING results (13 genetic loci,

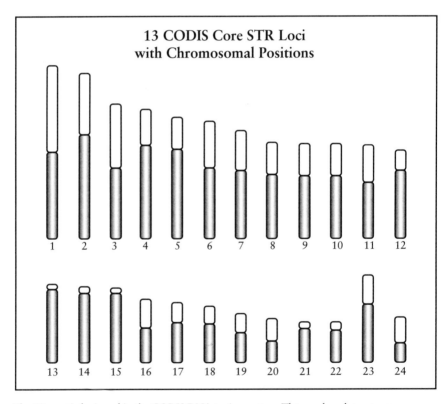

The 13 genetic loci used in the CODIS DNA typing system. The numbered structures are chromosomes and the shaded boxes indicate on which chromosome(s) the typed locations (loci) are found.

shown in the figure) and allows networked laboratories to share and compare DNA results. The system has proved to be particularly valuable in sex offense cases, in which DNA typing has been able to link suspects to multiple crimes within an area as well as across state lines. To date, several thousand cases have been solved by using CODIS, and the system is expanding to include countries outside the United States.

codominant In genetics, a situation in which an allele from both the mother and the father is expressed. For example, the ABO BLOOD GROUP SYSTEM is a genetic marker system with the types A, B, AB, and O. The A type is a result of AA alleles. The observable, measurable way in which any gene is expressed is called the PHENOTYPE, and so the phenotype of AA is A in this example (disregarding subtypes).

coelution In chromatography, the incomplete separation of two or more individual components in a mixture. Coeluting compounds have the same RETENTION TIME and thus emerge from the chromatographic column at the same time. Coelution complicates the identification of the coeluting species, but under some conditions, chromatographic DETECTORS such as MASS SPECTROMETERS (MS) can identify the individual components of a coeluting mixture.

coenzymes Chemical species that must bind to an enzyme to allow that enzyme to function and to facilitate the reaction in which it participates. Many vitamins are coenzymes.

COFSE The Council of Forensic Science Educators, a group of teachers and professors involved in teaching forensic science. COFSE is associated with the AMERICAN ACADEMY OF FORENSIC SCIENCES.

coincidental match A match or linkage, based on PHYSICAL EVIDENCE, that occurred by chance and not as a result of criminal activity or contact.

cold hit A term that refers to identification of a suspect or linked crimes by using DNA data and the CODIS database. The term generally refers to old violent crime cases such as rapes or murders that at first appear to be unrelated but that are later linked to each other and/or to a suspect on the basis of a DNA type in the database.

cold vapor deposition A technique used for the analysis of mercury by atomic ABSORPTION SPECTROPHOTOMETRY.

Coleoptera An order of insects (arthropods) that includes beetles, many of which are important in forensic entomology. The order contains several hundred thousand species. See also ENTOMOLOGY, FORENSIC.

collagen A tough fibrous protein that makes up structures such as the nose. Collagen fibers are also found in skin and provide the framework for other proteins and structures.

colloid (colloidal) A solution of a material in which it is suspended in the solvent as small particles rather than dissolved.

colloidal gold developer A PHYSICAL DEVELOPER used to visualize latent FINGERPRINTS. In a two-part procedure, the latent is first treated with the gold solution, then exposed to silver physical developer.

colloidal silver developer See SILVER PHYSICAL DEVELOPER.

colorimeter (colorimetry) SPECTROPHOTOMETRY conducted on colored samples. Visible light (colored light) makes up a small portion of the ELECTROMAGNETIC SPECTRUM, bordered on one side by ultraviolet (UV) radiation and on the other by infrared radiation (IR). By using BEER'S LAW, the amount of light absorbed can be related to concentration. Because the eye can be used as a detector, colorimetry was one of the first forms of spectrophotometry developed. In modern instruments, instrumental detectors replace the eye with

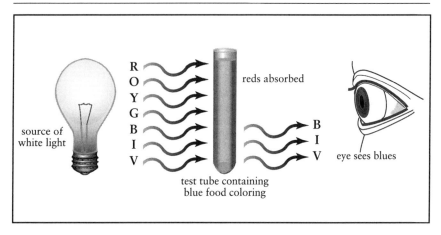

source of white light

R
O
Y
G
B
I
V

reds absorbed

test tube containing blue food coloring

B
I
V

eye sees blues

Simplified colorimetry. The color of the solution depends on which of the wavelengths are absorbed. The acronym ROY G. BIV is used to identify individual colors as *red*, *orange*, *yellow*, *green*, *blue*, *indigo*, and *violet*.

devices that convert light to electrical current that can be displayed or further manipulated.

color reversal A phenomenon that may occur when latent fingerprints are developed and the patterns take on a color opposite to the one expected.

color tests *See* PRESUMPTIVE TESTS.

combings A technique used to collect trace evidence, usually in sexual assault cases. When a rape has occurred, pubic combings are collected to find items such as pubic hairs of the perpetrator, fibers, or other transfer evidence that can be used to link the suspect to the victim. Other materials, including beards, can be combed as well, with the same intent.

combustion The chemical reaction broadly defined as burning that occurs when a hydrocarbon (a compound containing only carbon and hydrogen) or an oxygenated hydrocarbon is combined with oxygen at an elevated temperature. At the simplest level, a combustion reaction is one that takes place between a fuel and an oxidant. Heat is important, both

because it is a product of the reaction and because it ensures that the reaction has enough energy to be self-sustaining. The oxidant is usually (but not always) oxygen (O_2). The generic reaction for combustion of a hydrocarbon is given by

$$\text{Hydrocarbon (g)} + O_2\text{ (g)} \Rightarrow CO_2\text{ (g)} + H_2O\text{ (g)}$$

commingled remains Remains (human or other) that are mixed together; such mixing occurs after a mass disaster or in a mass grave.

common source Any evidence that can be shown to have come from one unique source or that can be shown to have originally been part of the same structure or unit is said to be evidence with a common source. Two bloodstains produced by one person have a common source, as do two different bullets fired from the same gun.

comparison microscope A microscope that consists of two COMPOUND MICROSCOPES linked together in such a way that two different objects can be examined side by side. The analyst looking through the viewer sees a circular area divided down the middle by a thin line, with the image from the left microscope stage on the left and the image from the right on the right of the viewer. There are two types of comparison microscopes, differentiated by the type of lighting used. For comparing objects that are opaque (BULLETS, CARTRIDGE CASES, and tools, for example), a vertical or reflected light source is needed; for objects that transmit light such as FIBERS, a transmission light source is employed.

compound   A chemical combination of two or more atoms in a stable form. A compound may be ionic, such as (NaCl), in which the bond between the two atoms is based on electrostatic attraction. Molecular compounds are held together by covalent bonds, which result when atoms share electrons. Water (H_2O) is a molecular compound.

compound microscope (biological microscope)   A microscope that consists of a light source, two lenses, a stage, and a barrel. The magnification is accomplished by two lenses, the objective lens (the one closer to the sample) and the ocular or eyepiece lens. The objective is often mounted on a platform called the nosepiece with two or more lenses that can be rotated into position above the sample. The view through the microscope can consist of one eyepiece (monocular) or two (binocular). Each lens has an associated magnification, total magnification is the product of the two. For example, if the ocular lens is selected as 10X (10 times magnification) and the objective lens is selected as 45X, the image of the object is magnified by a factor of 450.

computer models   Simulations created by software that model events or phenomena. Computer models are used for crime scene reconstruction and for investigation of automobile and other transportation accidents, to name just a few applications.

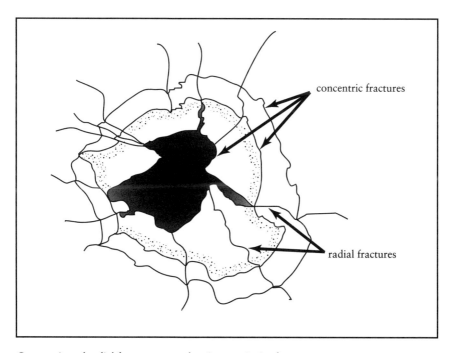

Concentric and radial fractures around an impact site in glass.

computing, forensic (forensic computing, computer forensics, cyberforensics) The application of the techniques of computer science and digital technology to electronic forms of evidence called digital evidence. Tasks of forensic computing include recovering data from disks and other storage media that have been erased, formatted, damaged, or tampered with; finding hidden, disguised, or encrypted information; and breaking passwords and other security devices used to hide evidence.

concentration The amount of a substance or ANALYTE that is contained in a sample. Concentration may be reported in many different units, such as parts per million or percentage.

concentric fracture (concentric crack) A crack found in GLASS or other media that is roughly circular at some distance from the impact; multiple concentric fractures at different distances from the impact are often observed. (See figure on page 43.) *See also* CONCHOIDAL LINES; RADIAL FRACTURES.

conchoidal lines (Wallner lines; ridges) Stress marks that are found in glass after an impact. A cross section of glass that has experienced an impact shows a series of curving lines that are parallel to the side where the force was applied and perpendicular to the side opposite the applied force. Thus, the pattern will be different in a radial crack versus a concentric one.

conclusive evidence Evidence that is nearly irrefutable, is not subject to multiple interpretations, and is powerful enough to override other evidence.

condenser The portion of a COMPOUND MICROSCOPE that collects light from the illumination source below the microscope stage and directs it through the specimen that is mounted on the stage and below the OBJECTIVE LENS.

conditional evidence Evidence that must be interpreted in light of other facts, considerations, or evidence; evidence that depends on outside factors.

condom A latex cap that is used to cover the penis during sexual intercourse to trap the ejaculate and prevent pregnancy. The composition of the condom and the lubricant used can be important evidence in sexual assault cases.

confidence interval A statistical quantity based on the standard deviation of a set of replicate measurements and that can be used to express the uncertainty of the

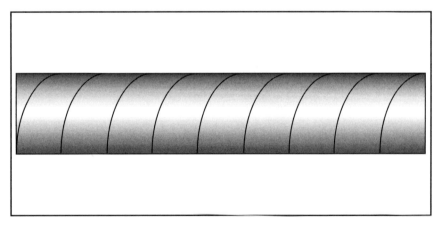

Conchoidal lines as they would appear in a cross-section of a fracture in glass. The direction of the lines depends on whether the fracture was conchoidal or radial.

result. For example, a sample found to contain cocaine might be analyzed three times with the following results: sample 1, 25.2 percent cocaine; sample 2, 25.0 percent; and sample 3, 26.8 percent. The average of these three values is 25.67, but since all three were different, reporting the result as a range would be more appropriate. Using tables and the standard deviation, the result could be reported with a 95 percent confidence interval, equal to 2.45 percent. Put another way, for the results presented here, there is a 95 percent confidence that the true percentage of the sample is between 23.2 and 28.1 percent. Using a confidence interval associates a range and a probability with a result.

conflagration A very large, highly destructive fire.

connecting stroke In handwriting, the pen (or pencil) stroke that connects the end of one letter to the beginning of another.

constructive interference Interference between two electromagnetic waves that results in reinforcement rather than canceling out (DESTRUCTIVE INTERFERENCE). Patterns of constructive and destructive interference are important in MICROSCOPY, optics, and many types of SPECTROPHOTOMETRY. *See also* MICHELSON INTERFEROMETER.

contact wound A gunshot wound produced when the barrel of a firearm is in direct contact with or only a few millimeters away from the skin surface. Contact wounds are characterized by a stellar or STELLATE PATTERN of skin rupture (caused by expanding gases) and a large degree of stippling produced by unburned powder. *See also* BULLET WOUNDS.

contamination The inadvertent or unintentional transfer of foreign materials or other alteration of evidence that can ruin the evidence or lead to incorrect interpretation of results. Contamination can be introduced at any stage from collection through the actual analysis and can lead to FALSE POSITIVES or FALSE NEGATIVES.

contrast The difference in intensity and colors of light between different media or across a boundary. For example, when a latent print is dusted with a white powder and lifted, for best visualization the lift should be placed on a black (or contrasting) background. In microscopy, degree of contrast is an important consideration and variable in many types of examination; PHASE CONTRAST MICROSCOPY can be used when differences in contrast between a medium and a sample are small.

controlled substances (Controlled Substances Act [CSA]) The federal law governing dangerous drugs, analogs, and raw ingredients (PRECURSORS) used in CLANDESTINE LAB synthesis. The CSA was passed as part of the Comprehensive Drug Abuse and Prevention Control Act in 1970. The act classifies drugs in one of five schedules (indicated with a roman numeral), which are based on accepted medical use and the danger of physical and psychological dependence. Each schedule also specifies penalties, ways the drug can be obtained, and permit requirement, when applicable.

control samples (controls) Samples with a known composition, identification, type, or source used for comparison or to ensure that an analysis is being properly performed and that the results are reliable. Controls are part of QUALITY ASSURANCE/QUALITY CONTROL (QA/QC) procedures that forensic labs follow to ensure the reliability of data and to eliminate incorrect results including FALSE POSITIVE and FALSE NEGATIVE results. Both positive controls and negative controls are used; a positive control demonstrates a positive result and a negative, the lack of a response.

contusion A bruise or wound that causes bleeding and resultant discoloration beneath the skin's surface.

contusion ring (marginal abrasion) A ringlike bruise often found around an entrance gunshot wound. It is caused by the stretching and abrasion of the skin that occur as the bullet penetrates it. *See also* BULLET WOUNDS.

cooling rate *See* BODY TEMPERATURE.

coomassie brilliant blue/commassie blue A common histological stain used to study cell structures and tissues and to aid in visualizing latent FINGERPRINTS.

copybook A book that is used to learn and practice techniques of handwriting and penmanship. It contains examples that the student mimics.

coroner From Latin *corona*, "crown," an elected or appointed official who is charged with determining the cause of death in cases in which the death appears to be the result of foul play, was unattended, or occurred under questionable or suspicious circumstances. The position of coroner was a remnant of Roman law instituted in England during the 12th century. The coroner system was common in the United States until 1877, when Massachusetts became the first state to abolish it in favor of a MEDICAL EXAMINER system. New York City followed in 1915, and many jurisdictions have since converted to MEs. The modern coroner is a judicial officer who is elected or appointed, whose tasks are mainly administrative. The job of the coroner is to determine the cause of death by using whatever resources are necessary, which can include ordering an autopsy by a pathologist or forensic pathologist.

corpus delecti Literally, "the body of the crime." This term can refer specifically to the thing or person on which a crime was perpetrated, such as a corpse, but more commonly it refers to the "foundation and substance" of a crime. According to *Black's Law Dictionary* (6th edition, 1990, West Publishing Co.), *corpus delecti* is defined as the criminal act along with the criminal agency associated with it. In modern usage, *corpus delecti* usually refers to the entire body of evidence that leads to the conclusion that a crime has been committed.

correction ribbon A ribbon used in typewriters to cover an error with a white-colored coating. In the past correction ribbons were important in questioned document examination because the ribbon often retains an image of the original letter that was covered. However, the decreasing use of typewriters means that correction ribbons are rarely encountered anymore.

corroborative reconstruction A crime scene or event reconstruction designed to corroborate or refute a given theory about what happened during the actual event.

cortex The central portion of a HAIR in which structures such as the MEDULLA, FIBRILS, CORTICAL FUSI, and pigments are found.

cortical fusi Vacuoles (air pockets) found in the CORTEX of HAIR.

countercoup An injury that occurs when a blow to the head is received. The COUP is the injury that occurs at the impact site; the countercoup is the injury that occurs directly opposite the impact. In head injuries, the degree of injury at these two locations can be useful in reconstructing the situation that created them. If the head is moving and strikes a stationary surface, such as when someone falls to the floor, the hemorrhage inside the skull is most pronounced on the side opposite the impact. If, on the other hand, the head is stationary and is hit by a weapon such as a baseball bat, the impact site (the coup) shows greater damage.

counterimmunoelectrophoresis *See* CROSSED-OVER ELECTROPHORESIS.

coup An injury that occurs when a blow to the head is received. The coup is the injury that occurs at the impact site; the COUNTERCOUP is the injury that occurs directly opposite the impact. In head injuries, the degree of injury at these two locations can be useful in reconstructing the situation that created them. If the head is moving and strikes a stationary surface, as when someone falls to the floor, the hemorrhage inside the skull is most pronounced on the side opposite the impact.

If, on the other hand, the head is stationary and is hit by a weapon such as a baseball bat, the impact site (the coup) shows greater damage.

courtroom testimony Evidence presented in a courtroom by a person (a witness) who is deemed competent and whose knowledge and/or opinions are relevant to the issue at hand.

court systems Within the United States, courts are organized on the basis of jurisdiction, which can be local (city or county), state, or federal. In the federal system, the district court acts as the gateway. If a jury is to be the TRIER-OF-FACT, the panel consists of 12 people and the verdict must be unanimous. The first level of appeal above the federal district court is found in circuit courts, which consist only of panels of judges that review district court decisions that are appealed on the basis of claims of procedural errors. The next higher level of appeal in the federal system is the U.S. Supreme Court. State courts are often modeled on the federal system but there are no uniform standards. County and city courts (often called municipal courts) are similarly diverse.

creatine and creatinine Creatinine is a natural component of urine, present as a result of muscle action. Creatine breaks down to creatinine, resulting in elevated levels of creatinine in the urine.

cremains The material left over when a body is burned or incinerated. In addition to ash, this can include teeth, bone bits, and metal such as screws used in medical procedures.

crime According to *Black's Law Dictionary* (6th edition, 1990, West Publishing Co.), a *crime* is defined simply as "a positive or negative act in violation of penal law; an offense against the state or the United States." A crime is therefore a violation of a code or written law that was created to protect society. Crimes are first divided on the basis of whether they are classified as felonies or misdemeanors. There are many types of crimes, including crimes against property (burglary, for example), crimes of omission (failure to stop and aid victims of an accident), crimes of passion, crimes of violence, and white-collar crimes such as fraud, bribery, and software piracy.

crime labs An informal term used to describe forensic laboratory facilities or laboratories associated with a law enforcement agency.

Crime Scene Investigation (CSI) *See* CRIME SCENES.

crime scene reconstruction A recreation of the events immediately before, during, and after the commission of a crime with the intent of determining what might or might not have happened. The reconstruction is primarily based on a combination of physical evidence, deduction, and witness testimonies. Reconstructions can determine a likely series of events or can refute them.

crime scenes Locations, indoors or out, where a crime was committed. Crimes may have more than one scene, such as when a person is killed and the body is buried in a remote location. In this example, the death scene is the primary crime scene; the burial site is the secondary scene.

criminal investigation The process of investigation of a criminal act carried out by law enforcement agencies, forensic scientists, courts, and allied personnel.

criminalistics Derived from *kriminalistik,* a word used by the Austrian HANS GROSS, a pioneer of the field, to describe the application of natural sciences to matters of law and law enforcement.

criminal law *See* CRIMINAL LAW VS. CIVIL LAW.

criminal law vs. civil law In criminal cases, the purported incident or action is a violation of a written penal code, the goal of which is the protection of society. On the other hand, civil law relates to disputes between individuals or parties rather than between the state and individuals or parties. In some states, separate courts handle the two types of cases, but usually one court hears both types.

Criminalistics Certification Study Committee (CCSC) A committee of forensic science professionals in the United States and Canada that was formed as a result of a grant from the National Institute of Justice (NIJ). The CCSC met from 1975 to 1979 and discussed the issue of certification of criminalists. Although their work did not result in a certification program, it formed the basis for certification programs that followed in the late 1980s (California Association of Criminalists [CAC]) and the early 1990s (AMERICAN BOARD OF CRIMINALISTS).

criminology A social science based on the study of criminals, crime, and the penal system. This term is often confused with CRIMINALISTICS, which involves the application of physical sciences (chemistry, geology, physics, and so on) and biological sciences to the analysis of physical evidence. The two terms are related but not interchangeable.

crossed-over electrophoresis A method used to determine the species of origin of a bloodstain by using gel ELECTROPHORESIS. Sample extract containing ANTIGENS is placed into a gel well nearest the CATHODE while ANTIBODIES for different species are placed in nearby wells closer to the ANODE. Each stain extract is tested against several species such as human, dog, cat, and whatever other species might have been the source of the stain. When power is applied, antigens and antibodies move rapidly toward each other. If the antibody reacts with the antigen (such as with human blood and antihuman antigen), a milky white precipitate forms in the gel. The advantages of crossed-over electrophoresis compared to other methods of species determination are its speed and its ability to do many sample pairings at once.

cross-examination In a courtroom proceeding, the process of questioning a witness who was called by the opposing party.

crystal tests (microcrystal, microcrystalline tests) PRESUMPTIVE TESTS (and in some cases specific tests) used as part of DRUG ANALYSIS. There are also crystal tests used to confirm the presence of blood through reactions with hemoglobin as well as tests for EXPLOSIVES. Crystal tests are performed by placing a small amount of the sample or sample extract on a microscope slide and adding the appropriate reagent. The shape of the crystals is examined by a COMPOUND MICROSCOPE or a polarizing light microscope.

crystal violet A histological stain used to study cell structures and tissues and to aid in visualizing latent fingerprints.

CSF1PO A gene locus classified as a SHORT TANDEM REPEAT (STR) that is typed in current deoxyribonucleic acid DNA TYPING procedures. It is one of the 13 loci included in the CODIS system.

CSI A television show broadcast on the CBS network that has popularized forensic science, particularly with younger audiences. In 2002, *CSI* (the acronym stands for *crime scene investigation*) was the top-rated television drama; it led to the creation of another series, *CSI Miami*.

CT scan Computed tomography scan, a medical imaging technique informally known as a "CAT" scan. CT is occasionally used as part of autopsy to image a body or remains.

Culliford, Bryan (1929–1997) A British forensic serologist best known for his book *The Examination and Typing of Bloodstains in the Crime Laboratory*, which was published in 1971 while he was with the Metropolitan Police Laboratory in London. The manual, which collected information on BLOOD GROUP typing and typing of ISOENZYME SYSTEMS, became a primary reference in forensic serology worldwide.

Until the recent ascension of DNA TYPING, typing of blood groups and isoenzyme systems was the primary tool available for classifying blood and identifying possible sources of stains.

cuticle The outermost part of a HAIR, consisting of scalelike structures that cover the CORTEX. The scale pattern of the cuticle provides important information for classification of a hair.

cutting agents (diluents) Materials used to dilute drugs such as COCAINE and HEROIN. Cutting agents, also known as diluents or adulterants, range from CAFFEINE to flour, starch, and sugars such as table sugar (sucrose), dextrose, mannitol, inositol, and fructose. Many can be identified microscopically, whereas others such as caffeine require chemical analysis for definitive identification.

CV (coefficient of variation/%RSD) A statistical quantity applied to a set of replicate measurements obtained by dividing the standard deviation by the mean and multiplying by 100. It is a measure of PRECISION or REPRODUCIBILITY; the smaller the %RSD, the closer the replicate measurements are to each other, that is, the smaller the spread of the individual data points.

cyanide A deadly poison that can kill by ingestion, inhalation, or absorption through the skin. The cyanide ion consists of a carbon atom and a nitrogen atom bonded by a strong triple bond and represented as CN⁻. The most common forms of cyanide encountered in forensic science and TOXICOLOGY are the powders sodium cyanide (NaCN) and potassium cyanide (KCN), both of which can react in the presence of carbon dioxide or acids to form the acid HCN, also known as hydrogen cyanide, hydrocyanic acid, or prussic acid.

cyanoacrylate Also known as Super Glue, a chemical (specifically cyanoacrylate ester) used to visualize FINGERPRINTS

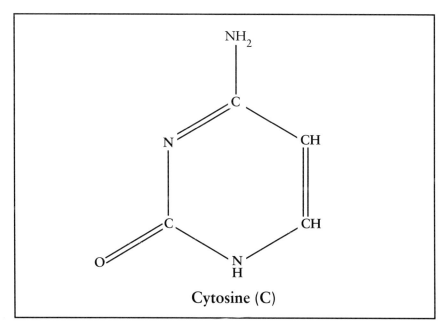

Cytosine (C)

The structure of cytosine (C), one of the four bases found in DNA and the complement of the base guanine (G).

on a variety of surfaces, including plastic, wood, leather, and human skin. Introduced in the early 1980s, cyanoacrylate fuming is easy to perform and has become one of the most common methods of developing latent fingerprints.

cystolithic hairs A microscopic feature found on the upper surface of the leaves of MARIJUANA (*Cannabis sativa*). Also known as bear claws because of their distinctive shape, cystolithic hairs have a relatively broad oval base supporting the clawlike structure that encases an aggregation of calcium carbonate ($CaCO_3$, the cystolith). Since it is found on the leaves, cystolithic hairs are indicative of marijuana and HASHISH, but since marijuana is not the only plant that has cystolithic hairs, they do not constitute a conclusive test.

cytosine (C) One of four NUCLEOTIDE bases that compose deoxyribonucleic acid (DNA) and ribonucleic acid (RNA). Because of its molecular structure, cytosine associates with guanine (G), and the two are referred to as complements of each other. (See figure on page 49.)

D13S317 A gene locus classified as a SHORT TANDEM REPEAT (STR) that is typed in current DNA TYPING procedures. It is one of the 13 loci included in the CODIS system.

D16S539 A gene locus classified as a SHORT TANDEM REPEAT (STR) that is typed in current DNA TYPING procedures. It is one of the 13 loci included in the CODIS system.

D18S51 A gene locus classified as a SHORT TANDEM REPEAT (STR) that is typed in current DNA TYPING procedures. It is one of the 13 loci included in the CODIS system.

D21S11 A gene locus classified as a SHORT TANDEM REPEAT (STR) that is typed in current DNA TYPING procedures. It is one of the 13 loci included in the CODIS system.

D3S1358 A gene locus classified as a SHORT TANDEM REPEAT (STR) that is typed in current DNA TYPING procedures. It is one of the 13 loci included in the CODIS system.

D5S818 A gene locus classified as a SHORT TANDEM REPEAT (STR) that is typed in current DNA TYPING procedures. It is one of the 13 loci included in the CODIS system.

D7S820 A gene locus classified as a SHORT TANDEM REPEAT (STR) that is typed in current DNA TYPING procedures. It is one of the 13 loci included in the CODIS system.

D8S1179 A gene locus classified as a SHORT TANDEM REPEAT (STR) that is typed in current DNA TYPING procedures. It is one of the 13 loci included in the CODIS system.

DAB Diaminobenzidine, a compound that has been used to develop latent fingerprints in blood.

Dacron A common type of polyester fiber.

dactyloscopy An older name for the general study of fingerprint and friction ridge patterns.

dAMP An abbreviation for the compound deoxyadenosine monophosphate, the NUCLEOTIDE that consists of the amino acid ADENINE, ribose (a sugar), and a phosphate group (PO_4^{3-}). The structure of dAMP is shown in the figure on page 52.

dandy roll A rolling mechanism used in the production of some types of papers. Pulp is placed on the roll and pressed to dry; occasionally marks that correspond to imperfections on the dandy roll are made in the paper.

Darvon (propoxypene) A prescription painkiller that was once widely abused.

date rape drugs (predator drugs) Drugs used to render victims unconscious or otherwise unable to resist sexual assaults. For this reason, these drugs are also called "predator drugs." Many drugs have been used in this way, most notably ALCOHOL, but recently several other drugs have received the attention of law enforcement and forensic chemists. Of these, two are currently of most concern, ROHYPNOL and GHB.

Deoxyadenosine Monophosphate (dAMP)

The structure of the nucleotide created when the adenine is bonded to the sugar ribose and a phosphate group. It is one of the four nucleotides found in DNA.

***Daubert* decision (1993)** A United States Supreme Court decision regarding the admissibility of scientific evidence. *Daubert v. Merrell Dow Pharmaceuticals* (113 S. Ct. 2786 [1993]) involved a case in which parents claimed that the birth defects of their children were the result of a morning sickness medication. The ruling of the court significantly altered the general approach to the admissibility of scientific evidence that had been set forth in a 1923 case known as the FRYE DECISION. In *Frye*, the court had ruled that in deciding whether to accept new or novel scientific methods and the accompanying testimony into evidence, the primary consideration was general acceptance by the scientific community. In the *Daubert* ruling, the court stated that it is the responsibility of the trial judge to determine whether scientific evidence is relevant and reliable. This role assigned to the judge is often referred to as that of "gatekeeper," and the court offered suggestions for making that determination, while leaving flexibility to the judges. In 1999, the court extended *Daubert* in the case of KUMHO TIRE V. CARMICHAEL to cover all expert testimony, not just that of scientific experts.

Daubert hearing   A hearing held by a court to determine the admissibility of scientific evidence and expert testimony using the standards set forth in the *DAUBERT* DECISION.

dCMP   An abbreviation for the compound deoxycytidine monophosphate, the NUCLEOTIDE that consists of the amino acid CYTOSINE, ribose (a sugar), and a phosphate group (PO_4^{3-}). Its structure is analogous to that shown for dAMP.

DEA (Drug Enforcement Administration)   Law enforcement agency administered by the DEPARTMENT OF JUSTICE responsible for investigation and enforcement of federal drug laws, primarily the CONTROLLED SUBSTANCES ACT. The DEA is also responsible for drug intelligence operations, seizure of assets related to drug trafficking, and coordination of federal, state, and local law enforcement activities related to drug crimes.

death   In a forensic context, death can be considered as two distinct events, starting with somatic death, in which the heart and breathing stop and brain activity ceases. Next occurs a process of cellular death and decay called AUTOLYSIS. At somatic death, body chemistry begins to change rapidly and certain enzymes that promote degradation of molecules such as proteins and carbohydrates become active. At this point, DECOMPOSITION has begun.

death: cause, manner, and mechanism   The cause of death is that event or process such as a disease or injury that led to the death. The mechanism of death is set in motion by the cause of death. For example, a stab wound would be a cause of death and the mechanism of that death could be internal bleeding. The manner of death is the category that describes the circumstances that led to infliction of the cause of death. The categories used are homicide (intentional killing by someone other than the victim), suicide (self-homicide), accidental, natural (disease), treatment induced (for example, death during surgery or drug interaction), and indeterminate. The acronym NASH is often used for possible manner of death; natural, accidental, suicidal, or homicidal.

death investigation   *See* CORONER; MEDICAL EXAMINER.

death investigator   A person, usually employed by the CORONER or MEDICAL EXAMINER, who performs the initial investigation of an unattended, sudden, unexpected, or otherwise suspicious death. The investigator's role is to determine whether further investigation is warranted.

deception analysis, forensic (FDA)   The application of psychological principles and techniques to determine the credibility of a person's words or actions as they apply to legal matters. Fundamentally, deception analysis is an attempt to determine whether a person is lying, be it in statements, actions, or symptoms. An example of the latter would be a person's pretending to have a mental or physical illness.

deciduous teeth   *See* PRIMARY DENTITION.

decomposition   The sequential process that occurs after death and ends with skeletonization. Although the phases of decomposition are known, environmental factors, principally temperature, dramatically affect its specific path and rate. As a result, the progress of decomposition is not alone sufficient or reliable for estimating the POSTMORTEM INTERVAL. Rather, it is one of many tools used to estimate the interval since death occurred. In these early stages of decomposition, microorganisms already present in the gut multiply and spread. The result is gas production that causes a bloating in the abdomen and then generalized bloating that can swell the body to twice or three times its original size before collapsing. The bloating can cause the eyes and tongue to swell and protrude as well. The appearance of the skin changes, passing through stages of green before reaching black, a stage appropriately named *black putrefaction.* Bacteria flourish in blood, which is an excellent medium for supporting them, accelerating the breakdown process. During this stage of decomposition, called putrefaction, characteristic odors of decomposition are given off. Stages of advanced decomposition can be reached in 12–18 hours in warm and humid environments, but in cold or freezing areas, the process slows dramatically, stretching into weeks, months, or even years, depending on temperature. Putrefaction gives way to dry decay or mummification, again depending on environmental conditions. Eventually, most or all tissue is dried and decays away, leaving only bones. When skeletonization is complete, the decomposition process is complete. *See also* ENTOMOLOGY, FORENSIC; TAPHONOMY.

dedicated dimensional standard A ruler, usually with both English and metric units, that is clearly labeled and used in photographs of physical evidence.

deductive reasoning A common form of logic applied in forensic science. In deductive reasoning, conclusions are based on evidence and facts that are already established. The following statement is an example of deductive reasoning: All animals that have spines are vertebrates. Humans are vertebrates, so it follows that humans have spines. Deductive logic and the complementary INDUCTIVE LOGIC are both part of the scientific method and are both utilized in forensic work.

defendant The accused in a criminal proceeding; in a civil case, the person or party being sued.

defense attorney The attorney who represents the defendant (suspect) in a criminal trial or the defendant in a civil action. In criminal cases, defendants may be appointed a lawyer from the Public Defender's office (a governmental agency) or may hire their own attorney. In high-profile cases, use of a team of defense lawyers, as in the O. J. SIMPSON case, is not uncommon. The defense lawyer has the right to question expert witnesses and to ensure their credibility and reliability through a process called VOIR DIRE.

defense experts Defendants in criminal and civil cases are entitled to have physical evidence analyzed by experts other than those employed by the agency responsible for their prosecution. Forensic labs strive to preserve at least half of any samples that they analyze for such a purpose. If that is not possible, they may be required to obtain the permission of the DISTRICT ATTORNEY (DA) before proceeding. Defense experts may work in a variety of organizations, including other government or private forensic labs, or may be consultants from a university or the private sector.

defense wounds Wounds found on a body or living person that resulted from defensive actions taken during a struggle. For example, when a victim is attacked by a person wielding a knife, one common defensive move is to grab the blade. This results in distinctive cuts on the hands and fingers that would be classified as defense wounds.

deflagration Very rapid burning with extreme heat; combustion that is one step away from DETONATION.

dehyrogenases A class of ENZYMES that catalyze biological oxidation/reduction (REDOX) reactions. Specifically, dehyrogenases catalyze the removal of a hydrogen atom from a molecule, in a process considered to be an OXIDATION.

delta A friction ridge feature found in fingerprints; the name originates from comparison of ridge patterns to flowing streams or rivers. As in a river, a delta is found wherever two lines diverge. The delta is the ridge point that is closest to the point of divergence. Occasionally, the letter delta (Δ) is also used to express a difference between two measurements; for example, the term ΔT could stand for the difference between two temperatures and would be calculated by subtracting one from the other: $\Delta T = T_2 - T_1$.

Demerol (meperidine) A synthetic opiate ALKALOID that is a narcotic with effects similar to those of MORPHINE. Demerol is widely used for pain relief in hospitals and is sometimes abused in ways other narcotics such as HEROIN are. Meperidine was first marketed as the product Demerol in 1939.

demography (demographics) The statistical study of human populations. In forensic science, demography is important in determining frequencies of blood types (BLOOD GROUP SYSTEMS and ISOENZYMES) and DNA types. In many systems, frequencies differ, for example, among Caucasians, African Americans, Hispanics, and Asians, and this difference must be taken into account when blood typing and DNA TYPING analyses are evaluated. The classes as written are the ones used most in forensic. It refers (usually) to just Americans.

demonstrative evidence Physical evidence (as opposed to testimony) that has bearing on the question at hand.

denaturation In proteins, the loss of three-dimensional shape and structure. This occurs when HYDROGEN BONDING interactions are weakened, altered, or destroyed by factors such as temperature or changes in pH. The cooking of an egg white causes denaturation and results in conversion of a liquid egg white into a solid. As part of DNA TYPING, heat or chemical compounds are used for denaturation, breaking the double helix into single strands for copying.

dendrimer A polymer with multiple branching that can encapsulate or enclosure other nanostructures such as small crystals or molecules. Dendrimers that incorporate particles such as CADMIUM SULFIDE are used to visualize latent fingerprints.

density The mass (m) of an object or substance divided by its volume (v) ($d = m/v$). The standard unit of density is grams per milliliter (g/mL). The density of water at 25°C, 1.00 g/mL, is used as a benchmark for comparison. Small amounts of dense materials (such as lead) are heavy, having a density of greater than 1.00, and sink when placed in water. Conversely, comparable amounts of low-density materials such as foam are light, have a density of less than 1.00, and float. In forensic science, density of materials such as SOIL samples and GLASS can be used to help classify them.

density gradient Glass or plastic tubes filled with liquids of known densities. These were at one time widely used to determine relative densities of evidence such as GLASS and SOIL.

dental profile A description of the likely appearance of a person's teeth and DENTITION based on evidence such as a bite mark. Dental profiles are provided by forensic ODONTOLOGISTS (dentists).

dental stone A plasterlike material used to create casts of three-dimensional impressions such as tire and shoe prints. It is used by dentists to make casts of a patient's DENTITION. Dental stone is sold as a powder and mixed with water to the consistency of thin pancake batter before being poured into a mold around the impression. It hardens quickly, within a few minutes, and creates a durable cast.

dentin (dentine) The substance that makes up most of the structure of a

TOOTH. The dentine covers the root and is itself encased in the enamel.

dentistry, forensic *See* ODONTOLOGY.

dentition The collective structure and appearance of a person's teeth including dental work such as fillings and crowns. *See also* ODONTOLOGY; TOOTH.

Department of Justice (DOJ) The agency that coordinates federal law enforcement. The Department of Justice houses the FEDERAL BUREAU OF INVESTIGATION and the DRUG ENFORCEMENT ADMINISTRATION, both of which play critical roles in forensic science. The department also administers grant programs that help fund research in forensic science.

deposition A sworn statement that is given under oath but not in a courtroom setting. The person documenting the deposition is an officer of the court, and the statement is taken by a court reporter. Occasionally, forensic scientists testify in a case by way of a deposition when traveling, scheduling, or other unavoidable conflicts prevent or prohibit them from appearing in open court. Attorneys or representatives for both the prosecution and the defense are usually present, allowing for questioning much as would take place in open court.

depressants Substances that depress the central nervous system (CNS) and can produce effects including loss of coordination, impairment of judgment, and sleep. ALCOHOL is a depressant, as are the BARBITURATES, tranquilizers, BENZODIAZEPINES such as diazepam (Valium), INHALANTS, and methaqualone (Quaalude).

depth of field In a microscopic examination, the distance or depth that the

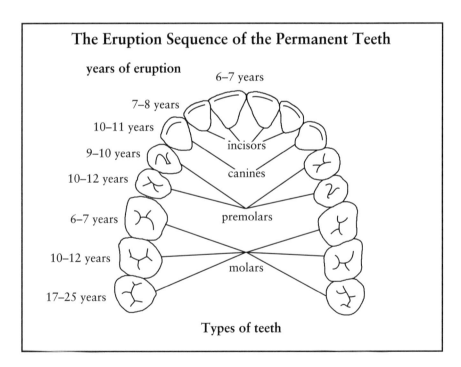

The Eruption Sequence of the Permanent Teeth

years of eruption

6–7 years

7–8 years

10–11 years

9–10 years

10–12 years

incisors

canines

6–7 years

premolars

10–12 years

molars

17–25 years

Types of teeth

Human dentition. All teeth are shown, along with identification of type and age when they normally appear (erupt). This information can be used to estimate the age of skeletal remains.

object being studied can be moved vertically without losing significant focus.

depth of focus In a microscopic examination, the depth into the object being studied to which focus can be maintained. As magnification increases, depth of focus decreases.

derivative/derivatization A new chemical compound made from an existing one, usually to facilitate chemical analysis and the process for creating it. Derivatization is used in certain types of chromatography such as GAS CHROMATOGRAPHY and HIGH-PERFORMANCE LIQUID CHROMATOGRAPHY.

dermal nitrate test A PRESUMPTIVE TEST for the presence of GUNSHOT RESIDUE (GSR), also called the paraffin test, that is no longer used. Whenever a person fires a gun, traces of residue are transferred to the hands. The dermal nitrate test was based on detecting the nitrate ion (NO_3^-), an ingredient in PROPELLANTS used in AMMUNITION. In one variant, the hands of the suspect were painted with hot wax (paraffin) that was allowed to dry. The cast was then removed and tested by using the reagent diphenylamine combined with sulfuric acid (H_2SO_4). Locations on the cast that showed a blue color indicated the possible presence of nitrate. In addition to gunshot residue, nitrates are common in many materials that may be found on the hands including tobacco, cosmetics, fertilizers, and urine. This large number of potential FALSE POSITIVE findings caused the test to be abandoned since it was not sufficiently specific to GSR.

dermis The inner layer of skin (the outer is the EPIDERMIS), which constitutes most of the mass of skin. The dermis is a dense layer of connective tissue that contains fibrous and elastic tissues and COLLAGEN. The dermis contains the majority of the sweat and oil glands as well.

design element The individual raised plugs or "islands" of rubber found on a tire. The design of a tire tread is made up of these individual design elements.

designer drugs Illegal drugs synthesized in CLANDESTINE LABS that are closely related to controlled substances in structure and effect. The first designer drugs were derivatives of methamphetamine and AMPHETAMINE and appeared in the 1970s. 3,4-Methylenedioxymethamphetamine (MDMA) is also known as ECSTASY; methylenedioxyamphetamine (MDA) is the amphetamine equivalent. Because of their similarity to existing illegal substances, designer drugs are also known as analogs.

destructive interference The interference of two waves of electromagnetic energy, which produces a canceling-out effect on the original waves. It is the opposite of CONSTRUCTIVE INTEREFERENCE. Patterns of constructive and destructive interference are important in MICROSCOPY, optics, and many types of SPECTROPHOTOMETRY.

destructive testing Testing or analysis that results in complete consumption or destruction of the original sample. Most tests performed in forensic chemistry and biology are destructive; however, since these tests typically require very little sample, untouched material remains for further analysis.

detection limit (limit of detection [LOD]; lower limit of detection [LLD]) In a chemical, biochemical, or biological analysis, the limit of sensitivity or detection of that technique. For example, an infrared spectrophotometer may require several nanograms (ng) of a substance to obtain usable data, whereas a GAS CHROMATOGRAPH/MASS SPECTROMETER can often detect femtograms (fg) of material. The LOD of a method takes into account all aspects, from sample extraction and preparation through to instrumental analysis. Thus, although an instrument or technique may have extremely good sensitivity, it is often the sample preparation that ultimately controls the LOD.

detectors Instrumental components that detect the presence of certain compounds introduced into the instrument.

Almost all detectors work by creating an electrical signal proportional to the detector response. For example, the detector in a SPECTROPHOTOMETRIC instrument converts electromagnetic energy to electrical current. Similarly, a FLAME IONIZATION DETECTOR (FID) used for GAS CHROMATOGRAPHY works by creating CHO⁺ ions in the gas stream and detecting the electrical current created when these charged species collide with a collector plate. Two criteria used to describe detectors are specificity (what does the detector respond to?) and sensitivity (how low a concentration can it detect?) An FID detector is selective for compounds that contain carbon–hydrogen bonds and can detect nanograms of material (1 ng = one-billionth gram).

deterrent (deterrent coating) Chemical treatments of propellants designed to control the rate of burning and performance of the ammunition. Different weapon sizes (calibers) and necessary projectile velocities, along with other considerations, determine optimal burn rate. Fundamentally, the role of the propellant is to burn rapidly to produce large quantities of hot, expanding gas that propel the projectile down and out of the barrel of the weapon at optimal speed. Excessively fast burning produces gases too quickly, whereas long burning continues even after the projectile has exited the barrel.

detonation An extremely rapid combustion reaction that produces a shock wave of hot expanding gases. It is a violent form of combustion and is the next step up from DEFLAGRATION.

detonation cord (det cord) A cord with a core that consists of RDX or PETN (HIGH EXPLOSIVES) wrapped in cotton or other fabric and sealed with a waterproof material such as plastic. The function of det cord is to set off other high explosives. The det cord is detonated by using a BLASTING CAP.

developers Materials (solids, liquids, or gases) that are used to visualize latent FIN-GERPRINTS or bloodstains. Example developers for fingerprints include black powder and NINHYDRIN; a well-known developer for bloodstains is LUMINOL.

dextrose An informal name for glucose ($C_6H_{12}O_6$); also called blood sugar.

DFO A developer for latent fingerprints. DFO (1,8-diazafluoren-9-one) serves as a substitute for NINHYDRIN. As ninhydrin does, DFO reacts with the AMINO ACIDS in the skin to create a red color. Additionally, the compound that is formed by this reaction fluoresces strongly when exposed to green light.

dGMP An abbreviation for the compound deoxyguanosine monophosphate, the NUCLEOTIDE that consists of the amino acid GUANINE, ribose (a sugar), and a phosphate group (PO_4^{3-}). Its structure is analogous to that shown for dAMP.

diaphysis The central portion (shaft) of a long BONE. *See also* EPIPHYSIS.

diastereoisomers Molecules that have the same formula, but different arrangements of the atoms in space. Stereoisomers have the same formulas and same geometric characteristics. Diastereoisomers are a special class of stereoisomers in which the mirror image of one cannot be superimposed on the other. Diastereoisomers have different physical and chemical properties and are sometime an issue in cases involving COCAINE. Cocaine has two enantiomers, *d*-cocaine and *l*-cocaine, which are mirror images of each other. There are three other such pairs of cocaine enantiomers, *d*- and *l*-pseudococaine, *d*- and *l*-allococaine, and *d*- and *l*-pseudoallococaine. These pairs are all diastereoisomers of cocaine.

diatomaceous earth (diatomite [DE]) A powdery material derived from limestone that contains skeletonized remains of microscopic DIATOMS, specifically their hard outer shell. DE is used as a filter medium such as in swimming pool filters and is sometimes used to insulate safes.

diatoms A form of algae that can be a source of TRACE EVIDENCE. Diatoms exist in fresh and salt water and thus can be important evidence in drowning cases, although use of diatom evidence is still somewhat controversial. If the heart is beating when a person enters or is forced into the water, the circulatory system can deliver the diatoms to parts of the body such as the liver, kidney, or bone marrow. Thus, finding diatoms in tissues far removed from the lungs can be interpreted to mean that the person was alive when he or she entered the water. Diatoms are characterized by a hard outer shell make of silicates.

diazepam (Valium) One of the most widely prescribed drugs in the world and one frequently abused. Valium is classified as a depressant and is a tranquilizer in the BENZODIAZEPINE family, which includes lorazepam (ATIVAN) and alprazolam (XANAX). Valium is manufactured as small pills and is used to treat anxiety without promoting sleep to the degree that BARBITURATES do.

dichroism A property of materials such as fibers in which the colors that are absorbed (and thus the colors that the material appears to be) is a function of direction. Dichroism is important in forensic MICROSCOPY and in the analysis of TRACE EVIDENCE and TRANSFER EVIDENCE. Specifically, in a dichroic material, the absorption pattern varies, depending on the direction or orientation of the material or on the orientation of polarized light that illuminates it. Although dichroism is often associated with colors and thus visible light, infrared absorption patterns can also be dichroic. To observe dichroism in a material, it is mounted on a microscope slide and examined by using polarized light (a POLARIZING LIGHT MICROSCOPE.) As the stage is rotated, the color of the material changes as the pattern of absorbance changes.

diffraction grating A device, also referred to as a reflection grating, used in many types of spectrophotometers to disperse electromagnetic radiation into component wavelengths (a wavelength filter). For colorimetry (spectrometry in the visible range), gratings have replaced prisms, which formerly served the same purpose, breaking white light into component colors. Gratings work by setting up patterns of CONSTRUCTIVE and DESTRUCTIVE INTERFERENCE of light that separate individual wavelengths and disperse them to different physical locations. Gratings are made of glass or metal-coated glass that has a series of fine grooves machined into the surface; its performance is defined by the grating equation $n\lambda = d(\sin\theta - \sin\phi)$, which means that for a given grating, for each angle if incidence (θ), constructive interference for a specific wavelength λ and reflection angles ϕ will occur. All other wavelengths will undergo destructive interference, and the result will be the physical dispersion of white light into component wavelengths. (See figure on page 60.) A similar phenomenon occurs when one holds a compact disk to the light and moves it around; the colors that are seen are the result of dispersion of white light on the surface of the disk.

diffusion The natural process of movement of materials from areas of high concentration to areas of lower concentration of that same material. For example, if a drop of food coloring is dropped into a glass of still water, it diffuses naturally until it is evenly distributed throughout the glass. This property is exploited in forensic techniques such as IMMUNODIFFUSION.

digestion A process used during sample preparations to break chemical bonds and to free the target ANALYTE or analytes from a complex matrix.

digital photography/digital imaging See PHOTOGRAPHY.

Dillie-Koppanyi test A PRESUMPTIVE TEST for BARBITURATES. It is a modification of an older test called the Zwikker test. The two-part test starts with the addition of a solution of cobalt acetate in methanol to the unknown sample (powdered), followed by the addition of a

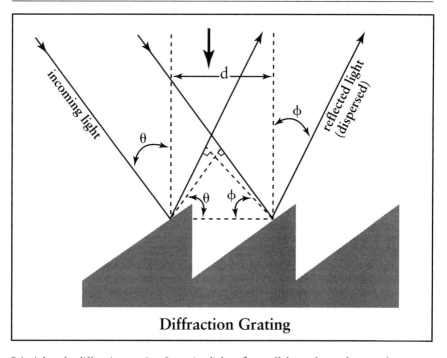

Diffraction Grating

Principles of a diffraction grating. Incoming light reflects off the surfaces of sawtooth grooves etched into the surface of the grating, creating zones where constructive and destructive interference occur, marked by the arrow.

methanolic solution of isopropylamine. A reddish violet color is indicative, but not definitive, of the presence of barbiturates.

diluents   *See* CUTTING AGENTS.

dinitrotoluene (DNT)   A high explosive that is used in dynamite made in Europe. DNT is either 2,4-DNT or 2,6-DNT depending on where the two nitro groups are attached to the toluene ring. DNT is also used as a standard in some types of explosive analyses.

dinucleotide repeat   A repeating pattern of two NUCLEOTIDES that are found side by side on DNA. An adenine (A), cytosine (C), and guanine (G) sequence such as ACGGACGG contains repeats of the dinucleotide pairs AC and GG.

diphenylamine test   A PRESUMPTIVE TEST for the presence of GUNSHOT RESIDUE. A solution of diphenylamine in sulfuric acid (H_2SO_4) is added to suspected residue; a blue color indicates the possible presence of nitrates (NO_3^- ions), which are ingredients in the PROPELLANTS used in AMMUNITION. The test is not specific, and the reagent reacts with nitrites (NO_2^- ions) and other oxidizing reagents. The reagent was used as part of the DERMAL NITRATE TEST, now abandoned because of its lack of specificity. The diphenylamine test can also be used in the analysis of paints, in which NITROCELLULOSE can be an ingredient.

Diptera   An order of insects that include tens of thousands of species of flies. Some, such as the blowfly, can be important indicators of the POSTMORTEM INTERVAL. *See also* ENTOMOLOGY, FORENSIC.

direct evidence Evidence that does not require any interpretation to reach a definitive conclusion, as opposed to CIRCUMSTANTIAL EVIDENCE. For example, if a suspect is found to possess a white powder and forensic analysis shows that the powder is COCAINE, that is direct evidence of a crime (possession of a controlled substance). Other types of evidence that are considered direct evidence include DNA TYPING results and FINGERPRINTS if linked to an individual. The term *direct evidence* is also used informally to indicate evidence that is considered to be very strong.

direct examination The first round of questions directed to a witness. The "direct," as it is often called, is conducted by the attorney who called the witness.

directionality The direction of travel of an object such as a bullet or a fluid such as blood that is or was at one time in motion. In BLOODSTAIN PATTERN analysis, the shape of a blood drop often indicates the directionality of the blood or the direction in which it was moving when it struck the surface.

directionality angle In BLOODSTAIN PATTERN analysis, the angle between a line drawn down the long axis of a blood drop and a reference point or line. For example, a reference line might be established on a floor along a wall, and the directionality angle of blood spots on the floor of the room could be determined by using that reference line to represent 0 degree.

disarticulation Literally, "disjointedness." In forensic applications, the term refers to a body that has been broken up or separated. As soft tissues such as ligaments and tendons decay or are torn, the joints come apart and the bones can become scattered, either by natural processes or by animals (scavengers). A corpse may also be purposely disarticu-lated to hide or otherwise dispose of the body.

discourse analysis An area generally classified under forensic LINGUISTICS devoted to courtroom transcriptions or other legal statements such as confessions for accuracy, perception, intent, and meaning. For example, the exclamation "Drop it!" might refer to a weapon or to a request for someone to change the topic of conversation. How such a statement was meant and how it was interpreted could be critical. Discourse analysis can also be important when translations are involved since the translator must often use individual judgment to select words or phrases in one language to express the meaning of another. Thus, two translators can take exactly the same statement and derive two different translations with differences that might seem subtle in one language but substantial in the other. Thus, word selection, sentence construction, and other linguistic elements become critical in conveying meaning.

discovery (disclosure) The pretrial process of revealing information to and among the prosecutor, the defendant, and the defense attorney. Although the rules of discovery vary by jurisdiction, in general, the defendant has a right to see any evidence incriminating to him or her, including reports and analysis performed in a forensic laboratory. In addition, the prosecution has an obligation to reveal to the defendant any evidence that is favorable to his or her case. In criminal cases, the defense may have a right to keep some test results confidential, but in civil cases, full disclosure/discovery is generally required of both parties. Many jurisdictions model their requirements on Federal Rules of Discovery.

discriminant function analysis (DFA) A form of multivariate statistical analysis (application of statistics to problems with two or more variables) that can be used to categorize items in terms of the value of those variables. For example, DFA could be used to categorize 100 different glass samples on the basis of color, density, texture, and other variables.

discrimination index (discrimination power) The ability of a given test to INDIVIDUALIZE evidence on the basis of probability. For example, in forensic biology and serology, the discrimination index is the probability that any two people selected at random in a given population will have *different* blood types. In the U.S. Caucasian population, approximately 42 percent have the ABO BLOOD GROUP type of A, 43 percent O, 12 percent B, and 3 percent AB. The discrimination index of the ABO blood group system is approximately 0.60, meaning that there is a 60 percent chance that any two people selected at random will have a different ABO type. Similarly, the isoenzyme system phosphogluco mutase (PGM) has three types (discounting subtypes)—1-1, 59 percent; 2-1, 35 percent; and 2-2, 6 percent—and a discrimination index of 0.52. Thus, PGM has a lower discrimination index and is less likely to distinguish two individuals selected at random. Another way to state this would be to say that PGM has a lower power of discrimination since nearly two-thirds of the population is PGM type 1-1. Discrimination indices can also be calculated for combined types.

disguised writing Handwriting that is purposely altered to make it difficult to identify the writer. This is differentiated from forgery, in which the writer is attempting to copy the handwriting of another. A familiar example of disguised writing is writing with the "wrong" hand. Ransom notes can fall into this category as well.

disorganized crime (disorganized criminal) A crime that shows a lack of planning and a spontaneous nature. A crime and/or victim over which the perpetrator had little control.

distal A point or direction that is farthest from the anchor or reference point. The distal end of a hair is the end that is farthest away from the root.

distance determination Although this can be a generic term for the process of measurement, in forensic science it most often means determining the distance between a firearm and a target when the weapon was discharged. Such determinations are useful in CRIME SCENE reconstruction and for support or refutation of different stories related to a shooting. Distance determinations are conducted by FIREARMS examiners, who must take into account environmental conditions and other variables when performing these tests.

distance wound A gunshot wound inflicted by a shot fired from farther away than approximately 18 inches. In contrast to a CONTACT WOUND, a distance wound has little or no stippling, has no loose powder, and lacks the stellar tearing pattern of the skin. *See also* BULLET WOUND.

district attorney (DA) The attorney appointed by a governmental entity (state, city, or county) who has responsibility for criminal prosecutions in his or her district. Forensic scientists, especially those employed by the same government entity, often work closely with the DA or one of the deputy DAs who handle cases.

D-loop A region of the DNA found in MITOCHONDRIAL DNA (mtDNA) that is typed in forensic mtDNA analysis. It is classified as a hypervariable region of DNA, making it a powerful tool for potential INDIVIDUALIZATION.

DMAC A compound with the chemical name *p*-dimehtylaminocinnamaldehyde that reacts with urea. Accordingly, it has been used as a reagent for a PRESUMPTIVE TEST for urine and for visualization of latent FINGERPRINTS.

DMORT Disaster Mortuary Operational Response Team. A team of professionals including forensic pathologists, anthropologists, and odontologists (forensic dentists) who respond to MASS DISASTERS such as airline crashes and terrorist attacks. DMORTs are organized by the federal government, specifically the Federal Emergency Management Agency (FEMA).

DNA (deoxyribonucleic acid) The molecule that makes up genes and carries inherited information. As shown in the figure, DNA is a POLYMER made up of a chain of NUCLEOTIDES, which consist of bases (cyclic molecules that contain nitrogen) connected to a sugar–phosphate backbone. The sugar in DNA is ribose, which assumes a pentagonal shape in the backbone structure. In DNA, there are four bases that can be incorporated into a nucleotide: ADENINE (A), THYMINE (T), CYTOSINE (C), and GUANINE (G). Through a process called HYDROGEN BONDING, the bases adenine and thymine are attracted to each other, as are cytosine and guanine.

This attraction ultimately leads to the double-helix shape of DNA.

DNA is composed of complementary strands of nucleotides in which the base pairs organize themselves opposite their partner. The interactions of the A with T and C with G pairs cause the two strands to twist around each other in a helical coil. The double-helix shape of DNA was first proposed in 1953 by Francis Crick (an Englishman) and James Watson (an American), although their work was the culmination of much research in the field. As shown in the illustration, DNA can be replicated by first "unzipping" the separate strands, a process that requires enzymes. Once they are separated, new complementary strands can be synthesized from nucleotides that arrange themselves

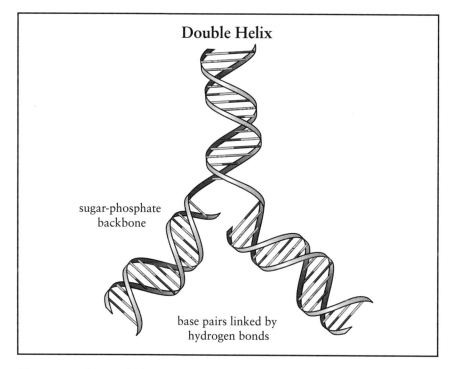

Double Helix

sugar-phosphate backbone

base pairs linked by hydrogen bonds

The structure of DNA. The backbone of the ladderlike structure is composed of the sugar-phosphate groups of the four nucleotides based on adenine (A), thymine (T), cytosine (C), and guanine (G). The rungs of the ladder are composed of complementary base pairs (A-T and C-G) held together by hydrogen bonding, three for C-G and two for A-T. These interactions lead to a twisting of the ladder structure to create the double-helix configuration of DNA.

in the same opposite pairings as in the original strand, creating two copies of the original or parent DNA molecule. In this way, each parent strand acts as a template for a new one.

DNA Advisory Board (DAB) A group formed as a result of passage of the DNA Identification Act of 1994. Members of the DAB were appointed by the FBI, and the standards they recommended took effect in 1998. These standards related to QUALITY ASSURANCE/QUALITY CONTROL for forensic DNA TYPING.

DNA fingerprint An informal term used to describe the combined set of a person's DNA types at the loci that are typed. Currently, this would refer to the 13 genetic loci typed for the CODIS system.

DNA probe A short segment of synthetic DNA used in DNA TYPING. The function of the probe is to find specific segments of DNA that will be amplified during POLYMERASE CHAIN REACTION (PCR) procedures. These probes are labeled by using radioactive or fluorescent tags and bind to the target DNA segment through complementary base pair interactions. For example, if the target sequence of NUCLEOTIDES were A-T-T, the probe would have a sequence of T-A-A. The primer acts to mark the boundary of the DNA region that will be amplified.

DNA typing (DNA analysis) A group of related procedures that have largely replaced traditional blood typing in forensic labs. DNA typing techniques were pioneered by molecular biologists and entered the forensic arena in the late 1980s and early 1990s. Since then, they have grown quickly into the tool of choice for the analysis of blood and body fluids. Rapid advances in the field continue, so DNA typing applications and techniques change and evolve. Smaller and smaller samples can be analyzed (currently as little as a billionth of a gram of DNA is needed), the DISCRIMINATING POWER (the ability of a typing test or tests to INDIVIDUALIZE a sample) is increasing. In addition to BLOOD and body

fluid stains, DNA typing can be used on a variety of different samples including HAIR, skin scrapings, and even dandruff. The first methods used were based on RESTRICTION FRAGMENT LENGTH POLYMORPHISMS, which have largely been replaced by methods based on SHORT TANDEM REPEATS (STRS) using a POLYMERASE CHAIN REACTION preparation. The FBI coordinates a database called CODIS, which contains types for 13 such genetic loci. DNA also exists outside the nucleus of the cell and can be typed by using MITOCHONDRIAL DNA typing techniques.

dNTP (dinucleotide triphosphates) A mixture of nucleotides that are added to the solution of DNA that is being amplified during the POLYMERASE CHAIN REACTION (PCR) step in DNA TYPING. For DNA amplification, the bases needed are ADENINE, THYMINE, CYTOSINE, and GUANINE, and the corresponding nucleotides (base plus sugar and phosphate) are deoxyadenosine monophosphate (dAMP), deoxythymidine monophosphate (dTMP), deoxycytidine monophosphate (dCMP), and dTMP.

dolophine See METHADONE.

dominant A genetic ALLELE that is expressed regardless of what the other allele is.

DOT number A number on tires that is specified by the U.S. Department of Transportation (DOT). The number can be used to determine when and where a tire was manufactured and by what company. Retread tires also carry a DOT number that provides information about the retread process.

double action A technique of firing a pistol (revolver or semiautomatic) in which the act of pulling the trigger raises the hammer and then drops it on the firing pin. In the case of a revolver, in double action pulling the trigger revolves the cylinder and places a fresh cartridge beneath the hammer. A double-action trigger pull is longer than a single-action trigger pull.

double based powder A type of SMOKELESS POWDER (propellant for ammunition) that consists of NITROCELLULOSE and NITROGLYCERIN along with various additives.

double helix The structure of the DNA MOLECULE. The helical shape is the result of HYDROGEN BONDING (a weak chemical interaction) between the BASE PAIRS adenine and thymine (AT) and cytosine and guanine (CG).

Doyle, Sir Arthur Conan (1859–1930) A British author famous for creating the immortal SHERLOCK HOLMES character, who has become a widely recognized symbol of scientific sleuthing. Many forensic scientists were inspired to enter their careers by his adventures, published between 1887 and 1927 in the form of short stories and novellas. Doyle, a physician, described many tests and techniques in his stories that would not become common tools of forensic science until years later. Pioneers in the field, particularly EDMOND LOCARD, praised Doyle's works and cited them as personal inspiration.

dpi Dots per inch, a measure of the resolution of devices such as computer printers (laser and ink-jet), computer monitors, and digital media such as photographs. The larger the number of dots, the greater the resolution.

DQA1 A genetic locus that codes for an antigen in the HUMAN LEUKOCYTE ANTIGEN system. The system is referred to as D1 alpha, D1A, or D1α. This was the first locus that was validated and applied in forensic science through a POLYMERASE CHAIN REACTION technique.

dram An older unit of fluid measurement used in pharmacies and drug dispensing (APOTHECARY UNIT). One dram contains 60 GRAINS and is equivalent to approximately four grams.

drawback Blood or another liquid substance on the barrel of the firearm that has been withdrawn deeper into the barrel.

DRE See DRUG RECOGNITION EXPERT.

drift In an instrument, a tendency for response, background, or other instrumental parameters to change slightly over time.

DRIFTS A specialized form of Fourier transform infrared spectroscopy (FTIR) that examines ELECTROMAGNETIC RADIATION (EMR) reflected from the surface of a sample rather than radiation that is absorbed. The initials stand for *diffuse reflectance infrared Fourier transform spectroscopy*. DRIFTS is a surface analysis technique similar to ATTENUATED TOTAL REFLECTANCE. DRIFTS techniques are more sensitive that FTIR, can operate over a larger concentration range, and thus can be very useful in forensic applications in TRACE EVIDENCE analysis. Sample preparation is easy, and both solids and liquids can be analyzed. The apparatus needed for DRIFTS can be obtained as an accessory for many FTIR instruments. Among other applications, DRIFTS has been used in the analysis of DRUGS, PAINTS, PLASTICS, and QUESTIONED DOCUMENTS.

drip pattern A bloodstain pattern created by blood that drops from a height to a surface such as a floor. A drip pattern can be used to determine whether the source was stationary or moving, and, if moving, the direction of motion. Occasionally the speed of movement can be estimated (walk versus run).

drop down A phenomenon that can occur during fires in which falling burning material drops down onto a surface that is not burning. For example, if a floor is burning and collapses into the basement, the fire can spread to the basement as a result.

drug A substance (not food) that when ingested into the body produces an effect on some function or functions in the body.

drug classification A system of categorizing illegal, abused, or controlled

drugs (listed in the CONTROLLED SUB-
STANCES ACT) on the basis of their physiological activity. The classes used and
examples are as follows:

1. Narcotics: substances that relieve pain
   (analgesics) and promote sleep. Contrary to popular belief, not all illegal
   drugs are narcotics, and this confusion
   has led to misclassification. Cocaine is
   often listed as or considered to be a
   narcotic even though by function, it is a
   central nervous system (CNS) stimulant. Narcotics abuse can lead to physical and psychological dependence
   (addiction). Heroin and the opiate
   ALKALOIDS are narcotics, as are synthet-
   ics such as OXYCODONE and meperidine (DEMEROL).
2. Depressants: substances that depress
   the CNS and can produce such effects
   as loss of coordination, impairment of
   judgment, and sleep. ALCOHOL is a
   depressant, as are the BARBITURATES,
   TRANQUILIZERS, BENZODIAZEPINES such
   as Valium, INHALANTS, and quaalude.
3. Stimulants: substances that stimulate the
   CNS, producing a feeling of wakefulness, decreased fatigue, decreased
   appetite, and general well-being.
   Cocaine, amphetamine, methamphetamine, methylphenidate (Ritalin), caffeine, and nicotine fall into this category.
   In higher doses, many stimulants can
   also act as hallucinogens.
4. Hallucinogens: substances that alter
   visual and auditory stimuli and produce hallucinations. LSD (LYSERGIC
   ACID DIETHYLAMIDE), MESCALINE,
   MARIJUANA, HASHISH, and ECSTASY are
   hallucinogens.

DRUGFIRE An automated computer
system for the identification of markings on
FIREARMS evidence. DRUGFIRE is a network of computer databases at local laboratories linked through the FBI and used to
determine whether the same weapon has
been used in more than one shooting in
which firearms evidence has been recovered. Although it emphasizes CARTRIDGE
CASINGS, STRIATIONS on BULLETS are also
included and can be searched. The BUREAU

OF ALCOHOL, TOBACCO, FIREARMS, AND
EXPLOSIVES maintains a similar system
called IBIS.

druggist's fold A folding pattern used on
paper that encloses small amounts of physical evidence such as a powder, hairs, or
fibers. The name originates from the
method druggists at one time used to dis-
pense small amounts of powders. To make
the fold, the paper is folded in thirds lengthwise, folding over the previous fold. The
paper is turned and the process repeated for
each remaining side. When it is completed,
the two free edges are tucked into each
other somewhat as on an envelope. The
packet is then sealed and initialed.

Drug Recognition Experts (DREs)
Police officers trained to evaluate drug
impairment and to determine the drug or
drugs a person has taken to become intoxicated. The Los Angeles Police Department
organized the first such drug evaluation
and classification (DEC) program in the
late 1980s. A DRE test is not a roadside
evaluation; it occurs after an arrest has
taken place and the intoxicated person has
been moved to a safe and controlled environment. The evaluation has 12 parts,
including a BLOOD ALCOHOL test, physical
examination and testing, and an interview
with the arresting officer. On the basis of
these tests, the DRE provides an opinion of
the type of drug involved (depressant,
stimulant, narcotic, hallucinogen, PHENCY-
CLIDINE, INHALANT, or MARIJUANA). The
opinion is confirmed by using toxicological
tests of urine, blood, or both.

Drunkometer A device developed in
the 1930s by Rolla Harger at Indiana
University. The device tested BREATH
ALCOHOL and related results to BLOOD
ALCOHOL and degree of intoxication.

dry origin impression An impression,
such as a SHOEPRINT or FINGERPRINT, that
is found or created in a dry material such
as dust or powder.

dTMP An abbreviation for the compound deoxythymidine monophosphate,

the NUCLEOTIDE that consists of the amino acid THYMINE, ribose (a sugar), and a phosphate group (PO_4^{3-}). Its structure is analogous to that shown for deoxycytidine monophosphate (dAMP).

Duffy system A blood group system that was rarely used in forensic serology before DNA TYPING. It contains two ANTIGENS and four types, the most common of which are Fy(a⁺b⁺) in Caucasians and Fy(a⁻b⁻) in African Americans.

Duquenois test (Duquenois-Levine test) A PRESUMPTIVE TEST for MARIJUANA and HASHISH. Also called the Duquenois-Levine test or the modified Duquenois-Levine test, it consists of three reagents that react with the active substance in marijuana, tetrahydrocannabinol (THC). The reagents are a 2 percent solution of vanillin and 1 percent acetaldehyde in ethanol, concentrated hydrochloric acid (HCl), and chloroform. Dried plant matter, seeds or seed extracts, or other material to be tested is placed into a test tube and the reagents are added in order as previously listed. The sample is shaken and a positive result is indicated by a purple color in the lower chloroform layer.

dyes (dyes and pigments) Compounds that are used as colorants. The term *dye* usually refers to an organic compound (synthetic or natural such as henna), whereas the term *pigment* refers to inorganic colorants such as zinc oxide, which is used as a sunscreen. In forensic science, dyes can be important in the analysis of FIBERS, INKS, and dyed HAIR. Fabrics and textiles are colored with one or more dyes, which penetrate the fibers and can be characterized by chemical and spectroscopic analysis.

dye staining The process of treating a latent fingerprint with a stain that has already been developed with cyanoacrylate (Super Glue). The stains commonly used, such as gentian violet, are also widely employed to stain biological samples such as cells.

dynamic load A load on a structure, component, or part that is not constant. An example of a dynamic load would be the force that the wind exerts on a tall building; the load changes with wind speed and, as is common with dynamic loading, produces a motion as the building sways.

dynamite A high EXPLOSIVE developed by Alfred Nobel in 1886 that consisted of nitroglycerin (NG) adsorbed onto an inert solid such as diatomaceous earth (DE) or a clay material. As a result, the usually unstable NG was rendered much more stable and safer to handle. Modern formulations often include sodium nitrate or ammonium nitrate ($NaNO_3$ or NH_4NO_3) as oxidizers, NITROCELLULOSE, sulfur, ETHYLENE GLYCOL DINITRATE, and other filler materials.

E

ear identification (ear print identification; otoscopy) Attempt to individualize a person on the basis of the shape and other characteristics of the ear. These characteristics can be studied by using photos or impressions. Such identification is not currently accepted as scientific evidence.

eccrine sweat The aqueous fluid excreted by the sweat (eccrine) glands in the skin. These pores are found on all skin surfaces; sebaceous glands (oil glands) are found in more limited areas. Pore location is part of the MINUTIAE of FINGERPRINTS, and the study of pore location as a tool for identification is referred to as poroscopy. Sweat is a complex mixture of inorganic and organic materials dissolved or suspended in water, which makes up about 98 percent of the volume. The amino acids found in sweat react with NINHYDRIN to produce a purple color and visualize a latent FINGERPRINT. Dissolved salts, sugars, and ammonia are also components of sweat. As a medium for toxicology, sweat can be used in some cases to detect drugs and metabolites and has the added advantage that it can be collected by using noninvasive procedures.

ECD *See* ELECTRON CAPTURE DETECTOR.

ecgonine methyl ester One of two primary metabolites of COCAINE that can be found in the blood and the urine after the drug has been ingested. The other important metabolite is BENZOYLECGONINE.

ecology, forensic The application of ecological principles and knowledge to law enforcement. This includes diverse areas such as knowledge of plants, pollen, and ecological SUCCESSION that can be used to identify CLANDESTINE GRAVES.

ecstasy A DESIGNER DRUG that is an analog of methamphetamine. The complete chemical name of the drug is 3,4-methylenedioxymethamphetamine, commonly abbreviated MDMA. Unlike other designer drugs, MDMA had a brief history of legitimate uses before it became popular as an illegal drug. Ecstasy use is increasing among adolescents and young adults, becoming a favorite party drug used at "raves." Ecstasy produces a sense of euphoria and heightened empathy and can lead to hallucinations. It is listed on Schedule I of the CONTROLLED SUBSTANCES ACT. Most of the drug is synthesized in CLANDESTINE LABORATORIES in Europe.

eddy diffusion A term used in the Van Deemter equation that describes the relationships among three fundamental terms in CHROMATOGRAPHY. Eddy diffusion is expressed as a numerical value that reflects the degree of band broadening (spreading) that occurs in a chromatography column when molecules of the analyte take differing (and thus longer or shorter) paths while traveling through it.

EDTA (ethylenediaminetetraacetic acid) A preservative used for liquid blood that will be subject to DNA TYPING or other serological analyses. EDTA is a common additive in food and consumer products such as shampoo. It works as a preservative by forming strong, water-soluble complexes with ions such as calcium and magnesium (Ca^{2+}, Mg^{2+}) and other ions that are needed by microbes. When the ions are bound up in EDTA complexes, they are unavailable to the microbes, and their growth is inhibited.

EDXRF (electron diffraction X-ray fluorescence) *See* SCANNING ELECTRON MICROSCOPY; X-RAY TECHNIQUES.

EGDN Ethylene glycol dinitrate, a compound classified as a high EXPLOSIVE and found in some formulations of dynamite.

ejector/ejector projector or pin The metal component that grips a used cartridge and ejects it. In an automatic or semiautomatic weapon, the ejector is operated by gas pressure created when the cartridge is fired. In revolvers, the spent cartridges must be manually ejected by using a pushing rod.

elastic matching (adaptive elastic string matching; string matching) A method (with many variations) used to compare digital representations of FINGER-PRINTS.

Elavil (amitriptyline) A tricyclic antidepressant that is in the same class of prescription drugs as fluoxetine (Prozac).

electrochemical detector An instrumental detector that can be used in HIGH-PERFORMANCE LIQUID CHROMATOGRAPHY, ION CHROMATOGRAPHY, and CAPILLARY ELECTROPHORESIS. Depending on type and configuration, such a detector may sense changes in conductivity, current, or voltage (potential). Detectors rely on the presence of ionic or charged species that allow electrical current to flow.

electrochemical etching A chemical technique that is used chemically to etch metal as when creating rifling or when attempting to restore an obliterated serial number. Reagents placed in contact with the metal surface cause an oxidation/reduction (REDOX, electron exchange) reaction that dissolves the metal in a controlled fashion.

electrode A metal surface at which an electron exchange can take place. An electrode may be passive or active in the elec-

trochemical reaction (redox, electron capture) that takes place. An inactive electrode merely conducts electrons.

electrofocusing *See* ISOELECTRIC FOCUSING.

electromagnetic energy and the electromagnetic spectrum EMR, including visible light, is composed of both electrical and magnetic energy and can be described as both a wave and a particle. When it is described as a wave, two terms are important: the wavelength (λ) and the frequency (ν). The idea can be visualized by imagining dropping a rock into a still pond. Energy ripples away from where the rock was dropped in waves that can be described by their wavelength (distance from crest to crest) and frequency (number of waves that pass a fixed point per second). Electromagnetic radiation that is high-energy has short wavelength and high frequency; low-energy EMR has long wavelength and low frequency. (See figure on page 70.)

When EMR is envisioned as a particle, the source of the energy can be thought of as a gun that "fires" discrete packets of energy, which are referred to as photons. The energy of a photon is described by the relationship $E = h\nu$, where h is constant (Planck's constant) and ν is the frequency. Thus, the wave model and particle model are related through the frequency. Instruments built to study the interaction of EMR and matter are called SPECTROPHOTOMETERS.

electron capture detector (ECD) A detector used in gas chromatography that is particularly sensitive to compounds containing chlorine, bromine, iodine, fluorine, and nitrogen groups. The detector contains the radioactive isotope nickel-63, which emits electrons. Molecules containing the atoms mentioned have high electron affinities and capture some of these electrons. The detector's response is registered as a decrease in the current (flow of electrons).

electronic transition The promotion of an electron in an atom or molecule

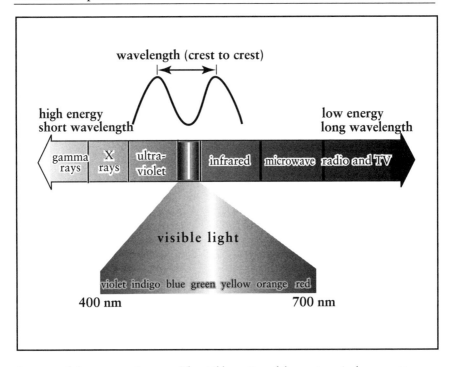

Spectrum of electromagnetic energy. The visible portion of the spectrum is shown; not to scale. The wavelength (λ) and frequency (ν) can be related to energy.

from a lower energy state to a higher energy state (EXCITED STATE). This most often is accomplished by the absorption of ELECTROMAGNETIC ENERGY such as ultraviolet (UV) light. It is the basis of many spectrophotometric techniques such as ATOMIC ABSORPTION, COLORIMETRY, and UV/VIS (ultraviolet/visible) SPECTROPHOTOMETRY. *See also* FLUORESCENCE.

electron microprobe (EMP) *See* SCANNING ELECTRON MICROSCOPY; X-RAY TECHNIQUES.

electron multiplier A device used in an instrument detector to amplify the detector's signal. This is accomplished by using electron-emissive materials that release large numbers of electrons for every single electron that collides with the material's surface. By facilitating a number of successive cascading collisions, an electron multiplier can multiply the initial response as much as a millionfold.

electroosmotic flow (EOF) The flow of a solution that occurs in CAPILLARY ZONE ELECTROPHORESIS. The mechanism is shown in the figure. Because the capillary is made of a type of glass, its surface has a net negative charge to which cations are attracted, beginning the process leading to EOF. The electroosmotic flow is sufficient to move all species—positively charged, negatively charged, and neutral—to the detector.

electropherogram The chart or graph produced by a CAPILLARY ELECTRO-PHORESIS instrument. It can be a printed sheet or a computer file. On the *x*-axis is given the RETENTION TIME and on the *y*-axis, the response of the detector at that time.

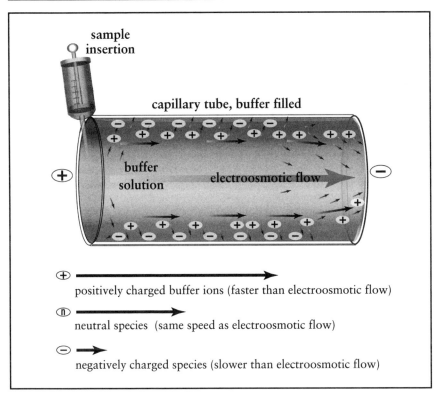

Electroosmotic flow in capillary electrophoresis. Positive ions are attracted to the negatively charged surface and form a layer. These ions are also surrounded by the solvent (water), so as they migrate toward the detector (negatively charged), the net effect is to drag solvent in that direction, creating the electroosmotic flow.

electrophoresis A technique of separating large charged molecules such as proteins on the basis of their mobility in an applied electrical field. In forensic science, electrophoresis was extensively used up until the mid-1990s for the typing of ISOENZYMES in BLOOD and BLOODSTAIN evidence. As shown in the figure on page 72, electrophoresis is carried out in a gel medium (or similar material such as polyacrylamide), which is attached to an electrical power supply. The ends of the gel are in electrical contact with an aqueous (water-based) buffer solution by means of a wet sponge or similar material. In the example shown, the sample is inserted into a slit at the extreme left-hand side of the gel. In the case of blood-

stained fabric, threads from the stain can be inserted directly. The power supply is activated, creating an electrical field with a positive and a negative end. The proteins, which usually carry a net negative charge, migrate to the positive pole since unlike charges attract. Different proteins in the samples are made up of different subunits and thus differ from each other in many characteristics, such as size, charge, and shape. Because of these differences and the structure of the gel, all components move at different speeds, resulting in separation.

electrospray A technique used to interface HIGH-PERFORMANCE LIQUID CHROMATOGRAPHY (HPLC) to MASS SPECTROMETRY. As

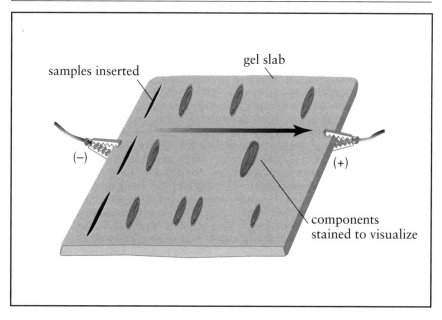

samples inserted

gel slab

(−)

(+)

components
stained to visualize

Gel electrophoresis, in which large molecules such as proteins are separated on the basis of their size and charge.

solvent and analyte emerge from the HPLC column, the electrospray device imparts surface charge to solvent droplets. The droplets are then dried, making the surface charges closer to each other as the droplets shrink. When the droplet gets so small that the charges are too close, the repulsion of the like charges causes the droplet to burst into a fine mist. As a result, most of the solvent is driven off before the material enters the mass spectrometer.

electrostatic detection apparatus (ESDA) A device used to visualize INDENTED WRITING. For example, if someone writes something on the top sheet of a notepad, the pressure exerted by the pen or pencil is often sufficient to indent the paper beneath. ESDA takes advantage of the same technology used in copying machines and laser printers to help visualize the indented writing. A sheet of thin plastic is placed over the paper that has the suspected indented writing, and both are placed into a vacuum chamber. A charge of static electric-ity is imparted to the plastic and then toner powder is applied to the surface of the plastic. An image of the indented writing is created on the plastic, with the added advantage that the original paper is not altered or damaged.

electrostatic lifting The process of using static electricity to lift or transfer a dry impression to a black background. A high-voltage source is used to create the charge in the dust and to attract it to the surface. This procedure is used for FIN-GERPRINTS, IMPRESSIONS, and QUESTIONED DOCUMENTS.

elemental analysis Chemical testing (using primarily INSTRUMENTAL ANALYSIS) aimed at detecting individual chemical elements. Both QUALITATIVE and QUANTI-TATIVE information can be obtained. The most common forensic applications of elemental analysis include analysis of GUNSHOT RESIDUE, in which elements such as lead, copper, barium, and anti-mony are of interest. Another example is the analysis of pigments in PAINT, in which elements such as iron and titanium are important. The techniques used for elemental analysis include ATOMIC ABSORPTION, inductively coupled plasma–mass spectrometry (ICP-MS), inductively coupled plasma–atomic emission spectrometry (ICP-AES), and X-RAY TECHNIQUES including those associated with SCANNING ELECTRON MICROSCOPES.

elimination prints FINGERPRINTS collected at a CRIME SCENE of all personnel who were in and at the scene and who might have inadvertently left prints on evidence. In addition to fingerprints, palm prints and footprints are often collected.

elimination samples (elimination evidence) Samples such as hair, fibers, fingerprints, or DNA that are collected from persons who may have contributed to a sample collected as evidence but had legitimate reasons for doing so.

ELISA (enzyme-linked immunosorbent assay) An IMMUNOASSAY technique used in toxicology to detect drugs and metabolites. Immunoassay relies on an antigen–antibody reaction between the drug being tested and an antibody specific for it. The antibody is attached to a solid surface such as the bottom of a plastic or glass well. A complex that consists of the drug and a label is added and the reaction occurs. As a result, the labeled drug is bound to the antibody. A sample that may contain the drug, such as urine, is added to the plastic well. If there is no drug or very little drug present, the labeled drug–antibody complex remains undisturbed. However, if there is a large concentration of the drug, this displaces the labeled drug from the antibodies, releasing the labeled drug into solution. The higher the drug concentration in the sample, the more is displaced. The amount of the displaced labeled drug is then measured. In ELISA, a chemical compound called a substrate is added to the solution in the well. The label is an enzyme that catalyzes a reaction in which the substrate is changed, forming a colored solution. The

deeper the color, the greater the concentration of the enzyme label present, suggesting a greater concentration of drug in the original sample.

eluate In CHROMATOGRAPHY, the fluid or gas that emerges from the end of the column.

eluent In CHROMATOGRAPHY, the mobile phase that is introduced into the column and used to move the sample molecules through it.

emission spectroscopy A form of SPECTROPHOTOMETRIC analysis. In forensic labs, the most common type of emission spectrometry is inductively coupled plasma–atomic emission spectrometry (ICP-AES), a type of ELEMENTAL ANALYSIS. Emission techniques are based on the analysis of the ELECTROMAGNETIC RADIATION emitted by atoms when they have been heated to extremely high temperatures. The energy created in these environments is used to promote electrons within atoms from a lower energy level to a higher energy level, a state called the EXCITED STATE. When these atoms relax out of the excited state and return to their original state (called the ground state), electromagnetic energy is released, usually in the visible or ultraviolet regions. By determining the wavelengths of the emitted energy and its intensity, it is possible to obtain QUALITATIVE and QUANTITATIVE information about the sample being analyzed. Early forms of emission spectrometry used flames or sparks to initiate the emission processes, but modern instrumentation relies on inductively coupled plasmas, which can reach much hotter temperatures, nearly 10,000°C.

EMIT (enzyme-multiplied immunoassay technique) An immunoassay technique that is similar to enzyme-linked immunosorbent assay (ELISA). In EMIT, the label catalyzes a common biological reaction, the conversion of nicotinamide adenine dinucleotide (NAD) to reduced NAD (NADH). The higher the level of

NADH detected, the greater the concentration of drug in the sample.

empirical evidence Evidence obtained from experimentation, analysis, and/or observation. Almost all forensic analyses yield empirical evidence.

enamel A type of paint or the outer layer of a TOOTH.

enantiomers A pair of molecules that are related to each other by being non-superimposable mirror images of each other. A common example of enantiomers are the right and left hands—they have the same arrangement of thumb and fingers but the mirror image of the right hand is not superimposable on the left hand. In forensic science, enantiomers are an important consideration in DRUG ANALYSIS, in which enantiomers and DIASTEREOISOMERS can be important, particularly in the case of COCAINE.

encapsulation A process used to cover tiny semiconductor crystals (nanoparticles) so that they can be used to visualize latent FINGERPRINTS.

encryption (encrypted evidence) In forensic computing, data or information that is scrambled, coded, or protected by a password to prevent access to it.

endogenous In forensic science, a term that generally refers to a material or compound that is synthesized or found naturally within the human body or in the bloodstream. It is the opposite of EXOGENOUS.

endothermic reaction A chemical reaction that absorbs heat from the surroundings. A familiar example is a portable "cold pack" that can provide instant cold for treating injuries or insect bites. When the pack is activated, it feels cold on the skin because the chemical reaction taking place is drawing heat out of the skin.

energy dispersive spectroscopy (EDS) See SCANNING ELECTRON MICROSCOPE.

engineering, forensic (forensic engineering) The application of engineering knowledge and techniques to legal matters and those that involve courts of law. Forensic engineering includes traffic accident reconstruction, failure analysis (as when buildings collapse), investigation of industrial accidents, product liability issues, and the investigation of transportation disasters such as airline crashes. Unlike professionals in other forensic disciplines, forensic engineers often find themselves involved in the investigation of incidents that are not crimes, although criminal activity may be a contributing factor. Forensic engineers try to understand why and how materials or machines failed and what can be done to prevent such failures in the future. A professional society devoted to forensic engineering, the National Academy of Forensic Engineers (NAFE), was formed in 1982.

enhancement The process of making an image, stain, or impression more visible by using chemicals, alternative or oblique lighting, photography, or other methods.

enthalpy (H) Heat energy stored in chemical bonds and released or absorbed during chemical reactions.

entomology, forensic The application of the study of arthropods such as flies and beetles to legal proceedings, both criminal and civil. Arthropods are animals with jointed legs, which include insects, arachnids (spiders), centipedes, millipedes, and crustaceans. Forensic entomology, principally related to insects, has become a common tool in death investigation and in determination of the interval since death occurred (POSTMORTEM INTERVAL, PMI), and/or establishment of the site of death. FLIES and BEETLES are among the most important insects in forensic entomology. The state of succession of insects on a body can be used to estimate how long it has been in a certain location. For example, flies arrive soon after death and lay eggs that hatch into maggots (see illustration on opposite page). Beetles then arrive

The Life Cycle of the Blowfly

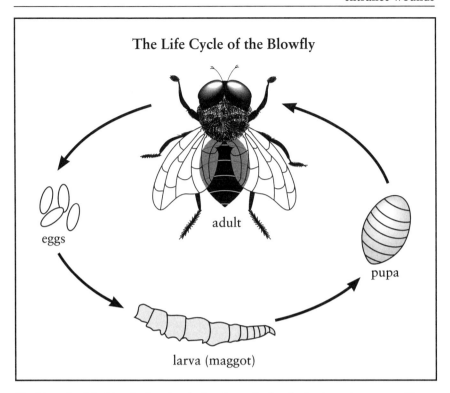

eggs

adult

pupa

larva (maggot)

The life cycle of the blowfly. In a day the female can lay hundreds of eggs in a corpse. The larval stage (maggot) lasts several weeks, providing food for other insects such as beetles.

to feed on the maggots even as the maggots progress through their life cycle. Thus, a decomposing body becomes the basis of a small ecosystem, the state of which can be used to determine the PMI. *See also* SUCCESSION PATTERNS.

entomotoxicology The insects found at a death scene are chemically analyzed to determine what substances they ingested while feeding on a body. Preparation of insects for chemical analysis uses techniques similar to those developed for the analysis of hair and fingernails, since the chemical composition of insect shells is similar to the composition of these materials. Entomotoxicology can be valuable in cases when a body is discovered so long after death that the blood and tissues needed for standard toxicological analyses have disappeared.

entrance wounds Wounds created where a bullet enters the body. When a bullet strikes flesh, the skin is first stretched, then broken as the projectile penetrates. As the bullet enters, material on its surface such as dirt and dust, lubricants, powder and primer residue, and lead are wiped onto the skin in a pattern called bullet wipe. The bullet also scrapes off skin cells, creating an injury called a CONTUSION RING. These features may be obscured or altered by the presence of clothing, and in some cases the bullet wipe pattern may obscure the contusion ring. The shape of the bullet wipe and contusion ring can provide clues about angles and relative positions of the shooter and victim. Beyond the bullet wipe and contusion ring there is a dispersed deposit of material (GUNSHOT RESIDUE) that contains flakes of unburned powder and other residues. The concentration of these residues and the extent of their spreading depend primarily on the distance between the shooter and the victim. *See also* BULLET WOUNDS.

entrophy (S) The inherent disorder of a system, chemical or otherwise.

environmental forensics The study of environmental data that is involved in legal proceedings. *Environmental Forensics* is also the title of a journal in the field that began publishing in 2000. Often the goals of litigation are to determine responsibility for environmental contamination and to decide which parties will be responsible for the cleanup or other remedial action. Environmental cases are not typically handled by forensic laboratories but rather by laboratories and private consultants working in the environmental field. The ENVIRON-MENTAL PROTECTION AGENCY, along with various state and local agencies, are responsible for enforcement actions.

enzyme A biological CATALYST that facilitates and speeds a chemical reaction that would otherwise take place very slowly. Enzymes are protein molecules that are used in many biochemical procedures in forensic science; their naming conventions describe enzyme functions. For example, a dehydrogenase enzyme catalyzes the removal of hydrogen from molecules.

enzyme kinetics The study of the speed or rate of chemical reactions (reaction kinetics) that are catalyzed by ENZYMES.

EPA (Environmental Protection Agency) The agency charged with overseeing environmental affairs, regulation, and enforcement at the federal level. The EPA has investigative authority and as a result requires scientific analysis of environmental samples and evidence. Currently, the EPA's Office of Criminal Enforcement, Forensics, and Training (OCEFT) oversees criminal investigations, compliance issues, and forensic analyses.

ephedrine Along with pseudoephedrine, a common ingredient in over-the-counter (OTC) drugs used to treat colds and flu, specifically the symptom of congestion. Ephedrine and pseudoephedrine occur naturally in plants in the Ephedra family, and ephedrine is a central nervous system (CNS) stimulant. Tablets containing ephedrine and/or caffeine, purporting to be amphetamine, are occasionally sold on the street. Ephedrine can also serve as a precursor in the clandestine synthesis of methamphetamine.

epidemiology The study of disease patterns in large populations. Epidemiology is an important part of ENVIRONMENTAL FORENSICS, in which disease patterns related to pollution are a critical concern.

epidermis The outer portion of the skin that is made up of several distinct cell layers beginning with the basal cell layer and progressing to the outer layer, where all epidermal cells are eventually sloughed off.

epi illumination (episcopic illumination) A lighting technique used in microscopy in which the specimen being studied is viewed by REFLECTED LIGHT versus TRANSMITTED LIGHT. For example, a thin section of tissue could be mounted on a slide and observed by directing light from the bottom of the microscope through the sample given that the sample is thin enough to transmit light. The epi illumination originates above or from the sides of the object being viewed, and the light is reflected rather than transmitted.

epiphysial union The point at which the growth center of one long BONE (epiphysis) is connected to another.

epiphysis Centers of bone growth found at the ends or edges of bones. An epiphysis is an OSSIFICATION CENTER and during normal development fuses with other ossification centers to form a single bone structure and an epiphysial union. *See also* DIAPHYSIS.

equivocal death A death in which the manner (natural, accidental, suicidal, or

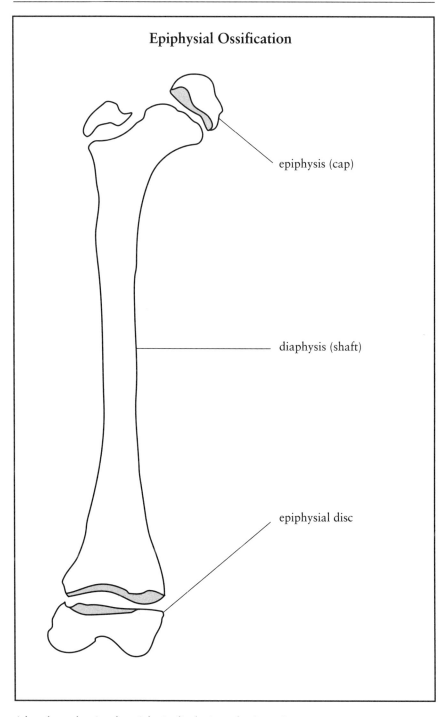

Epiphysial Ossification

epiphysis (cap)

diaphysis (shaft)

epiphysial disc

A long bone showing the epiphysis, diaphysis, and union point.

homicidal) cannot be determined; the cause of death is indeterminate.

erasure   The purposeful removal of letters, numbers, words, phrases, or other notations from a document. The erasure can be abrasive (such as from rubbing with an eraser or wetting and rubbing), chemical (such as bleaching), cut out (using a razor or other sharp knife), or lifted (using a CORRECTION RIBBON on a typewriter). Erasures are not the same as OBLITERATIONS, in which existing writing is overwritten or hidden but not removed.

Erlich's test   A PRESUMPTIVE TEST used in DRUG ANALYSIS to detect lysergic acid diethylamide (LSD) and other related ergot ALKALOIDS. The reagent contains *p*-dimethylaminobenzaldehye (*p*-DMAB) in a solution of sulfuric and hydrochloric acids and is often referred to as the *p*-DMAB test or by the older term *Van Urk test*.

erythrocyte acid phosphatase (EAP or ACP)   A polymorphic ISOENZYME system with six common PHENOTYPES present in blood. Before widespread adoption of DNA TYPING, this and other isoenzyme systems were typed in bloodstains in attempts to individualize the stain to the extent possible. Typing is performed by using gel ELECTROPHORESIS.

ESD (esterase D)   A polymorphic ISOENZYME system with five common PHENOTYPES. Before widespread adoption of DNA TYPING, this and other isoenzyme systems were typed in bloodstains in attempts to individualize the stain to the extent possible. Typing is performed by gel ELECTROPHORESIS.

ESDA   Electrostatic detection apparatus, a device used to visualize and create a copy of indented writing.

etchant   A chemical or mix of chemicals used to dissolve metal, usually during the course of a SERIAL NUMBER RESTORATION. Acids are common etchants, but not the only type available; for example, solutions of strong base such as sodium hydroxide (NaOH) are also used, depending on the metal to be etched.

ethanol   Ethyl alcohol, chemical formula CH_3CH_2OH. The type of alcohol that causes intoxication and that is included in alcoholic beverages.

ethics and accountability   Several forensic science professional organizations have codes of ethics, but there are no uniform national standards. The AMERICAN ACADEMY OF FORENSIC SCIENCES, the AMERICAN BOARD OF CRIMINALISTICS, and the CALIFORNIA ASSOCIATION OF CRIMINALISTS all have ethical guidelines that vary in extent and detail. Common themes in these ethical guidelines include prohibitions against misrepresenting or exaggerating one's scientific qualifications, use of the best available scientific methods of analysis, use of procedures to ensure the quality and reliability of data produced, and obligation to report individuals who violate the guidelines. Ethical guidelines are important in any profession, but even more so in forensic science given the nature of the work and the integral relationship with the justice system. Ethical codes, enforced by professionals working in the field, ensure that the work done is trustworthy, reliable, and impartial.

eukaryotic cells   Cells that have a nucleus and that divide by mitosis and meiosis. Human, animal, plant, and fungi cells are eukaryotic.

evidence   *See* PHYSICAL EVIDENCE.

evidence collection technician   A person who responds to a crime scene and is responsible for documentation, collection, and delivery of evidence.

evidentiary fact   A fact directly ascertained from the evidence, but not related directly to the matter in question. For example, consider the case of a murder involving a gun. If a suspect is identified and forensic analysis that is performed shows that there was GUNSHOT RESIDUE on the suspect's hands, that finding is an evidentiary fact. By itself, it does not prove that the suspect committed the murder: the matter in question. Additional information and evidence are required to establish whether or not the suspect is innocent or guilty of the crime.

exchange principle *See* LOCARD'S EXCHANGE PRINCIPLE.

excitation (excited state) Excitation occurs when an atom, ion, or molecule absorbs electromagnetic energy and is promoted to a higher energy state called the EXCITED STATE. Excited states are unstable and eventually the excited species will give off the extra energy and return to the normal GROUND STATE. In forensic science, excitation is often exploited as a means to generate FLUORESCENCE; as when LUMINOL is used to visualize bloodstains. In this example, the fluorescent light is produced by an electronic transition. Fluorescence is also part of many methods used to visualize latent FINGERPRINTS.

exclusionary evidence Evidence that excludes or eliminates a person or disproves a possible scenario. For example, if semen involved in a rape case is found to be of a type that does not match that of a suspect, that finding is considered exclusionary evidence since it eliminates that person as a possible source.

exemplar Examples of handwriting collected for comparison to suspect writing in a QUESTIONED DOCUMENTS case. Exemplars can be requested as part of the investigation or may be collected from other writing by the person whose handwriting is in question. A signature on an old credit card slip or check would be an example of this type of exemplar. The term is also used occasionally to refer broadly to standards used in comparisons, such as of fibers or hairs.

exit wound The wound produced when a bullet or bullet fragments exit the body. Unlike an entrance wound produced by a gunshot, an exit wound does not show characteristics such as bullet wipe or stippling.

exogenous In forensic science, a term that generally refers to a material or compound that is not synthesized or found naturally within the human body or in the bloodstream. It is the opposite of ENDOGENOUS.

exon A sequence of BASE PAIRS in DNA that contain the code that specifies the amino acid sequence for a protein.

exothermic reaction A chemical reaction that releases heat to the surroundings. A burning flame is an example of an exothermic reaction.

expert witness A person who is accepted by the TRIER OF FACT as having specialized knowledge that is relevant to the case and beyond the expertise of the average person. A key component of forensic scientists' job is to offer expert testimony in cases that involve evidence that they have analyzed in the lab. This communication to a court constitutes the "forensic" portion of their work, and the laboratory analysis and interpretation constitute the "science" portion. The expertise of a witness is judged on the basis of education, training, and experience and must pass any challenges offered by the defense.

explainable differences Differences observed when comparing samples that can be explained and are not significant to the comparison. For example, if a window was broken and several fragments were compared, there might be different patterns of scratching and weathering on the individual fragments even though they originated from the same source. However, these differences are explainable and would not be interpreted to indicate that the fragments were from different windows.

explosion A process similar to COMBUSTION in that a solid or liquid is converted to gases in a reaction that is exothermic (heat-producing). A source of fuel (the material that decomposes during the explosion) is required, as well as an oxidant. What distinguishes an explosion from combustion is the speed at which the reaction occurs. Explosions create huge amounts of gas that can travel at speeds of nearly 7,000 miles per hour, and this shock wave or blast effect can do enormous damage to anything in its path. A by-product of the blast effect is an extremely loud noise created by the pressure wave.

explosives Chemical compounds or mixtures that decompose rapidly to produce heat and gas in the form of an EXPLOSION. LOW EXPLOSIVES burn very quickly and must be in a confined space in order to explode; examples include BLACK POWDER and SMOKELESS POWDER (used as PROPELLANTS in AMMUNITION). Another low explosive, made infamous by the Oklahoma City bombing on April 19, 1995, is composed of ammonium nitrate and 6 percent fuel oil (ANFO). HIGH EXPLOSIVES are further divided into primary and secondary explosives. Primary high explosives are shock- and/or heat-sensitive and are often used as PRIMERS that ignite secondary high explosives. Secondary high explosives are much more stable and are usually detonated by the shock generated from a primary explosive. High explosives decompose at a much faster rate than low explosives, and their detonations generate shattering power; produce smaller, sharper fragments; and generally leave minimal residue. Examples include NITROGLYCERIN ("nitro" or NG), trinitrotoluene (TNT), HMX, RDX, TETRYL, and PETN. The terms *plastic explosive* and *plastique* usually refer to RDX mixtures that are moldable; C4 is a complex that is 90 percent RDX.

exsanguination Bleeding to death.

external reflectance A general class of techniques used in INFRARED SPECTROSCOPY, with forensic applications in drug analysis and TRACE EVIDENCE analysis. Traditional spectroscopy is based on measuring the degree of absorbance of a sample. For example, an infrared (IR) absorbance spectrum of a drug is obtained by exposing a sample to successive wavelengths of energy in the IR portion of the electromagnetic spectrum and plotting the amount of energy absorbed by the sample at each wavelength. External reflectance techniques measure energy that is reflected by the sample rather than absorbed by it.

external standard A type of instrument CALIBRATION used in techniques such as ATOMIC ABSORPTION SPECTROPHOTOMETRY and GAS CHROMATOGRAPHY. A CALIBRATION CURVE is created by using standards that are made in media other than the one the sample itself is in. For example, if a toxicologist uses an external standard calibration to determine the amount of a drug in a blood sample, the calibration curve is made by using that drug in a different medium such as an organic solvent rather than blood.

extinction angle In POLARIZING LIGHT MICROSCOPY, the angle at which a polarizer has to be turned to make the object being observed appear dark.

extraction In a chemical or biochemical analysis, the process of preparing the sample such that the compounds or components of interest are separated from any materials that may complicate or interfere with the analysis of those samples. For example, to perform DNA TYPING on a bloodstain, it is necessary to extract the DNA from the blood and from the material on which it is deposited. In drug analysis, extractions are used to separate the suspected drug from any cutting agents that might be present. Each type of analysis has a specific extraction procedure. Once the sample has been extracted, it is then ready for further analysis as appropriate.

extractor In automatic and semiautomatic FIREARMS, a metal piece that extracts a spent cartridge from the chamber of the weapon. Some extractors leave marks on cases as a result.

extractor marks (ejector marks) Markings created on CARTRIDGE CASINGS that can be useful in FIREARMS analysis and identification. These impressions are created by the metal-to-metal contact between the cartridge case and the extractor and ejector mechanisms used in the weapon. The extractor mechanism removes a cartridge from the chamber; the ejector throws away the cartridge once it is extracted. REVOLVERS do not have ejectors, but automatic and semiautomatic weapons (pistols and rifles) do, and as a result the cartridge cases used in such weapons are designed differently than AMMUNITION used in revolvers. Ejector marks can be studied by using a comparison microscope and may be useful in individualizing cartridge cases.

extruded powder   A type of SMOKELESS POWDER that is made by creating dough-like material that is forced through a mold (extruded) and cut to size.

extrusion marks   *See* EJECTOR MARKS.

eyepiece (ocular)   The lens of a microscope that is found in the top portion where the observer positions his or her eye to view a specimen. If the microscope has a single eyepiece, it is monocular; if it has two, it is a binocular microscope. The purpose of the ocular lens is to magnify the REAL IMAGE created by the OBJECTIVE LENS, the lens closer to the specimen. The eyepiece lens typically has a magnification of 10×.

eyewitness testimony   Testimony provided by a person who clearly observed the event or person in question. It arises from the memory of the witness.

fabric impressions/fabric prints A type of IMPRESSION EVIDENCE that can be produced when fabric comes in contact with a surface capable of retaining the pattern. A fabric impression can be studied for class characteristics such as weave, and if the fabric has unique features such as tears, wrinkles, creases, or wear patterns, it is sometimes possible to INDIVIDUALIZE the impression and match it back to a COMMON SOURCE. Fabric impressions can also be left by contact of the cloth with a dusty or bloody surface, in which case the evidence is treated much as a latent FINGERPRINT is. Glove prints are another example of a fabric impression that has the potential to be very useful, depending on the level of detail retained in the print.

facial reconstruction A group of techniques used to assist in the identification of badly decomposed or skeletal remains. A two-dimensional reconstruction is simply a drawing (or computer-generated drawing) of the face of the deceased; a three-dimensional reconstruction is a sculpted likeness built upon a skull or portions of it. Facial reconstruction is usually a joint effort of forensic ANTHROPOLOGISTS and forensic artists. Forensic sculptors are forensic artists who specialize in three-dimensional reconstructions.

facsimile (fax) A machine that transmits and receives documents transmitted over phone lines. The term *fax* also refers to the printed message itself, which may be created by a thermal imaging system, an ink-jet printer, or a laser printer. Each sheet printed by a fax machine has a header that identifies it; printouts used as evidence are examined by techniques applied to other computer printouts.

factor B A statistical value defined by Sir Francis GALTON and used in the equations he developed to express the uniqueness of a given FINGERPRINT. Factor B represents the probability that two fingerprints will have the same general pattern such as WHORL or ARCH.

factor C A statistical value defined by Sir Francis GALTON and used in the equations he developed to express the uniqueness of a given FINGERPRINT. Factor C represents the probability that two fingerprints will have the same number of ridges passing through the 24 regions he defined as composing a fingerprint.

failure analysis In the forensic context, a technique used in forensic ENGINEERING as part of the investigation of building collapses, airplane crashes, or other events that involved failure of some kind such as mechanical, electrical, or materials. For example, if a bridge collapses, various components used in construction such as bolts and cables are subject to failure analysis as part of an investigation.

false acceptance rate (FAR) In automated FINGERPRINT processing and classification, a term that expresses the fraction of fingerprint comparisons (for example, comparison of a submitted print to one stored in the database) that are falsely accepted as a match. The FAR is needed to gauge the overall accuracy of an automated classification and matching system such as AFIS.

false classification rate (FCR) In automated FINGERPRINT processing and classification, a term that expresses the fraction of fingerprints that at initial classification are assigned to the incorrect class. The FCR is needed to gauge the overall accuracy of an automated classification and matching system such as AFIS.

false exclusion During examination of class characteristics of an item or items, an incorrect elimination of one from its correct class. A false exclusion also occurs when an item is incorrectly eliminated in an attempt to find a COMMON SOURCE. For example, consider two bullets that were fired from the same gun and subsequently examined. If one bullet had been damaged after firing and some of the characteristic marks obliterated or altered, the FIREARMS examiner might incorrectly conclude that the two bullets were not fired from the same weapon, a false exclusion.

false inclusion During examination of class characteristics of an item or items, an inclusion of one in a class to which it does not belong. A false inclusion also occurs when an item is incorrectly included in an attempt to find a COMMON SOURCE. For example, if two bloodstains were incorrectly typed and shown to have the same DNA type although in fact they did not, this would be a false inclusion and an incorrect assignment of a common source.

false negative A type of error that occurs when a test or analysis produces incomplete or negative results that should have been positive and/or definitive. For example, consider a urine sample submitted by someone who has recently ingested a drug. If the instrumentation used in the testing is not functioning properly and fails to detect the drug, a false negative result is obtained. Another type of false negative can occur when typing blood or making other attempts to determine whether two pieces of evidence have a COMMON SOURCE. For example, if DNA TYPING does not match a suspect to a bloodstain even if the blood really was from that person, that is an example of a false negative. As a part of laboratory QUALITY ASSURANCE/QUALITY CONTROL (QA/QC), procedures are designed to minimize false negatives and to detect them when they occur.

false positive A type of error that occurs when a test or analysis produces a positive result or incorrect result that it should not have produced. False positives are not uncommon with presumptive tests for drugs and blood. For example, many of the reagents used to test stains for the presence of blood give positive results with plant matter or chemical oxidants. The potential of these substances to cause false positives is the reason the tests are classified as presumptive rather than conclusive or definitive. Another type of false positive can occur in an attempt to assign two pieces of evidence to a COMMON SOURCE, as when a bloodstain is subjected to DNA TYPING in an attempt to link it to a specific suspect. If the typing produces a match even when the suspect was not the true source of the stain, this is an example of a false positive. As with false negatives, the sample itself and/or laboratory procedures can produce false positives, and QUALITY ASSURANCE/QUALITY CONTROL using CONTROL SAMPLES and other protocols helps to reduce or eliminate them when they do occur.

false reject rate (FRR) In automated FINGERPRINT processing and classification, a term that expresses the fraction of fingerprint comparisons (for example, comparison of a submitted print to one stored in the database) that are falsely rejected as matches. The FRR is needed to gauge the overall accuracy of an automated classification and matching system such as AFIS.

falsifiability One of the hallmarks of scientific theories and a key consideration in separating science from PSEUDOSCIENCE. A scientific theory or explanation is purposely designed so there is a way to test it and thus to prove it false. Scientific theories are stated this way either by using a specific mathematical formula or by wording the theory in such a way that it can be verified by experimentation or observation. The idea of falsifiability is central to forensic testing. For example, if an analyst performs a DNA analysis on a bloodstain and determines the specific type, that result (the identification of a specific type) can be tested by another analyst's performing a separate and independent test. If the results can be objectively tested, they can also be proved to be wrong.

fast blue B   A dye commonly used in a presumptive test for SEMEN.

fatty acids   Long-chained organic molecules that are produced when a fat or oil reacts with a strong base (alkaline material) such as sodium hydroxide (lye, NaOH). Naturally occurring fatty acids include palmitic acid and oleic acid. Simple soap is a mixture of fatty acids originally made by combining animal fat with wood ash, which contains hydroxides.

Faulds, Henry (1843–1930)   A Scottish physician who, while working in Japan, became interested in fingerprints. He wrote a famous letter, published in the respected periodical *Nature* in October 1880, that made the first known mention of the value of fingerprints found at a crime scene.

feathers   A widespread component of many consumer products and thus a form of TRACE EVIDENCE. Feathers and down, the fine underfeathers of birds such as geese and ducks, are used in bedding, sleeping bags, pillows, mattresses, coats, and vests, and it is not unusual for feathers or portions of them to escape from an article and to become TRANSFER EVIDENCE. Parts of feathers are also encountered as a component of DUST, one of the easiest materials to transfer. Chicken, duck, turkey, and pigeon feathers make up the majority of feathers encountered in forensic cases, but others are occasionally seen.

feature extraction   In database searches such as those for fingerprints or shoe prints, mathematical algorithms that are used to identify visual features of the pattern in question. In fingerprints, for example, a feature algorithm would be used to find ridge patterns.

fecal matter (feces)   Excrement. Until the advent of DNA TYPING, feces were not considered to be a particularly valuable type of biological evidence. Feces are most commonly encountered in burglary, sodomy, and homosexual assault cases, and a test for the compound urobilin is used to confirm the presence of fecal material. Simple examination of the contents is sometimes useful to determine the origin of the material; the presence of hair and undigested food matter is suggestive of animal rather than human origin. Because fecal matter contains a great deal of cellular matter, DNA TYPING can be utilized in some cases.

Federal Bureau of Investigation (FBI)   The investigative arm of the United States DEPARTMENT OF JUSTICE and home to the largest forensic laboratory and forensic research center in the world. The bureau maintains an extensive website at www.fbi.gov. Among the most important aspects of the FBI Laboratory Division are the numerous databases and reference collections maintained and coordinated there. They include the DRUGFIRE firearms system, the Combined DNA Index System (CODIS), the Integrated Automated Fingerprint Identification System (IAFIS), and reference collections of glass, footwear tread patterns, tire tread patterns, explosives, paints, inks, as well as numerous collections relating to questioned documents. Collation and exploitation of this type of database are emerging as key components of the future of forensic science.

Federal Rules of Evidence   A set of rules describing the admissibility of scientific evidence and expert witness testimony. The rules, embodied in Article VII, "Opinions and Expert Testimony," form the basis of admissibility of such evidence in many jurisdictions outside the federal system. Rule 702, "Testimony by Experts" states:

> If scientific, technical, or other specialized knowledge will assist the TRIER OF FACT to understand the evidence or to determine a fact in issue, a witness qualified as an expert by knowledge, skill, experience, training, or education, may

testify thereto in the form of an opinion or otherwise.

In 1923, the FRYE DECISION established a standard of admissibility that could be summarized by the term *general acceptance,* meaning that if the scientific technique in question were generally accepted in the scientific community, then it was worthy of consideration in the court. However, as scientific methods became more specialized, the limitations of this standard became apparent. In a crucial ruling in 1993 (the DAUBERT DECISION), the Supreme Court decided that the standards set forth in the Rules of Evidence were more flexible and that general acceptance was not an absolute requirement. The court assigned to the trial judge the role of determining which scientific evidence and expert testimony are relevant to a particular case and whether or not that evidence and expert testimony should be admitted.

Fehlings solution A presumptive test for reducing sugars such as fructose, which can be used as a CUTTING AGENT (diluent) for drugs. It is composed of Cu^{2+} cations in a sodium tartrate solution.

felony A crime deemed serious enough to be punished by one or more years in jail or by death.

fentanyl A powerful NARCOTIC analgesic that is 100 times more powerful than MORPHINE, but shorter-acting; it has been marketed under the name Sublimaze. Because it is such a potent drug, low dosages are taken, complicating the task of the forensic CHEMIST or TOXICOLOGIST who must identify a sample or detect its presence in urine or blood. The chemical structure of fentanyl has been the basis of several illegally synthesized analogs, or DESIGNER DRUGS. Fentanyl and its analogs have been available on the street since the late 1970s.

ferric chloride A PRESUMPTIVE TEST used in drug analysis. The reagent is prepared by dissolving ferric chloride ($FeCl_3$) at a concentration of 10 percent in water. It is used to detect the presence of MOR-PHINE, which causes the solution to turn a blue-green color progressing to a green.

fibers Tendrils of material that may be natural or synthetic, animal, vegetable, or mineral in origin. Fibers can be first classified as natural or artificial. Natural fibers include those of mineral origin (glass wool or asbestos), vegetable origin (cotton and linen), and animal origin (wool). HAIR is a specialized fiber of animal origin, which is addressed in a separate entry. Cotton, the most common vegetable fiber, is composed of cellulose, as are all plant fibers. The artificial or synthetic fibers encompass those that are derived from natural fibers and those that are completely synthetic. The Federal Trade Commission (FTC) has developed a list of generic names for families and subclasses of synthetic fibers, and the different versions made by different manufacturers are identified by their trade names. For example, one of these families is spandex, and a specific brand of spandex is Lycra, which is used in sportswear and swimsuits. Because fibers are mass-produced and used in large quantities, it is rarely if ever possible to individualize them and to link any two fibers conclusively to one COMMON SOURCE.

fibrinogen *See* BLOOD.

filament (headlights) Coiled electrical wires in the exterior lights of cars that produce light when current is passed through them. The condition of filaments, particularly the presence of GLASS on them, can provide information about whether the lamp was on or off at the time of impact.

film A celluloidlike material that contains light-sensitive compounds that darken when exposed to light. The oldest films consisted of an emulsion of silver chloride (AgCl), a compound that turns dark purple when exposed to light. Color films have additional layers of different photosensitive materials and more complex chemical structure.

fingernails and fingernail scrapings Fingernails are a tissue that is similar to hair and as such can prove useful for forensic analysis. In addition, scrapings taken from underneath the fingernails of the victims of such crimes as homicide and sexual assault may often contain minute but valuable traces resulting from the assault including BLOOD, skin, HAIRS, and FIBERS. DNA TYPING techniques have been successfully applied to matter recovered from fingernail scrapings as well as the fingernails themselves. In addition, broken fingernails can serve as transfer evidence, and PHYSICAL MATCHING techniques can sometimes be used to link a fingernail fragment to a fingernail.

fingerprint development/fingerprint enhancement The process of applying different materials and reagents to LATENT FINGERPRINTS to assist in visualizing and documenting them. A variety of techniques and procedures are available, including FINGERPRINT POWDERS, chemical developers and stains, CHEMICAL FUMING, PHYSICAL DEVELOPERS, CYANOACRYLATE (Super Glue), NINHYDRIN and related compounds, and semiconductor NANOPARTICLES containing materials such as CADMIUM SULFIDE.

fingerprint powders Powders that are dusted onto LATENT FINGERPRINTS to improve visualization. They work by adhering to the oily components found in the fingerprint residue. Powders are selected by color to maximize contrast with the background. If a print were on a white surface such as a countertop, a gray or black powder would be selected. Additionally, powders may be magnetic and applied with a magnetic brush, which makes the dusting process easier. Powders that have other desirable properties such as FLUORESCENCE are also available.

fingerprint region The region in an INFRARED (IR) SPECTRUM that spans the WAVE NUMBER region between approximately 1,500 and 700 centimeters. This region of the spectrum shows a complex pattern of radiation absorbance that is considered to be sufficient in almost every case to identify a chemical compound

definitively. For such identification, the sample that is analyzed must be pure.

fingerprints An impression, either two- or three-dimensional, produced by contact of a finger with a substrate. A two-dimensional fingerprint would be one deposited on a flat surface, such as in dust on a windowsill. A three-dimensional print, or plastic print, would be found in material such as putty. LATENT FINGERPRINTS, such as those on a surface such as glass or paper, consist of the residues produced by sweat glands (ECCRINE SWEAT glands) along with oils, fats, ions, and amino acids.

Fingerprints are defined by the patterns of FRICTION RIDGES on the fingers. As a mark of individuality, fingerprints have a long history. Ancient cultures such as the Babylonian and Chinese used them as signatures, although it is not known whether the ancients recognized that fingerprints were unique to each individual. Modern interest in fingerprints as an aid to law enforcement traces back to the middle of the 19th century. Pioneers in the field include the Europeans SIR JOHN HERSCHEL, HENRY FAULD, SIR FRANCIS GALTON (a cousin to Charles Darwin), and SIR EDWARD RICHARD HENRY and the Argentine JUAN VUCETICH.

As most forms of evidence do, fingerprints have CLASS CHARACTERISTICS that can be used to divide fingerprint patterns into categories. The patterns of fingerprints are of four basic types, whorls, tents, arches, and loops, each of which can be further subdivided as shown in the accompanying figure. Systems of fingerprint classification are based on the presence or absence of these features and are used to categorize similar patterns into smaller groups that can then be searched and compared for the MINUTIAE that are used to make individual identifications. The Federal Bureau of Investigation (FBI) classification system, an extension of the Henry system, is the most widely used in the United States and is called a 10-finger system since prints from all fingers are used.

Fingerprints are formed in utero and do not change except for expansion due to

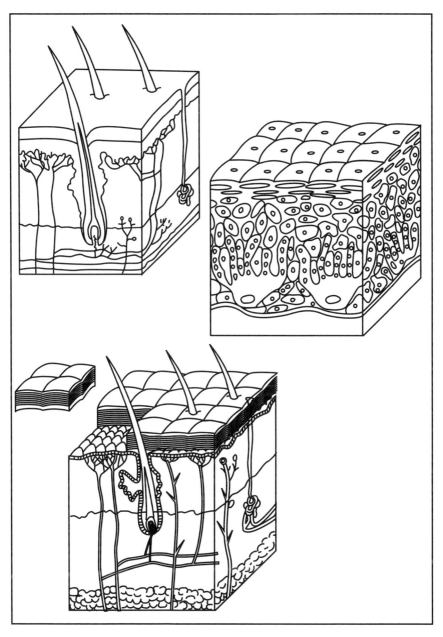

Cross-sectional view of skin showing the dermis, epidermis, eccrine sweat glands, ridge characteristics, and minutiae. The total thickness of the layers shown is approximately two millimeters.

growth. They are also unique—even identical twins do not have the same fingerprints. Because fingerprints can be initially subdivided by class characteristics, it is possible to develop sophisticated and efficient systems for classifying, filing, storing,

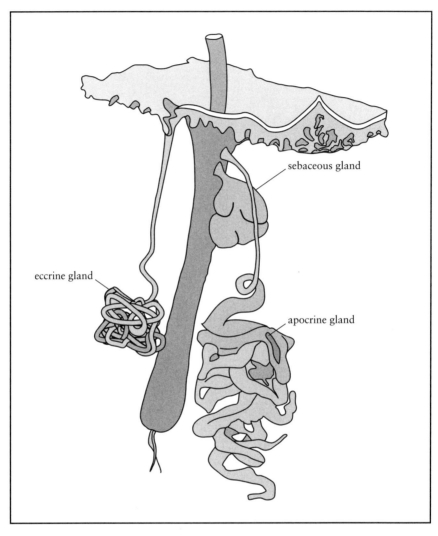

sebaceous gland

eccrine gland

apocrine gland

Detailed view showing the major secretory glands in skin, here associated with a hair follicle. Not all glands would be found on every portion of the body. The secretions from these glands are what create the latent print.

and identifying them. As a result of recent advances in digital imaging, computer databases, and software, fingerprints can be cataloged and searched electronically, providing the fingerprint examiner with a small collection of individuals who might have left the print wherever it was found. In the late 1960s, the FBI began research into the use of digital and computer technologies to assist in fingerprint classification and identification, and the first operable reader appeared in 1972. This technology continued (and continues) to evolve to the current system of automated fingerprint identification (AFIS).

Numerous methods exist for visualizing and preserving LATENT FINGERPRINTS, all based on chemical reactions with components deposited with the print. The process starts and ends with photographic documentation, but between these actions, there are many options available, depending on the type of surface the print was deposited on. Options include FINGER-PRINT POWDERS, chemical developers and stains, CHEMICAL FUMING, PHYSICAL DEVELOPERS, CYANOACRYLATE (Super-Glue), NINHYDRIN and related compounds, and semiconductor NANOPARTI-CLES containing materials such as CADMIUM SULFIDE.

firearms Weapons that exploit expanding gases created by a PROPELLANT to expel a projectile from the barrel at high speed. Modern firearms include pistols or handguns, rifles, SHOTGUNS, machine guns, automatic weapons, homemade firearms, and hobby guns, all of which are encountered by firearms examiners. Pistols are smaller guns designed to be fired with one hand and include the revolver and semiautomatic pistols, often erroneously called automatics. True automatics fire continuously as long as the trigger is pulled. Semiautomatic guns exploit gas pressure and springs to eject the spent cartridge, load a new one, and cock the weapon for the next shot, but a separate trigger pull is required to fire the next cartridge. Homemade weapons are usually small handguns that carry a few shots or one shot and have been referred to as "zip guns" and "Saturday night specials." Hobbyists such as those who reenact Civil War battles use smooth bore guns with round projectiles (ball shot) and BLACK POWDER.

In many laboratories, firearms examiners also study TOOLMARKS, GUNSHOT RESIDUE, and chemical characteristics of bullets and do SERIAL NUMBER RESTORA-TION and make DISTANCE DETERMINA-TIONS of how far a shooter was from the victim or target. Although the term *forensic ballistics* is sometimes used to describe this work, it is not a technically correct description. Ballistics is the study of the motion and trajectory of projectiles, whereas firearms analysis focuses on the study of BULLETS, CARTRIDGE CASES, and other materials associated with firearms as physical evidence.

Tasks of firearms investigator are similar to those of other forensic analysts in that they seek CLASS CHARACTERISTICS and other characteristics that INDIVIDU-ALIZE evidence by linking a specific weapon to a specific bullet or cartridge. Modern guns all have rifled barrels, meaning that a series of grooves (LANDS AND GROOVES) are machined into the barrel in a spiraling pattern. When the bullet is forced over these lands and grooves, spin is imparted to the bullet, stabilizing the trajectory and greatly increasing accuracy over that of smooth-bore weapons. This contact creates distinctive STRIATION patterns on the bullet that can later be matched to the weapon. The striations are usually unique to one gun, determined by the way the gun is manufactured and used. The primary tool of firearms examination is the COM-PARISON MICROSCOPE. The pioneers of firearm analysis include VICTOR BALT-HAZARD and CALVIN GODDARD.

fire investigation Similar to the investigation of ARSON, and occasionally used synonymously with the term *arson investigation*. Although arson and fire investigators often work side by side, the primary difference between the two jobs arises from the ultimate cause of the fire. If arson is involved, a criminal act has occurred, and that is the responsibility of the arson investigator. However, most fires are not the result of arson. These fires are the subject of fire investigation. Fire investigation is often considered to be part of or closely related to forensic ENGINEERING.

fire load The amount of flammable material that is available in a given place such as a structure or an aircraft. Fire load is typically reported in British thermal units per square foot (BTU/ft^2).

fire patterns The patterns left on surfaces by a fire, burning, smoke, or soot. Such patterns are used in fire and ARSON investigation to determine the point(s) of origin of a fire.

fire point The temperature at which a liquid accelerant produces a sufficient amount of vapor to sustain combustion. It is sometimes referred to as the IGNITION TEMPERATURE.

firing pin A small metal projection that strikes the PRIMER of a CARTRIDGE that is loaded into the breech of a firearm. The impact causes the primer to explode, thereby igniting the PROPELLANT. The impact of the firing pin on the primer creates FIRING PIN IMPRESSIONS.

firing pin impressions (or indentations) Small marks created in the PRIMER by the impact of the firing pin. The action of pulling the trigger of a firearm causes the firing pin to strike the primer, which consists of a tiny amount of a shock-sensitive explosive. Ignition of the primer in turn ignites the PROPELLANT, causing a rapid buildup of gases behind the bullet. The pressure drives the bullet down the barrel and out toward the target. The surface of the primer struck by the firing pin is relatively soft metal, which can pick up the pattern on the surface of the firing pin. Either as a result of the original machining or through use and wear, the markings on the surface of the firing pin become unique.

first-order reaction/first-order process A chemical reaction in which the rate of the reaction depends on only one reactant.

first responder The first person to arrive at a scene other than witnesses. The first responder is often a police officer, firefighter, or member of an emergency medical team.

FISH An acronym with two different meanings in the forensic context. A Forensic Information System for Handwriting is a database of handwriting samples used in QUESTIONED DOCUMENT analysis. Letters or other writings involved in cases (bank robbery notes, and so on) are scanned into a database for future reference and comparison. If a new document that is submitted shows similarities to others already in the database, a document examiner can be alerted so that further study can be made. In this sense, FISH is similar to the DRUGFIRE system for cartridge casings and bullets. The FISH system was developed by the United States SECRET SERVICE. The other term that has the acronym *FISH* is *fluorescent in situ hybridization,* a method that can be used to determine the sex of cells.

Fish and Wildlife Service (wildlife forensics) The federal law enforcement agency responsible for protection of wildlife and prosecution of crimes such as poaching. In 1989, the agency opened the first and currently the only forensic laboratory dedicated to the investigation of crimes related to wildlife, a specialty referred to as wildlife forensics. The laboratory, which is located in Ashland, Oregon, serves the national and international communities. The lab specializes in the identification of animals by species and subspecies on the basis of bones, biological fluids, or parts of the animal submitted as evidence. Cause of death is determined to the extent possible, and PHYSICAL EVIDENCE such as FINGERPRINTS and FIREARMS is tested.

fixation The process of making something permanent such as a photographic image, developed or enhanced fingerprint, or stained specimen on a microscope slide. Fixation is usually a chemical process. The term is also used in forensic psychology, psychiatry, and other behavioral sciences to describe a person's focus on or obsession with another person or an object.

flame ionization detector (FID) A type of detector used in GAS CHROMATOGRAPHY (GC), which is frequently used in the analysis of ARSON evidence. An FID is sensitive to compounds containing bonds between carbon and hydrogen, making it an ideal detector for the analysis of hydrocarbons such as gasoline and other common ACCELERANTS. Gas emerging from the outlet of the GC instrument is mixed with flammable hydrogen gas and ignited such that the flame burns constantly whenever the detector is on. Compounds are separated in the GC and enter into the flame, where the hydrocarbon molecules burn to form CHO⁺ ions. Because ions carry a positive electrical charge, they are drawn toward the collector, which has a negative charge. The greater the number of positive ions collected, the greater the detector response.

flammability   The ability of a solid, liquid, or gas to ignite and to burn rapidly. The United States Department of Transportation defines a *flammable liquid* as one that has a flash point of 100°F (38°C).

flammable liquids   *See* ACCELERANTS.

flammable range   The concentration range of a flammable liquid in air that will burn. It lies between the LOWER EXPLOSIVE LIMIT and the UPPER EXPLOSIVE LIMIT. The flammable range for gasoline is 1.3 percent to 6.0 percent concentration in air. Below 1.3 percent, the mixture of gasoline vapors and air is too lean to burn, and above 6 percent, it is too rich.

flanking region   A section or region of DNA that is located directly beside a region of interest. DNA primers bind to flanking regions during the thermal cycling amplification step of DNA TYPING.

flashback   In a fire set by using a flammable liquid, the movement of a fire back to the container. For example, if an arsonist leaves a trail of gasoline leading to a gas can and then lights the end of the gasoline trail, flashback occurs as the flame travels back to the gasoline can.

flashover   A phenomenon that can occur if a fire is contained to a room or other area and becomes hot enough to raise all materials in that room to ignition temperature at nearly the same time. Flashover results as these materials catch fire, causing rapid spread.

flash point   The temperature at which a liquid gives off enough vapor to form an ignitable mixture. For gasoline, the flash point is –50°F. The National Fire Protection Association (NFPA) defines a flammable liquid as one with a flash point of less than 140°F.

flexion creases   The pattern of creases, folds, and lines in the palm of the hand that exist at the points where the hand flexes.

flintlock   A type of firearm used into the 19th century in which flint striking a surface created the spark that ignited the PROPELLANT.

floater   An informal term sometimes used to refer to a body recovered from water.

float glass   Glass that is made by pouring molten glass over liquid tin (Sn), where it is allowed to cool and solidify, producing a flat glass. The infusion of small amounts of tin into the glass can be useful in identifying a sample as float glass since the residual tin fluoresces when exposed to ultraviolet (UV) light or black light. However, since only one side of the glass was exposed to the tin, only one side shows such behavior.

Florence test   *See* CHOLINE TEST.

flow pattern   The shape and direction of flow of wet blood, which are dictated by gravity. On a vertical wall or other surface, blood flows downward. Movement of a surface can also influence the flow pattern. For example, if liquid blood is on a tabletop and the table is upended, the blood flows according to gravity; however, if the blood is dry, no additional flow occurs. This type of information can be useful in crime scene reconstruction.

fluorescein   A reagent used as a PRESUMPTIVE TEST for blood that works similarly to LUMINOL. It is used most commonly to visualize stains not visible to the eye, such as when a scene has been cleaned. Fluorescein is used in conjunction with an ALTERNATE LIGHT SOURCE that produces fluorescence.

fluorescence When atoms or molecules absorb ELECTROMAGNETIC RADIATION of the appropriate wavelength, the energy that is absorbed can be used to promote the electrons into what is called an EXCITED STATE. When the electrons decay back to their original ground state, energy is emitted. In the case of fluorescence, the energy is emitted in the form of electromagnetic energy that has the same wavelength as or a longer wavelength (lower energy) than the excitation energy. This emission of energy is called fluorescence, and it ceases as soon as the excitation energy source is turned off. If the emission continues after the source is turned off, the phenomenon is called PHOSPHORES-CENCE, familiar to many in the form of glow-in-the-dark watches. Fluorescence can be used to classify and identify many materials of interest to the forensic scientist and can be performed by techniques called SPECTROFLUOROMETRY or use of fluorescence microscopes.

fluorescent polarization immunoassay *See* FPIA.

fluorophore A species that is capable of fluorescence when exposed to electromagnetic radiation. It may be a chemical functional group, a molecule, or a compound that is inserted into another.

fluoroscopy A real-time X-ray technique that is used in airport security scanners and as a first step in some autopsies, especially those associated with mass dis-

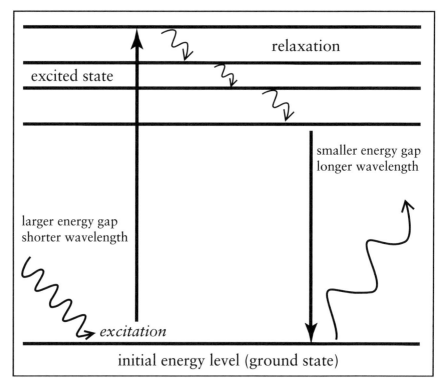

Fluorescence. Energetic radiation, such as in the ultraviolet (UV) range, is absorbed, promoting an electron in the molecule to an unstable excited state. Collisions or other nonradiative processes dissipate some of the energy, and then the electron returns to the ground state, emitting light of lower energy and a longer wavelength.

asters or deaths in which hazardous articles or materials might be found on the remains.

flurazepam A member of the BENZODIAZEPINE family of drugs marketed under the name of Dalmane. It is prescribed to treat sleeping disorders and insomnia.

focal length With a simple convex (converging) LENS such as is found in microscopes and cameras, the focal length is the distance between the principal focus point and the lens as measured from the geometric center of the lens.

follicle The small cavity in the skin in which a HAIR is anchored. The hair grows out of the hair follicle.

font In typewriters and computer printouts, a set of characters with common design elements, size, and typeface. Typewriters had Pica and Elite fonts; with the advent of computers, thousands of fonts are available. This book is printed in Times New Roman font.

Food and Drug Administration (FDA) The U.S. federal agency that regulates the nation's food and drug supply for purity, quality, and effectiveness. The FDA maintains a Forensic Chemistry Center in Cincinnati, which was established in 1989; most of its work relates to product tampering and contamination. The lab also works on cases involving the sale and use of drugs that are not approved in the United States, counterfeit food and drug products, and fraudulent or deceptive labeling. The importance of forensic capabilities in tampering first drew the public's attention in the early 1980s, when the first cases involving cyanide-tainted Tylenol occurred in Chicago. As a result, the Forensic Chemistry Center has developed numerous rapid screening techniques for many poisons that can be added to food and pharmaceutical products.

footprints (footwear impressions) See SHOE PRINTS.

footwear database A database maintained by the FBI that contains the sole patterns of thousands of shoes.

forcing cone A cone-shaped flare in the barrel of a revolver at the breech end (firing pin end). The function of the forcing cone is to facilitate the feeding of the cartridge into proper seating position before firing.

FORDISC A computer program written by Drs. Richard Jantz and Steven Ousley that is widely used in forensic anthropology. The program takes skeletal measurements as inputs and assists the anthropologist in classifying remains by race and sex.

forensic In its simplest and oldest meaning, *forensic* relates to public debates (in a "forum"). The term is still used in this context to describe debating clubs and societies in some high schools and universities. However, the common modern use of the term is related to the legal system, a formalized public debate epitomized in the courtroom. Thus, FORENSIC SCIENCE can be broadly considered to be the application of science to a legal context.

forensic light source See ALTERNATE LIGHT SOURCE.

forensic pathologist A medical doctor who specializes in the study of the causes of death. After obtaining an M.D. degree, a candidate completes a residency in pathology followed by one to two years of studying forensic pathology. Forensic pathologists are certified by the American Board of Pathology (ABP).

forensic science In the broadest sense, the application of the techniques of science to legal matters, both criminal and civil. Forensic science includes a number of disciplines and subdisciplines such as forensic anthropology, engineering, and medicine. There is some debate as to how the terms CRIMINALISTICS and *forensic science* relate to each other; usually criminalistics is considered to be the largest subdivision of forensic science, which encompasses the analysis of physical evidence. In this scheme, criminalistics includes the traditional divisions found in forensic science laboratories such as SEROLOGY and DNA (BIOLOGY), CHEMISTRY, and TRACE EVIDENCE.

Forensic Science Service (FSS) The main provider of forensic services in the United Kingdom (England, Wales, Scotland, and Northern Ireland) which is somewhat analogous to the FBI laboratory in the United States. The FSS, part of the Home Office of the UK government, was established in 1991 and consolidated several forensic laboratories throughout England and Wales. The Metropolitan Police Laboratory (city of London) was merged into the FSS in 1996.

forensics *See* FORENSIC SCIENCE.

forensic system for handwriting (FISH) *See* FISH.

forgery As defined by *Black's Law Dictionary, forgery* is the "false making or material altering of a document with intent to defraud. . . . A signature of a person that is made without the person's consent and without the person authorizing it." At its core, forgery is an attempt fraudulently to imitate an original, be it a signature, lottery ticket, will, painting, sculpture, historical document, or ancient artifact. In forensic science, evidence associated with forgery usually falls within the purview of QUESTIONED DOCUMENTS, but many cases of forgery involve, for example, artworks and archaeological artifacts.

formula weight The average mass of a compound given in units of grams per mole. The compound may be IONIC, such as NaCl, or MOLECULAR, such as H_2O. Masses of the individual atoms are obtained from the Periodic Table of Elements. The formula weight is called an average weight because the masses of the individual atoms are given as weighted averages that incorporate all of the naturally occurring isotopes. *See* APPENDIX III.

forward spatter Blood that is spattered in the forward direction by an impact. This blood travels in the same direction as the projectile, weapon, or force that created the injury and contains greater amounts of material than any BACK SPATTER that occurs.

Fourier transform/Fourier transform infrared spectroscopy (FTIR) A mathematical series that is used to express a curve as a sum of sine and cosine terms. The Fourier transform, coupled to a device called a MICHELSON INTERFEROMETER, is the basis of Fourier transform infrared spectroscopy (FTIR). Mirror movements create a series of waves that undergo periodic constructive interference and destructive interference before reaching the sample. Although computationally intensive and complex, application of the Fourier transform allows the resulting signal curve to be decomposed into signals at individual wavelengths and an INFRARED SPECTRUM. FTIR offers many advantages over traditional instruments including greater speed and sensitivity. FTIR is the predominant type of INFRARED SPECTROPHOTOMETRY used in forensic science.

FPIA Fluorescent polarization IMMUNOASSAY, an immunological technique that uses a fluorescent label that interacts with POLARIZED LIGHT in different ways, depending on whether it is bound to the antibody or free in solution.

fragmentation patterns In forensic science, a term that usually refers to the way in which a molecule is fragmented in a mass spectrometer. When molecules of the same compound are introduced into a MASS SPECTROMETER, they are fragmented by collisions with electrons and the MASS SPECTRUM records the mass of the fragments versus their abundance. Under the same instrumental conditions, a compound should always produce comparable fragmentation patterns. Mass spectra are invaluable for compound identification in drug analysis, toxicology, and other forensic chemistry applications.

fraud A willful attempt by a person or entity to deceive another person or entity by lying, concealing the truth, misrepresenting, making false statements or representations, or misusing authority or perceived authority. Shredding of legal documents and willful tax evasion are examples of fraud, and many fraud cases involve QUESTIONED DOCUMENTS.

free base A term that usually refers to the basic form of drugs, specifically COCAINE, giving rise to the term *free basing*. Many drugs can exist in both an acidic form (acidic salt) and a basic form; the appearance and physiological effects are quite different. Cocaine hydrochloride is a sparkling white powder that is usually snorted or dissolved in water and injected. Cocaine free base is a sticky, resinous material that can be smoked, causing an almost instantaneous effect. Conversion of a drug from one form to the other requires only simple chemical reactions and can be accomplished by using reagents as simple as baking powder or sodium hydroxide (NaOH, lye).

free energy *See* GIBBS FREE ENERGY.

freehand simulation (drawing) In the forgery of handwriting or a signature, an attempt to copy the writing without tracing, that is, by copying or drawing freehand. The quality of the forged signature may appear good, but forgery is often betrayed by signs of slow penmanship and lifting characteristics not found in a genuine signature.

freon Usually, an informal term for compounds containing chlorofluorocarbons (CFCs) that were once widely used as propellants for sprays such as NINHYDRIN. CFCs are no longer available as propellants because of their propensity for ozone depletion in the stratosphere, and substitute propellants are now used.

frequency A term that has two meanings in forensic science. The *frequency* of electromagnetic energy is defined as the number of waves that pass a given point in space per second. The higher the frequency, the greater the energy. Frequency is related to wavelength and the speed of light in a vacuum by the equation $c = \lambda \nu$ where c represents the speed of light, λ is the wavelength, and ν is the frequency. The units of frequency are hertz per second[-1]. The other common usage of *frequency* in forensic science is related to DNA TYPING. Here the term refers to frequencies of expression of a genetic allele, which are determined from population studies. These are also referred to as FREQUENCY ESTIMATES.

frequency estimates Estimates of the number of people in a specific population who have a given genetic characteristic that are based on previous knowledge, data, and testing. For example, in the ABO BLOOD GROUP SYSTEM, approximately 42 percent of the Caucasian population is type A; this figure is a frequency estimate. The term is now used primarily in relation to DNA TYPING, but the idea can be extended to any kind of evidence that exists within known classes or subdivisions.

friction ridge/friction ridge skin Skin surface that has ridges that provide friction to increase gripping power. In humans, these surfaces are the soles of feet, palms of hands, and fingers. These form patterns such as whorl patterns.

Froehde's reagent A presumptive test in drug analysis that is most often used to detect the presence of LSD (LYSERGIC ACID DIETHYLAMIDE). It consists of a solution of molybdic acid and sulfuric acid, which turns an olive green to blue-green color in the presence of LSD. The test is somewhat limited in that LSD is typically found in very small amounts (microgram quantities) or less on the individual blotter papers on which it is usually sold. Froehde's reagent can also be used to indicate the presence of heroin (purple to olive green color), and mescaline (yellow to greenish color).

***Frye* decision/*Frye* standard** A court ruling handed down in 1923 that has greatly influenced the ways scientific evidence and expert witness testimony are admitted and used. The ruling was a rejection of the validity of the POLYGRAPH (LIE DETECTOR) test, which stated in part:

> Just when a scientific principle or discovery crosses the line between the experimental and demonstrable stages is difficult to define. Somewhere in the twilight zone the evidential force of the principle must be recognized, and while courts will go a long way in admitting expert testimony deduced from a well-recognized scientific principle or discovery, the thing from which the deduction is made must be sufficiently established to have gained general acceptance in the particular field in which it belongs.

This ruling led to a criterion referred to as general acceptance that governed the admissibility of scientific evidence in many jurisdictions. General acceptance is also referred to as the *Frye* standard.

FTIR *See* INFRARED SPECTROSCOPY AND MICROSCOPY.

fuel cell A device that extracts electrical energy from reagents via an oxidation/reduction (REDOX) process. Fuel cells are used to power some portable BREATH ALCOHOL measurement instruments.

fungus/fungi A class of plants that do not have chlorophyll and that reproduce by spreading spores. Mushrooms, molds, mildew, and yeasts are all examples of fungus. In forensic science, fungi can play an important role in DECOMPOSITION and forensic BOTANY.

G

Gabor filters A mathematical algorithm used to enhance digital images of FINGERPRINTS. These filters treat the patterns of ridges and valleys in localized areas of a fingerprint image as a wave pattern (a sine wave) that can be treated with filters designed for signal processing.

gait measurement The measurement of the length of a person's typical stride, which can be important in areas such as forensic BIOMECHANICS.

Galton model The first attempt to address statistically the rates of occurrence of different FINGERPRINT patterns. It was set forth by FRANCIS GALTON in 1892 and, although never widely adopted, initiated the continuing discussion concerning the probabilities associated with a given fingerprint.

Galton ridges A term sometimes used to describe the ridges in a FINGERPRINT.

Galton, Sir Francis (1822–1911) An English researcher in heredity and a pioneer of early fingerprint studies. He was also the cousin of Charles Darwin. He is credited with developing the first classification system for fingerprints, which was adopted by the British government as an adjunct to the BERTILLON system of body measurements and photographs that was then the primary method of identification of criminals. In 1892, he published the influential book *Finger Prints,* which helped put fingerprinting at the forefront of criminal identification. It is still considered to be one of the primary references in the field. The book was notable for stating the fundamental principles that fingerprints are unique and unchanging. Galton also was the first proponent of classification using the basic patterns of the loop, arch, and whorl. In the United States, the term GALTON RIDGES is used to describe one of the features found in fingerprints.

gamma radiation Electromagnetic radiation given off during the radioactive decay of an atomic nucleus. Unlike alpha and beta radiation, gamma radiation is energy and not a particle and thus has high penetrating power. Gamma radiation is utilized in NEUTRON ACTIVATION ANALYSIS.

gamma ray spectroscopy *See* NEUTRON ACTIVATION ANALYSIS.

gas chromatography (GC) An instrumental technique used forensically in drug analysis, ARSON, TOXICOLOGY, and analysis of other ORGANIC COMPOUNDS. GC exploits the fundamental process common to all types of CHROMATOGRAPHY: separation based on selective partitioning of compounds between different phases of materials. Here, one phase is an inert gas (helium [He], hydrogen [H_2], or nitrogen [N_2]) that is referred to as the mobile phase (or carrier gas), and the other is a waxy material (called the stationary phase) that is coated on a solid support material found within the chromatographic column. In older GC systems, the stationary phase was coated on tiny beads and packed into glass columns with diameter about the same as that of a pencil and lengths of six to 12 feet, wound into a coil. The heated gas flowed over the beads, allowing contact between sample molecules in the gaseous mobile phase and the stationary phase. Called packed column chromatographs, these instruments were widely used for drug, toxicology, and

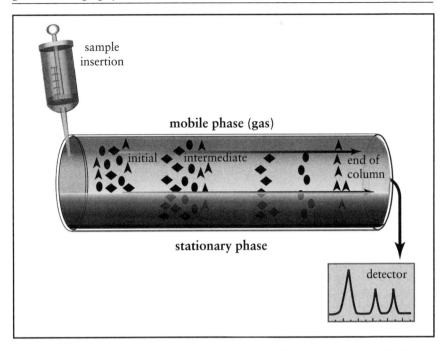

Depiction of the separation process in gas chromatography. Refer to the text for a detailed discussion.

arson analysis. Around the mid-1980s, column chromatography began to give way to capillary column GC, in which the liquid phase is coated onto the inner walls of a thin capillary tube (about the diameter of a thin spaghetti noodle) that can be anywhere from 15 to 100 meters long, also wound into a coil. Capillary column chromatography represented a significant advance in the field and greatly improved the ability of columns to separate the multiple components found in complex drug and arson samples. However, a few applications still require packed columns.

The purpose of the gas chromatograph is to separate mixtures into individual components that can be detected and measured one at a time. A plot of the detector output, called a CHROMATOGRAM, charts the detectors' response as a function of time, showing the separate components. The separation is based on differences in affinities for the two phases. As shown in the figure, the sample is introduced into the GC column by way of a heated injector, which volatilizes all three components and introduces them into the gas flowing over the stationary phase. In this example, the compound represented by the arrowheads has the least affinity for the stationary phase. As a result, it moves ahead of the other two components and reaches the detector first. The compound symbolized by the diamond has the greatest affinity for the stationary phase and spends the most time associated with it. As a result, this compound is the last to reach the detector. Separation has been achieved on the basis of the different affinities of the three types of molecules found in the sample. In reality, complex mixtures cannot always be completely separated, and some compounds emerge from the column simultaneously. This effect, called COELUTION, can often be overcome by using detectors such as MASS SPECTROMETERS.

A number of detectors are available for use in gas chromatography. In forensic applications, the most commonly used are MS, FLAME IONIZATION, and NITROGEN-PHOSPHORUS detectors. The mass spectrometer is the most commonly used of the three, principally because it can provide definitive identification of compounds (in almost all cases) along with quantitative information.

gasoline *See* ACCELERANTS.

gastric contents Contents of the stomach (sometimes including the upper intestine) recovered from a body during autopsy. They can be used for toxicological studies or for determination of what was last eaten and approximately when; however, the usefulness of gastric contents is often exaggerated in popular literature and fiction.

gauge A device used to identify the size of a SHOTGUN barrel. *Gauge* originally referred to the number of lead pellets having the same diameter as the barrel that would be required to obtain one pound of lead. The larger the diameter of the barrel, the larger the pellets, and the fewer would be needed to make a pound. Thus, a 12-gauge shotgun has a larger barrel diameter (0.730 inch) than a 16-gauge (0.670 inch). There is one exception, the 0.410 gauge shotgun, in which the barrel diameter is 0.410 inch. This convention is similar to the caliber naming convention used for handguns and rifles.

Gaussian error Error that is as often positive as negative, and generally small. Such errors can also be referred to as RANDOM ERRORS. These errors are usually associated with an instrument or device of some type.

GC/MS (GC-MS; GCMS; GCMSD)
Acronyms used for a GAS CHROMATOGRAPH (GC) coupled to a MASS SPECTROMETER (MS). GC/MS systems are fundamental in many toxicology and forensic chemical analyses. The gas chromatograph separates a complex mixture into individual components (in most cases), and the mass spectrometer identifies and quantitates these components.

gel A supporting medium for ELECTROPHORESIS. Gels can be made of agarose, a derivative of seaweed, or POLYACRYLAMIDE. Gels are structured materials that can absorb large amounts of water and thus electrolytes and ions that are dissolved in the water. The presence of water and electrolytes allows the gel to conduct electricity.

gel diffusion A phenomenon that is exploited in forensic biology. Because gels contain large amounts of water within a structured framework, molecules can diffuse through gel, but at a much slower rate than they would diffuse through water alone. Techniques that exploit gel diffusion include OUCHTERLONY test diffusion and other tests for determining species.

gel electrophoresis A technique used to separate large molecules such as proteins by using a gel (AGAROSE or POLYACRYLAMIDE), a BUFFER solution, and an applied electrical field. It is used in typing of ISOENZYME systems and in DNA TYPING.

gender determination *See* SEX DETERMINATION.

gene *See* GENETICS.

gene frequency The percentage or relative proportion present in a population of alleles for a given genetic trait, GENETIC MARKER type, or band found in DNA TYPING. Knowledge of gene frequency is necessary when determining the commonness of a given type; the HARDY WEINBERG law can be used to estimate frequencies.

general acceptance *See* FRYE DECISION.

generalist The issue of whether a forensic scientist should be a generalist well versed in most aspects of the field or a dedicated specialist in one area has been debated since the earliest days of forensic science. On one extreme are those who argue that the discipline has advanced to the stage that specialization not only is the norm, but is required given the depth and technical complexity of each area (FIRE-ARMS, QUESTIONED DOCUMENTS, and so on). The opposing viewpoint holds that forensic scientists should be knowledgeable across a broad spectrum of areas, even if they do not work in those areas on a daily basis. Pioneers such as HANS GROSS and PAUL KIRK held this opinion, and it was Kirk that expressed the view that a CRIMINALIST is in fact an expert in comparison and INDIVIDUALIZATION rather than a specialist in any one forensic area.

General Knowledge Exam (GKE) A certification examination offered by the AMERICAN BOARD OF CRIMINALISTICS (ABC) introduced in 1993. The initial test is written and covers a wide range of topics in forensic science and CRIMINALISTICS; passing it gives the analyst Diplomate status, designated as *D-ABC*. To advance to Fellow status (*F-ABC*), the analyst must pass a second written examination in a specialty area as well as a laboratory proficiency test. Specialty tests are given in DNA TYPING, TRACE EVIDENCE ANALYSIS, and drug analysis. The first test is broadly based in criminalistics; its emphasis on the forensic scientist as a GENERALIST supports the notion that such a person should be familiar with other aspects of the field even if he or she does not work directly with them.

Genesolv 2000 The trade name of a solvent manufactured by Honeywell consisting of dichlorofluoroethane. This compound is considered less damaging to the ozone layer than traditional solvents such as the freon family and has been used as a solvent for NINHYDRIN.

genetic engineering The purposeful manipulation of the genetic code of an organism by insertion or deletion of genetic material.

genetic linkage *See* LINKAGE.

genetic marker systems Inherited characteristics that show variation within a population. Such a variation is called a POLYMORPHISM ("many forms") and can

be illustrated by the most well known genetic marker system in blood, the ABO BLOOD GROUP. A person's ABO blood type is inherited from the parents and can be type A, B, AB, and O. Thus, ABO is a genetic marker system that is polymorphic with four types in the population. Genetic marker systems exist in BLOOD and can also be typed, under certain circumstances, in body fluids such as semen or saliva. To be useful in forensic work, a genetic marker system should have several variants that subdivide the population into several smaller groups, the more groups, the better. Furthermore, the system must be robust and must resist degradation long enough to be typed in bloodstains that may be old and/or in very poor condition. This last requirement limited the widespread typing and use of many genetic marker systems. Although fairly easy to type in fresh whole blood, many of the systems are fragile and could not be routinely and reliably typed in stains.

Blood can be subdivided into different fractions, all of which contain genetic marker systems. After centrifuging, blood separates into serum and the cellular components. Serum, a yellowish liquid, contains serum blood group systems such as haptoglobin (Hp) and group-specific component (Gc) that are polymorphic. Within the cellular component, white blood cells (leukocytes) contain the HUMAN LEUKO-CYTE ANTIGEN (HLA) system, includes many individual factors and types. Both the serum blood group systems and the HLA system were difficult to type in stains and were not routinely used in forensic casework.

Unlike red blood cells, the white blood cells have a nucleus, which is the source of DNA used in DNA TYPING. The 13 loci that are usually typed in current practice can also be classified as genetic markers since they are inherited and polymorphic. Red blood cells are the richest source of non-DNA genetic marker systems that were once widely used in forensic SEROL-OGY. These cells (erythrocytes) have on their surface the antigens that make up blood group systems such as ABO and Rh. Within the cell are found the ISOENZYME

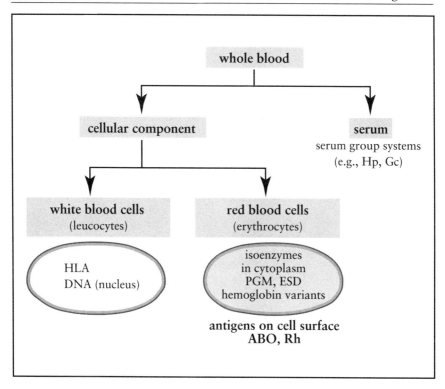

The genetic markers in blood and the locations where they are found. The plasma portion of blood retains the serum proteins such as haptoglobin; the remainder is found in the cellular component. DNA is found in the white blood cells, as is the human leukocyte antigen (HLA) system. The isoenzymes such as PGM and blood group systems are associated with the red blood cells.

SYSTEMS such as phosphoglucomutase (PGM) and esterase D (ESD) as well as variations of the hemoglobin molecule. The ABO blood group and isoenzymes were the most used in casework.

genetics The study of heredity, the fundamental unit of which is the gene. Genes are encoded in DNA, and the place in which a particular gene is found is called its locus. Multiple sites are called loci. Different forms of the same gene are called alleles. Human genes are found on 46 chromosomes organized into 23 pairs. Within those pairs, one chromosome originated from the mother, the other from the father. Many of the genetic character-

istics of interest in forensic science are codominant, meaning that both the allele from the mother and that from the father are expressed. For example, the ABO BLOOD GROUP SYSTEM is a genetic marker system with the types A, B, AB, and O. The A type is a result of AA alleles. The observable, measurable way in which any gene is expressed is called the PHENOTYPE, and so the phenotype of AA is A in this example (disregarding subtypes). When such a variation occurs at a given locus, it is referred to as a POLYMORPHISM ("many forms") since variants exist within the population. POPULATION GENETICS, the study of the relative frequencies of different types or variations within a given population (Caucasian, African American, and so on) is of critical importance in interpreting the result of forensic DNA or genetic marker analysis.

genetics, forensic Broadly speaking, the application of GENETICS to forensic science. Most often this term refers to POPULATION GENETICS, FREQUENCY ESTIMATES, and issues related to DNA TYPING.

genome The complete complement of an organism's or a species's DNA. For example, the human genome is all of the DNA (including variants) found within the DNA of every person. The term generally refers to DNA found in the cell nucleus as opposed to MITOCHONDRIAL DNA. A person's genome is considered to be all the DNA found within his or her body.

genotype The actual alleles found at a gene locus, as compared to the phenotype, which is the way the trait is expressed. For example, the ABO BLOOD GROUP SYSTEM is a genetic marker system with the types A, B, AB, and O. The A type is a result of AA alleles, and the genotype is AA. The observable, measurable way in which any gene is expressed is called the PHENOTYPE, and so the phenotype of this person with the AA genotype is A, since he or she has A antigens on the surface of the red blood cells and has type A blood (disregarding subtypes). Similarly, a person whose genotype is AO has a phenotype of A.

gentian violet A histological stain used for staining of tissues and cells and for visualization of latent FINGERPRINTS.

geographical profiling A technique that uses locations of crime scenes in an attempt to link serial crimes such as rape or murder. It is based on the theory that crime scene location is not entirely random and that in linked crimes, it may be possible to glean information from the locations themselves. The process often uses GIS (Geographic Information Systems) software and accompanying databases to provide investigative information.

geology, forensic Primarily the study of SOILS, which are a common type of TRANSFER EVIDENCE. HANS GROSS and EDMUND LOCARD, pioneers of forensic science, were among the first to recognize the value of soil evidence in linking an individual to a particular place. Early work and explorations of forensic geology were reported in the middle of the 19th century, but the first documented cases are generally credited to Georg Popp of Frankfurt, Germany. In the early 1900s, Popp used soil analysis, and particularly evidence found in layers of soil, to link suspects to crime scenes. This process is still one of the primary roles of the forensic geologist. The analysis of soil is similar to the analysis of GLASS in that it relies heavily on the measurement of physical and optical properties. Perhaps the most important tool for the forensic geologist is MICROSCOPY using STEREOMICROSCOPES, POLARIZING LIGHT MICROSCOPES, and SCANNING ELECTRON MICROSCOPY/ENERGY DISPERSIVE X-RAY INSTRUMENTS.

geometrical isomers Two or more chemical compounds with the same chemical formula that differ in the arrangement of the atoms or groups around a ring or a double bond. The prefixes *cis-* and *trans-* are used to differentiate geometrical isomers.

geophysical methods Instruments and techniques used to measure the characteristics and properties of the Earth, principally beneath the surface. An example of a forensic use of a geophysical technique would be employment of ground-penetrating radar to search for CLANDESTINE GRAVES. Other geophysical methods use metal detectors, seismographs, and magnetometers.

GHB Gamma hydroxybutyrate, a substance classified as a DATE RAPE DRUG. It was at one time used as a bodybuilding supplement but was pulled from stores in 1990. Easily synthesized, it has been abused in much the same way as Rohypnol. Low doses relieve tension and promote relaxation, but higher doses produce sleep (sometimes suddenly) and nausea, and alcohol enhances these effects. In recognition of its dangers, GHB was

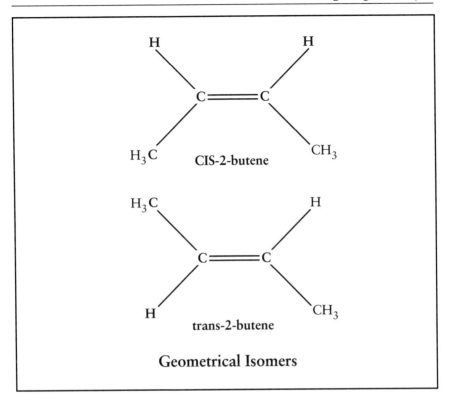

CIS-2-butene

trans-2-butene

Geometrical Isomers

Two different structures (geometrical isomers) possible for the organic compound 2-butene (C_4H_8). These are geometrical isomers in that they differ only in the arrangement of the methyl group ($-CH_3$) around the double bond. In the cis configuration, the two methyl groups are on the same side of the double bond, and in the trans configuration, they are on opposite sides.

placed on Schedule I of the CONTROLLED SUBSTANCES ACT in 2000.

Gibb's free energy (G) A measurement of the energy in a system that includes terms for ENTHALPY (H), temperature (T), and ENTROPY (S, disorder). The free energy change in any process, such as a chemical reaction, determines whether that reaction is spontaneous or not and is calculated as $\Delta G = \Delta H - T\Delta S$. If the sign of ΔG is negative, the reaction is spontaneous; if it is positive, it is not. Another way to state the last part of this is to say that if the sign of ΔG is positive, the reverse process is spontaneous.

GIS Geographic Information System software, which fundamentally is mapping software combined with database capabilities. Such software can be used for crime mapping and for GEOGRAPHICAL PROFILING.

glass/glass analysis Glass is an amorphous solid lacking the rigid, ordered crystal structure of materials such as table salt. In the strictest sense, glass is considered a viscous solid or a supercooled liquid rather than a true solid. Because of its lack of order and pattern at the molecular level, glass breaks in random patterns and has other unique properties that make it a valuable and widely used material. The most common ingredient in most glass is quartz sand, SiO_2. Soda lime glass, of which windows are usually made, has additives of Na_2O, CaO, MgO, and Al_2O_3, all metal oxides. Different additives are used to impart different physical characteristics to the glass. Colored glasses are made by adding oxides from the transition metal family. Red glass is made by adding cadmium or selenium; adding cobalt compounds can make blue. Leaded glass used in fine crystal contains high concentrations of lead oxide (PbO). Tempered or safety glass is created during the manufacturing process by subjecting the glass to thermal cycling, which creates a glass that breaks into small squares rather than into sharp shards. Windshields are yet another type of specialty glass made by sandwiching glass on both sides of a plastic sheeting material. Most glass made today is called float glass because of the way it is manufactured. Earlier processes used blowers or rollers to mold molten glass into flat sheets. Float glass is made by pouring the molten material onto a bath of melted tin, producing a very smooth surface that needs little or no additional polishing.

Broken glass is frequently encountered as evidence in burglaries or violent crimes. The process of breaking glass creates a shower of particles in both the forward and the reverse direction; particles may be transferred to the person breaking the glass if he or she is standing close by. In many (but not all) cases, it is possible to determine the direction from which the glass was broken. An impact in glass produces two kinds of fractures, RADIAL (radiating out from the point of impact) and CONCENTRIC (forming a circle around the impact). The patterns found in the cross sections of these fractures can indicate the direction from which the force was applied. The lines created in the glass, called ridges, CONCHOIDAL lines, or Wallner lines, show distinctive patterns that depend on whether the fracture from which they are collected is concentric or radial. If there are multiple impacts in the same pane of glass, the sequence of events can be deduced from the fracture patterns. Existing fractures from the first impact act as barriers to fractures created by the second impact, and these abrupt terminations are easily identifiable. Shattering of glass by heat creates a distinctive, different fracture pattern characterized by wavy smooth cracks.

Currently, the only way to link two fragments of glass to a common source is by the process of PHYSICAL MATCHING. Since glass breaks randomly, each breakage is unique. As a result, if a fragment of glass such as a piece of a broken window or headlight can be fitted back into the original as if it is a puzzle piece, that fit individualizes the glass and proves that it can only be from that source. The two most common physical parameters used to characterize glass samples are measurements of the REFRACTIVE INDEX and DENSITY. The FBI maintains a database of several hundred glasses showing the frequency of refractive indices and densities of all those cataloged, and these frequencies can be used to estimate whether a certain type of glass is common. Chemical characterization of glasses is not widely used in forensic science for a variety of reasons. Glass samples often display significant internal variations, meaning that chemical composition, particularly at the trace level, can vary within the same sample of glass. Instrumentation needed is often expensive and requires specialized training and a great deal of time, further limiting usefulness. Finally, many methods of chemical analysis require that the sample be dissolved in acids, and such destructive analyses are often not feasible or appropriate.

glove prints Impressions created by the patterns in gloves, deposited in a manner analogous to the way in which a FINGERPRINT is deposited. Gloves made of wool or other FIBERS can leave FABRIC IMPRESSIONS, whereas leather can leave a type of skin impression because it is a hide.

glowing combustion Combustion that occurs on a surface but without flame, such as that of charcoal in the embers of a fire. Cigarettes burn by glowing combustion.

glucose A sugar, sometimes referred to as "blood sugar," grape sugar, or dextrose, with the molecular formula $C_6H_{12}O_6$. It is the primary energy source for the human body and is classified as a monosaccharide. Glucose can also serve as the base unit in glucose polymers such as starch and cellulose. There are two optical isomers of glucose, D-glucose and L-glucose, and the molecule can exist as a straight chain or as a ring, which also has two forms, depending on the orientation of the –OH group at carbon number 1.

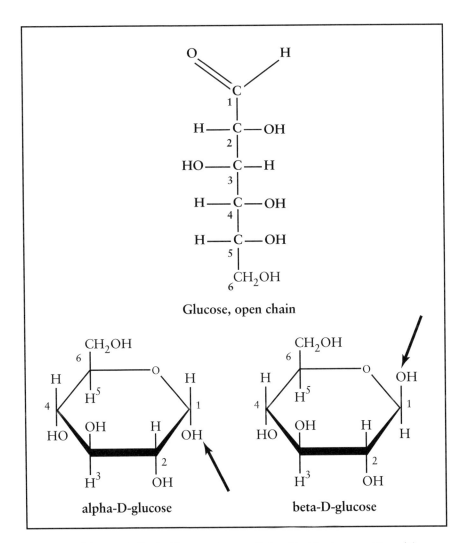

Structures of glucose, $C_6H_{12}O_6$. The two rings are distinguished by the opposition of the hydroxyl (OH) group on carbon number 1; if it is pointed up, it is the beta (β) form, and if downward, it is the alpha (α) form. The dextrorotatory (D) optical isomer is shown.

glue lifts Usually, an informal term describing the development and recovery of latent FINGERPRINTS using CYANOACRYLATE (Super Glue).

glycerides Fats (lipids) that contain a glycerol ($C_3H_8O_3$) skeleton. There are three locations on the molecule that can react with a fatty acid, the three –OH groups. If one reacts, the product is a monoglyceride; if two react, a diglyceride; and if three react, a triglyceride. Triglycerides are the primary form in which excess fat is stored in the body.

glyoxalase I (GLO I) An ISOENZYME system with three common types, 1, 2-1, and 2.

Goddard, Calvin (1891–1955) An American widely credited with establishing scientific examination of FIREARMS evidence in the United States. Goddard was a retired army physician and gun enthusiast who had risen to the directorship of the Johns Hopkins Hospital in Baltimore, Maryland. In 1925, he joined the Bureau of Forensic Ballistics, a private organization founded by Charles E. Waite. A few years earlier, Waite had been involved in the STIELOW CASE, in which firearms evidence

was used to reverse the conviction of an innocent man. Goddard worked with others at the bureau and quickly became a vocal advocate of the use of the forensic COMPARISON MICROSCOPE in firearms casework. Waite died in 1926, and Goddard became the director of the bureau; during this time he wrote several articles on firearms examination, including an article that appeared in *Popular Science Monthly* in 1927. Also in 1927, Goddard became involved in the *SACCO-VANZETTI CASE,* in which his examination of bullets and guns conclusively linked Sacco's gun to a fatal bullet, a finding that was upheld during a reexamination several years after Goddard's death.

As a result of his growing reputation, Goddard was called to examine evidence of the infamous Saint Valentine's Day Massacre in Chicago in 1929. During the coroner's grand jury inquest, he was able to show that all the murdered men had been killed by two Thompson submachine guns. Some of the jurors in that case later raised money to establish a forensic laboratory at Northwestern University in Evanston, Illinois, called the Scientific Crime Detection Laboratory. Goddard was appointed as the first director and stayed until 1932. The laboratory moved to Chicago and became the Chicago Police

Glycerol

Structure of glycerol, backbone of the mono-, di-, and triglycerides.

Department laboratory in 1938. Goddard also assisted the FBI in establishing a firearms analysis capability at their new lab, inaugurated in 1932.

gold bromide and gold chloride Reagents used in drug analysis. Gold bromide (AuBr) and gold chloride (AuCl) along with the platinum analogs (PtBr and PtCl) are used in CRYSTAL TESTS to make microcrystals that are characteristic of a given drug. Gold chloride is used for COCAINE, and a gold chloride volatility test is used to form crystals with the AMPHETAMINES. A crystal test can also be used as part of the analysis of heroin by using platinic chloride.

good laboratory practice (GLP) A series of guidelines for laboratory analysis designed to ensure the goodness and reliability of the results produced. Originally designed for use in environmental, food, and pharmaceutical analyses, the concept of GLP has expanded to cover any type of laboratory analysis. Although specific GLP guidelines vary among disciplines and laboratories, they generally cover aspects such as personnel training and testing, laboratory procedures and protocols, documentation, facilities and equipment, calibration and standardization, and QUALITY ASSURANCE/QUALITY CONTROL. GLP for forensic science laboratories is addressed through analyst and laboratory CERTIFICATION procedures.

grain An older unit of mass that is still encountered occasionally in pharmaceutical applications (APOTHECARY UNIT). One grain equals 0.0648 gram or one gram is equivalent to 15.4 grains.

Gram negative A classification of bacteria based on the inability of the bacteria to hold a stain of crystal violet when rinsed with alcohol or acetone. The test was named after the Danish physician who developed it.

Gram positive A classification of bacteria based on the ability of the bacteria to retain a stain of crystal violet even when

rinsed with alcohol or acetone. The test was named after the Danish physician who developed it.

grand jury An inquisitional judicial proceeding used at both the state and federal levels. The job of the grand jury is to hear evidence and decide whether it is sufficient to indict someone. If so, the grand jury returns an indictment and criminal charges can be brought against the person in question. At the state level, a grand jury indictment is often required for capital crimes and felonies, but the standards vary among the different jurisdictions. An indictment does not mean that the person is guilty, but only that there is sufficient evidence to prove that a crime did occur and that the person may have committed that crime. Forensic scientists can be called to testify before grand juries just as they can in any other court proceeding. The grand jury traces its origins to English common law.

graphite furnace atomic absorption/ electrothermal atomic absorption ATOMIC ABSORPTION (AA) spectrophotometry that exploits a small graphite furnace rather than a flame to heat and atomize a sample. A small amount of the sample is placed on a shelf inside a tiny graphite tube. Graphite conducts electricity; directing a high current through a furnace produces rapid heating that drives off solvent and then vaporizes and atomizes a sample.

graphology The analysis of handwriting for the purpose of identifying personality and psychological traits of the writer. Graphology is a PSEUDOSCIENCE and is not to be confused with forensic handwriting analysis conducted by a QUESTIONED DOCUMENTS examiner. In such cases, the examiner is evaluating writing samples for purposes such as detecting forgeries or determining whether the same person was responsible for different writing samples. This type of handwriting examination is systematic and based on sound principles and procedures subject to constant review and improvement. Graphology is not part of forensic document examination and is not considered as admissible or reliable in courts.

grating, diffraction *See* DIFFRACTION GRATING.

gray wipe *See* BULLET WIPE.

grid search/grid method A method of searching a crime scene in which the area to be searched is divided into roughly square sections that are searched individually.

Griess test A PRESUMPTIVE TEST used to detect the residuals of GUNPOWDER or EXPLOSIVES. The test detects the nitrite ion (NO_2^-) and can be modified also to detect the nitrate ion (NO_3^-). Reagents used in the test are naphthylamine in methanol, sulfanilic acid in acetic acid, and zinc metal. Nitrates and nitrites are produced whenever a gun is fired, and nitrate is a breakdown product of nitroglycerin, an ingredient found in many explosives and in SMOKELESS POWDER used as a PROPEL-LANT in AMMUNITION. Thus, this test was at one time widely used in attempts to determine whether a suspect had recently fired a gun or had handled explosives containing nitroglycerin. The test is also used in DISTANCE DETERMINATION, in which GUNSHOT RESIDUE can be visualized by using the test. Finally, the reagents can be used as a developer for THIN LAYER CHRO-MATOGRAPHY applied to explosives.

groove A term that has two principal uses in forensic science. It is used in FIREARMS analysis, in which LANDS AND GROOVES (RIFLING) machined into the barrel of a rifle or pistol impart spin to the bullet. The pattern of striations produced on the bullet by the lands and grooves is characteristic of that weapon. Second, in a tire, grooves surround the DESIGN ELE-MENTS and are part of the tread design.

Gross, Hans (1847–1915) Gross, an examining magistrate in Graz, Austria, is credited with coining the term CRIMINAL-ISTICS to describe what was then an infant science. Gross viewed forensic science

holistically and believed that experts in diverse fields should contribute to the analysis of physical evidence and solving crimes. He understood the value of bio-logical evidence, SOIL, DUST, and many other types of transfer and TRACE EVI-DENCE. In 1893 he published the first text-book in forensic science, translated into English under the title of *Criminal Investi-gation,* and he started a journal called *Kriminologie,* which is still published. Gross was a pioneer in the field, and he exerted a strong influence on his contem-poraries including other pioneers such as Edmund LOCARD.

ground state A state of an atom, mole-cule, or electron in which it is in its lowest energy state. This is in contrast to the EXCITED STATE, which develops if the atom or molecule absorbs ELECTROMAG-NETIC ENERGY. Ground and excited states are fundamental to SPECTROSCOPY.

grouping The process of typing blood or body fluids, for example, finding blood groups. This is an older term primarily associated with forensic SEROLOGY and typing of the ABO BLOOD GROUP SYSTEM and ISOENZYMES.

group-specific component (Gc) A GENETIC MARKER SYSTEM found in the serum of BLOOD. Gc is produced by the liver and shows variations in the popula-tions, with the primary types 1, 2-1, and 2. Type 1 is the most common; about 51 percent of the population have it. Type 2-1 is found in about 41 percent; the remainder are type 2. Typing can be accomplished by using ELECTROPHORESIS and ISOELECTRIC FOCUSING. With the ascendance of DNA TYPING techniques, typing of Gc in bloodstains is rarely needed or used.

guaiacum test A PRESUMPTIVE TEST for blood, one of the first developed, which is no longer used. It relied on guaiacum (a resin isolated from trees) in combination with hydrogen peroxide. If a stain turned blue when treated with these reagents, it was considered a positive result indicative of presence of blood. However, as with all presumptive tests, the results are not conclusive; nor do they prove that the blood detected is human.

guanine (G)   One of four NUCLEOTIDE bases that compose DNA and ribonucleic acid (RNA). Because of its molecular structure, guanine associates with cytosine (C), and the two are referred to as complements of each other.

guard column   A small column placed in front of the analytical column of a HIGH-PERFORMANCE LIQUID CHROMATO-GRAPH. The function of the guard column is to filter out large particulates and other contaminants that might damage or shorten the life of the analytical column.

Gun Control Act   An act passed in 1968 after the assassinations of John F. Kennedy, Robert Kennedy, and the Reverend Martin Luther King Jr. The act required better and more detailed record keeping for gun sales and limited the ability of convicted felons and other groups to buy guns. Mail ordering of rifles and shotguns was banned.

guncotton   NITROCELLULOSE (NC) that is 13.3 percent nitrogen by weight.

gunpowder   A term most often used to describe the PROPELLANTS used in AMMUNITION. There are two kinds of powder available, both classified as low explosives, and the term *gunpowder* has been applied to both. BLACK POWDER, the original gunpowder, consists of 75 percent potassium nitrate (KNO_3 or saltpeter), 10 percent charcoal, and 15 percent sulfur. Because of the smoke produced, its use in firearms ended in the 1800s, but black powder is still available for hobby guns and is used in applications such as fuses. SMOKELESS POWDER, which is used in modern FIREARMS, consists of nitrated cotton or

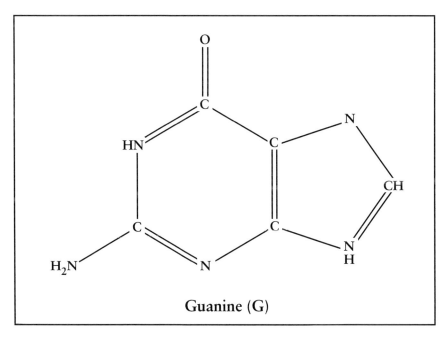

Guanine (G)

The structure of guanine (G), one of the four bases found in DNA and the complement of the base cytosine (C).

NITROCELLULOSE (single base) or nitroglycerin combined with nitrocellulose (double base). The word *gunpowder* usually refers to smokeless powder because it is the modern standard, but the term can be and often is used to describe black powder, particularly in a historical context.

guns *See* FIREARMS.

gunshot residue (GSR) The residue that escapes from a gun when it is fired. GSR is considered a type of TRANSFER EVIDENCE and can be detected by using chemical tests and INSTRUMENTAL ANALYSES. The residue is mostly from the primer, which is part of the CARTRIDGE containing the BULLET and PROPELLANT in modern AMMUNITION. Particles of unburned powder are also part of residue. The most common elements found in GSR are lead (Pb), antimony (Sb), and barium (Ba), and this combination of elements is telling. When a particle contains these three elements, it is almost certainly gunshot residue. Other elements that may be found include copper (Cu), aluminum (Al), iron (Fe), and zinc (Zn). Chemicals in the propellant (usually SMOKELESS POWDER) in GSR are nitrate ions (NO_3^-), nitrite ions (NO_2^-), NITROCELLULOSE, and nitroglycerin. During the firing process, compounds of metal elements and atoms of carbon, nitrogen, oxygen, and hydrogen can form as a result of the combustion. Thus, GSR is chemically complex, particularly at the trace level. The size of the GSR particles and their chemical makeup vary, depending on the type of weapon, powder, primers, and projectiles used. The particles themselves as well as the individual elements and the chemical compounds in GSR can be detected by chemical, microscopic, and instrumental methods of analysis. Presumptive tests for GSR that have been and are used include the GRIESS TEST, DIPHENYLAMINE TEST, DERMAL NITRATE TEST, WALKER TEST, and sodium RHODIZONATE TEST.

Gupta model A mathematical and statistical model proposed in 1968 that describes the probabilities of finding certain features of latent FINGERPRINTS. *See also* MINUTIAE.

gypsum A calcium sulfate ($CaSO_4$) powder derived from a mineral, which is used in Sheetrock, a common building material.

hair/hairs   Hair is an animal fiber characterized by a scaly cuticle that is easily recognizable when viewed under a microscope. An experienced forensic examiner can distinguish animal hair from human, but it is rarely possible to link a questioned hair to an individual definitively unless DNA TYPING can be performed.

Thus, for hair evidence, the primary method (and often the sole method) of analysis is visual examination. Hair is produced below the surface of the skin in the follicle, and the base of the hair comprises living tissue. As the hair grows and reaches the surface, the cells die and become keratinzed. Keratin is a strong

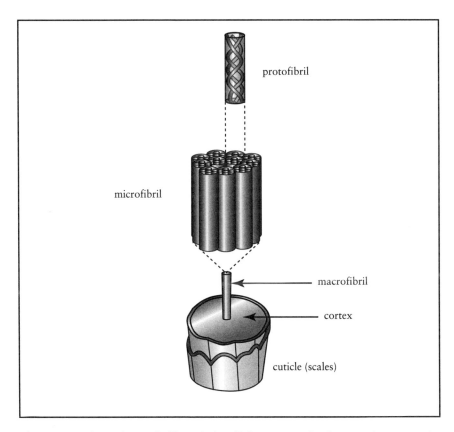

The structure of a single strand of hair; the bundled structure makes it extremely strong and resilient.

protein that is found in high concentrations in the fingernails and imparts to the hair durability and resistance to degradation. Inside the follicle is the root, where the living portions are found, along with the bulb and sheath tissue. The follicle is associated with an oil gland called the sebaceous gland.

Hair possesses scales that protrude from the outer cuticle, with the unattached "flap" of the scale pointing toward the tip of the hair. On the surface, the scales form an irregular pattern most often compared to shingles on a roof. The "shine" of hair is related to the position of the cuticles and how tightly they adhere to the cuticle. Inside the cuticle (the cortex), hair consists of fibrils of protein arranged in a stacked arrangement of an individual protein strand enclosed in a protofibril, which in turn is encased in a microfibril and then a macrofibril. A tubular structure called the medulla runs down the middle of the hair; it may be thick, thin, continuous, or discontinuous. Pigment granules are responsible for color and are scattered throughout the cortex. Except in true blonds and redheads, there are only three colors of pigment found in human hair: yellow, black, and brown. The cortical fusi are void spaces within the cortex that are concentrated nearer the root end.

Individual hair growth occurs in three stages (ANAGEN, CATAGEN, and TELOGEN) independently of neighboring follicles. The structure and appearance of the root (characteristics of its MORPHOLOGY) can be used to determine at what phase a hair was lost. Likewise, if a hair is forcibly removed, as is often the case in violent crimes, portions of the follicular tissue adhere to it and can be identified microscopically. Hair differs in appearance on different parts of the body, an important consideration since hair comparisons must be based on hairs from the same area.

half-life ($t_{1/2}$) The amount of time required for one-half of the original amount of material to decay or otherwise change form. In forensic science, this term is most often used to describe radioactive decay, or the amount of time that drugs and metabolites remain in blood and urine.

hallucinogens See DRUG CLASSIFICATION.

hammer forging A method of creating the land and groove rifling in FIREARMS. The barrel is shaped by hammering it down over a mold, thereby transferring the land and groove pattern to the metal.

handguns A classification of FIREARMS that includes REVOLVERS, PISTOLS, and homemade guns that are specifically designed to be fired by one hand. This is in contrast to shoulder guns such as RIFLES and SHOTGUNS. Of all firearms, handguns are the ones most often used in crimes.

handwriting Writing that is produced directly and manually by a person using a writing instrument such as a pen or pencil. It can be cursive or printed.

hanging Death by means of ASPHYXIA (loss of oxygen to the brain) caused by a noose tightened by the body weight of the victim. The pressure of the noose quickly cuts off the blood flow to the brain, leading to the rapid onset of unconsciousness. The airway also closes, causing death in a few minutes. This effect can be achieved without complete suspension, so people can die of hanging while kneeling or otherwise still having partial contact with the ground.

haploid cell A cell such as a sperm or egg cell that contains only half of the organism's genetic material or one set of chromosomes.

haptoglobin (Hp) A serum blood group system and GENETIC MARKER SYSTEM. Before the ascendance of DNA TYPING, this was one of the systems that could be typed in bloodstains using electrophoresis. There are three types of haptoglobin, of which 2-1 is the most common, found in approximately 40–50 percent of the population, depending on race. Type 1 is shown by about 20–30 percent and type 2 by 20–35 percent.

Hardy Weinberg law A relationship used by forensic biologists to estimate the FREQUENCIES of alleles (variants of a gene) and the resulting distribution of types within a population. It provides a statistical method to test frequencies and to provide estimates of the DISCRIMINATING POWER of a given type. A coin flip provides a simple example.

The process of consecutively flipping a coin twice mimics the combination of two genes (one from a mother and one from a father) to create a variant at the site of the gene (its locus). For any one toss, the probability of obtaining a head (H) is 0.50, meaning that 50 percent of the time, heads should appear. This is assigned to a variable labeled p. The variable q, the frequency of tails (T), is also 0.50, and $p + q$ must always equal 1. In terms of the Hardy Weinberg law, this is expressed as $(p + q)^2$, and then by using a technique called a binomial expansion, the expression $p^2 + 2pq + q^2$ is obtained. This is the equation that can be used to estimate frequencies. If a coin is tossed twice, there are three possible outcomes: two heads (HH), one of each (HT), and two tails (TT). The equation can be used to estimate how often each combination is expected to occur. HH would correspond to p^2 or 0.50 × 0.50 = 0.25 or 25 percent. This means that 25 percent of the time, when a coin is flipped twice, a combination of HH should occur, and the same is true for TT. For one of each, the predicted frequency is 2 × 0.50 × 0.50 ($2pq$) or 0.50, meaning that half the time, two flips will result in one of each. These are the frequencies expected, assuming the coin tosses are completely random and the coin is not altered to favor heads or tails. This simple example can be extended to GENETIC MARKER SYSTEMS and DNA TYPING, except that population data are available for the actual frequency of types. A test called the CHI-SQUARED TEST is used to determine whether this difference is significant.

hashish A potent derivative of MARIJUANA (*Cannibis sativa*) that has a high concentration of the psychoactive ingredient tetrahydrocannabinol (THC). Hashish is a tarry substance that varies in color from dark green to almost black. It is the resin of the marijuana plant, which is often prepared by extracting the flowering tops with an alcohol. Like marijuana, hashish is classified as a hallucinogen and is listed on Schedule I of the CONTROLLED SUBSTANCES ACT.

headlamps The headlights of a vehicle. In those lamps that use a filament, the condition of that filament can be useful in accident investigations. Since the filaments carry electrical current to generate light, they become very hot. If the sealing glass or plastic that encloses it is broken while the filament is on and hot, a coating of oxidation can be observed (visually or microscopically) on the filament. If the light is off and the filament is cold, no such oxidation is seen. Also, if the filament is encased in a glass headlamp, shards of glass can come in contact with the filament. If the lamp is on and is hot, the glass can melt and form beads on the filament that can be observed with a microscope.

headspace The void or empty space above a solid or liquid held in a sealed container. In a sealed bottle of soda, it is the empty space above the line of the liquid.

headspace analysis A technique used in ARSON evidence analysis, forensic TOXICOLOGY, and other specialized chemical analyses. The headspace is the air or other gas above a sample (solid or liquid) that is enclosed in a sealed container. Fire debris, for example, is usually collected in paint cans, filling a portion but not all of the can's volume. The lid is sealed, creating an enclosed air space above the fire debris. Any materials that are volatile enough (evaporate at a low enough temperature) collect in this headspace. By withdrawing a sample of the headspace, volatile components from the debris can be identified by various analytical techniques. In an arson case, the components of interest are any ACCELERANTS such as gasoline used to start and sustain a fire. Headspaces can be cold, or the container may be heated to increase volatilization. The headspace can also be dynamic, meaning that it is periodically swept out of the container and directed over a material that can trap the components contained in it. This technique is also known as purge-and-trap. In the analysis of fire debris, several variations of headspace sampling have been or are used, including cold headspace, heated headspace with collection on a charcoal trap, collection of headspace vapors on a charcoal strip, and dynamic headspace with a charcoal trap.

hearsay Testimonial evidence from a witness that relates to events of which the witness has no personal knowledge. Rather, the witness is testifying about something he or she heard another person say while that other person was not under oath. In a courtroom, if a witness is on the stand and asked a question such as "What did Mr. Jones say about that?" or "What did you hear Mr. Jones say about that?" the response the witness gives is hearsay evidence. Hearsay evidence is usually inadmissible, but there are numerous exceptions to the hearsay rule.

heavy metals In forensic science, a term that usually refers to metals that are also POISONS. In this context, the metals of interest include ARSENIC, LEAD, MERCURY, cadmium (Cd), bismuth (Bi), antimony (Sb), and thallium (Tl). Of these, arsenic is the best known as a poison—accidental, suicidal, or homicidal. Strictly speaking, it is classified as a metalloid or semimetal in terms of chemical properties and behavior. Antimony is also found in gunshot residue, and thallium is used in the electronic industry.

heel The edge of a BULLET or the rear portion of a SHOE IMPRESSION.

hemagglutination Also called agglutination, a clumping reaction that occurs when red blood cells are mixed with ANTISERA to the ANTIGENS found on their surface. An example is the agglutination reaction used to determine a person's ABO BLOOD GROUP type.

Hemastix A test strip designed to test for blood in urine that is employed as a PRESUMPTIVE TEST for blood.

hematin/hematin test (Teichman test) A confirmatory test for blood based on the formation of distinctive hematin crystals that are viewed under a microscope. The test was developed in 1853 by Ludwig Teichmann and its name is sometimes spelled as *Teichmann*. The Teichman test is one of a group of confirmatory tests for blood referred to as CRYSTAL, microcrystal, or microcrystalline tests. Crystal tests require a larger sample than do the PRESUMPTIVE TESTS for blood and are susceptible to interference.

heme The nonprotein part of the HEMOGLOBIN molecule, the part that carries the oxygen molecule. Each heme group can complex one O_2 molecule, for a total of four O_2 molecules per hemoglobin. The heme group can act as a catalyst for oxidation/reduction (REDOX) reactions; this is the basis of PRESUMPTIVE TESTS for blood.

hemochromogen/hemochromogen test (Takayama test) A confirmatory test for blood based on the formation of distinctive hemochromogen crystals that are viewed under a microscope. The test was developed in 1912 by Masao Takayama in Japan and has evolved into the most commonly used of the so-called microcrystalline tests for blood. The test requires a larger sample than do the PRESUMPTIVE TESTS for blood and is susceptible to interference.

hemoglobin (Hb) The predominant protein in red BLOOD cells, which transports oxygen from the lungs to the tissues. The hemoglobin molecule is made up of four protein subunits (the globin portion), two alpha (α) subunits, and two beta (β) subunits (see figure on page 116); for this reason, hemoglobin is classified as a tetramer. Each of the four subunits possesses a HEME unit with an iron ion (Fe^{2+}) at the center (see figure on opposite page); it is the heme units that bind to the oxygen, four per hemoglobin molecule. PRESUMPTIVE TESTS and microcrystalline con-

Heme Group

The structure of the heme group in hemoglobin. Oxygen (O_2) complexes with the iron cation and is delivered to tissues in that form. Each hemoglobin molecule has four heme groups.

firmatory tests for blood used in forensic science rely on reactions with the hemoglobin molecule. There are also variants of hemoglobin that can be typed by using ELECTROPHORESIS and ISOELECTRIC FOCUSING techniques.

hemp A slang term for marijuana, the term also is used to describe the tough fibrous tissues found in the marijuana plant that can be used to make rope, fibers, and clothing.

Henry classification system/Henry system A FINGERPRINT classification system

developed by Sir Edward HENRY. Based on a bin system, it classified sets of all 10 fingerprints into smaller categories that made it easier to store and catalog large numbers of fingerprints. A modified Henry system is still used in the United States and Europe, although initial searching is now done by computers such as those that are components of the AFIS system.

Henry, Sir Edward (1851–1931) An English police officer who, like GALTON and VUCETICH, was a key figure in the early use of FINGERPRINTS in criminal investigations. He first employed them for

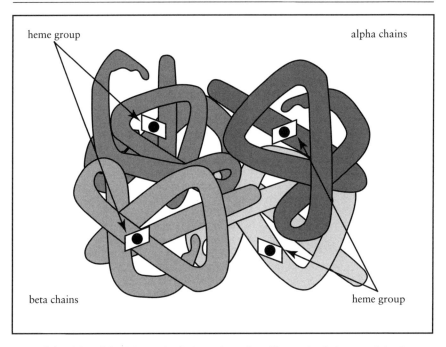

Hemoglobin. Two alpha (α) protein chains and two beta (β) protein chains containing heme groups are combined into a complex three-dimensional structure.

identification purposes while he was the inspector-general of police in Bengal, India. Later, he became commissioner of the Metropolitan Police in London (Scotland Yard), and in 1901, his fingerprint classification system was adopted. Modifications of the Henry system are still used throughout Europe and in the United States.

Henry model A statistical approach to describing the individuality of a fingerprint pattern developed by Sir Edward HENRY and proposed in 1900.

Henry's law An expression that describes the solubility of a gas in a liquid as a function of pressure. The equation is written as

$$K_H = \frac{[A]}{P_A}$$

in which the K_H is called the Henry's law constant, [A] represents the concentration

of gas A dissolved in a liquid, and P_A represents the pressure of that gas above the liquid in question. At a set temperature, the value of K must remain the same; that means that if the pressure of a gas increases above a liquid (P_A increases), then the concentration of that gas dissolved in the liquid ([A]) must also increase. The gas above the surface of the beverage has an elevated pressure of CO_2 relative to the atmosphere. This can be illustrated by the everyday situation of carbonation (CO_2 gas) in beverages such as soda pop, sparkling wines, and champagne. These beverages are bottled under elevated pressure, allowing more carbon dioxide to dissolve in them. When the cap is opened, the familiar hissing sound indicates that the pressure is equalizing and pressure over the liquid is falling back to normal atmospheric levels. In terms of the equation, P_A is decreasing. At the same time, bubbles of carbon dioxide form inside the liquid and escape, since at the lower pressure, it is less soluble ([A] is also decreasing).

heroin A derivative of MORPHINE that is highly addictive and widely abused. Also called diacetylmorphine, diamorphine, and acetomorphine, heroin is easily synthesized from morphine by the addition of acetyl chloride or acetic anhydride. The morphine in turn is obtained from OPIUM poppies, so heroin is classified as one of the opiate ALKALOIDS. It is most commonly found in the form of a hydrochloride salt with a white or off-white color; however, a brownish black resinous form known as "black tar" originating in Mexico is also seen. Heroin is water-soluble and so is usually injected, although snorting and smoking are possible. Because of its potency, heroin is both physically and psychologically addictive and has no acceptable medical use. It is listed on Schedule I of the CONTROLLED SUBSTANCES ACT.

Herschel, Sir William J. (1833–1917) An Englishman who, along with HENRY FAULDS, is credited with advancing the use of FINGERPRINTS as a means of criminal identification. Herschel was stationed in India in 1853 and used fingerprints as a form of signature in contracts with nonliterate Indians; he also studied ridge patterns and noted that they were unchanging. Hershel developed a classification system that it was never widely adopted.

hesitation marks/wounds Wounds found on the body of suicide victims that indicate earlier unsuccessful suicide attempts. For example, a person who commits suicide by slitting the wrists may have multiple shallow cuts or slices on the wrist from the first tentative cuts.

heterogeneous Lacking a uniform composition throughout a substance, mixture, or material. The composition varies, depending on where it is observed, tested, or sampled.

heterozygote/heterozygous Having two different alleles at a genetic locus. For example, for the ABO BLOOD GROUP SYS-TEM a heterozygote would have a GENOTYPE of AB, as well as the PHENOTYPE AB.

HETP Height equivalent of a theoretical plate, a number that provides a way of comparing the separation power of CHROMATOGRAPHIC columns: higher numbers represent greater separation ability for compounds amenable to that column. Such columns do not contain physical plates; rather, the terminology arises from the analogy of a chromatographic separation to the separation of components of crude oil by distillation (a process called "cracking"). In such distillations, the greater the number of plates in the cracking tower, the more refined the separation of the volatile components of crude oil.

hexane A solvent used in chromatography and in some types of extractions in ARSON analysis. Hexane is a straight-chain HYDROCARBON (properly referred to as *n*-hexane) with six carbons and 14 hydrogens all bonded with single bonds. In practice, a mixture of the various straight-chained and branched hexanes (all C_6H_{14}) is often used for chemical extractions and is referred to as hexanes.

high explosive A category of explosive materials divided into two classes, primary and secondary explosives. Primary high explosives are shock- and/or heat-sensitive and are often used as primers that ignite secondary high explosives. PRIMERS in ammunition and blasting caps contain primary high explosives. Secondary high explosives are much more stable and are usually detonated by the shock generated from a primary explosive. High explosives decompose at a much faster rate than low explosives, and detonations generate shattering power; produce smaller, sharper fragments; and generally leave minimal residue. High explosives include the famous NITROGLYCERIN (NG) ("nitro"), which was invented in 1847; trinitrotoluene (TNT); HMX; RDX; TETRYL; and PETN.

high-performance liquid chromatography (HPLC) An instrumental system based on chromatography that is widely used in forensic science. The *HP* of the acronym sometimes designates high pressure (versus high performance), but it refers to the same analytical system. As is all CHROMATOGRAPHY, HPLC is based on selective PARTITIONING of the molecules of interest between two different phases. Here, the mobile phase is a solvent or solvent mix that flows under high pressure over beads coated with the solid stationary phase. While traveling through the column, molecules in the sample partition selectively between the mobile phase and the stationary phase. Those that interact more with the stationary phase lag behind those molecules that partition preferentially with the mobile phase. As a result, the sample introduced at the front of the column emerges in separate bands (called peaks); the bands that emerge first are the components that interacted least with the stationary phase and as a result moved more quickly through the column. The components that emerge last are the ones that interacted most with the stationary phase and thus moved more slowly through the column. A detector is placed at the end of the column to identify the components that elute. Occasionally, the eluting solvent is collected at specific times correlating to specific components. This arrangement provides a pure or nearly pure sample of the component of interest. This technique is referred to as PREPARATIVE CHROMATOGRAPHY.

Many detectors are available for HPLC. The simplest and least expensive is the REFRACTIVE INDEX (RI) detector. Although this detector is a universal detector—meaning it responds to any compound that elutes—it does not respond well to very low concentrations and as a result is not widely used. Detectors based on the absorption of light in the ultraviolet (UV) and visible (VIS) ranges (UV/VIS detectors and UV/VIS SPECTROPHOTOMETERS) are most commonly used, because they respond to a wide variety of compounds of forensic interest with good to excellent sensitivity. The photodiode array detector (PDA) is especially useful since it can produce not only a peak-based output (a CHROMATOGRAM) but also a UV/VIS scan of every component. In many ways, the ideal detector for HPLC is a MASS SPECTROMETER (MS), which provides both quantitative information and in most cases a definitive identification of each component (qualitative information). However, HPLC-MS systems are relatively complex and expensive and are not readily available in all labs. Other detectors that are sometimes used include fluorescence detectors (which are very sensitive) and electrochemical detectors.

high-risk victim   A person such as a drug user or prostitute who is, by the nature of his or her lifestyle, at a greater risk of being a victim of a violent crime than others.

high-velocity impact spatter   A BLOODSTAIN PATTERN created by a high-velocity impact on a person's body. High-velocity spatter patterns are created by gunshot wounds and by some striking wounds caused by great speed and force. High-velocity spatter patterns are characterized by fine tiny droplets that may form a mist. However, not all stains that are very small are the result of high-velocity impact.

HLA   *See* HUMAN LEUKOCYTE ANTIGEN.

HLA-DQA1   *See* DNA TYPING.

HMX   A high explosive that is formed as a by-product during the synthesis of another high explosive, RDX. The chemical name for HMX is cyclotetramethylenetetranitramine, and the meaning of the acronym *HMX* is unclear.

hollow cathode lamp (HCL)   A monochromatic (single-wavelength) light source used in atomic absorption. The lamp consists of the same metal that the analysis seeks to detect; thus, if the TARGET ANALYTE is barium in a test for GUNSHOT RESIDUE, the HCL selected is composed of barium.

hollow point   A type of bullet in which a portion of the nose has been removed,

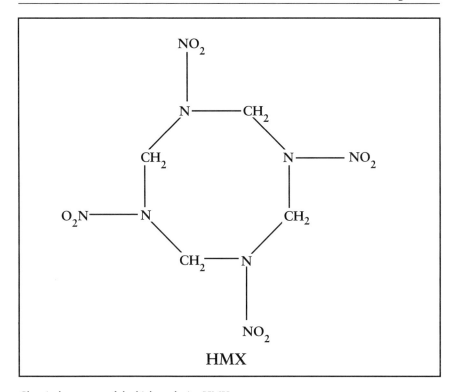

HMX

Chemical structure of the high explosive HMX.

leaving a hollow cavity. Upon impact, the gas trapped in the cavity is severely compressed, resulting in a mushrooming effect, expanding the surface area of the nose significantly. Consequently, the bullet can do greater damage to the tissue it passes through.

Holmes, Sherlock The famous fictional detective created by SIR ARTHUR CONAN DOYLE. Over the course of 40 years (1887–1927), Doyle published 56 short stories and four novellas about Holmes and his good-humored assistant Watson as they solved crimes using logic and science. In his career, Holmes delved into many areas of forensic science including forensic BIOLOGY, TRACE EVIDENCE and TRANSFER EVIDENCE, FIREARMS, and QUESTIONED DOCUMENTS using fictional techniques that in many cases accurately predicted later developments in the field. Doyle's stories

influenced and inspired many of the pioneers of forensic science including HANS GROSS and EDMUND LOCARD.

Home Office The branch of the government in the United Kingdom (UK; England, Northern Ireland, Wales, and Scotland) that houses the FORENSIC SCIENCE SERVICE. The FSS is the main provider of forensic services in the UK and is loosely analogous to the FBI laboratory in the United States.

homicide A purposeful killing undertaken with the intention of causing the death of the victim.

homogeneous A homogeneous substance, mixture, or material that has a uniform composition throughout. It is the same everywhere it is observed, tested, or sampled.

homologous A series of compounds that differ in a specific way. The term is most often used to describe molecules within the same functional group such as the alkanes methane (CH_4), ethane (C_2H_6), and propane (C_3H_8).

homozygote/homozygous Having two of the same alleles at a genetic locus. For example, for the ABO BLOOD GROUP SYSTEM, a homozygote has a GENOTYPE of AA.

hook cutter A cutting tool and method that is used in some manufacturing processes to cut the lands and grooves (RIFLING) into the barrel of a firearm.

Hoover, J. Edgar (1895–1972) Hoover, best known for his 48-year tenure as head of the FBI, was an early champion of forensic science. In 1926, the Bureau of Investigation, as it was then called, assumed the primary responsibility for fingerprint cards and identifications for law enforcement agencies throughout the country, a responsibility that still represents one of its largest single tasks. In 1932, Hoover oversaw the establishment of a technical laboratory. Early examinations focused on FIREARMS and QUESTIONED DOCUMENTS. The laboratory moved to the Department of Justice building in Washington, D.C. Although not the first forensic lab in the United States, the Technical Laboratory, as it was called then, was the largest and evolved into the single most influential source of forensic analysis and information in the world. The organization of the lab served as a model for many other state and local facilities formed in subsequent years.

Hoover was born in Washington, D.C., and received a law degree in 1917, after which he began working at the Department of Justice; within seven years, he was head of the agency that would eventually become the FBI. High-profile escapades of special agents during the gangster eras and World War II made Hoover famous, well respected, and powerful. He used that power to expand the size and budget of the FBI and was known as a strident anti-communist. Although controversial in his later years, he was instrumental in making the FBI a key force in American law enforcement and forensic science.

hot stage An accessory used with a microscope for the analysis of FIBERS and GLASS. In glass analysis, a piece of glass is immersed in oil with a known REFRACTIVE INDEX. The hot stage gradually heats the oil, changing the refractive index while the analyst observes a feature known as the BECKE LINE. The Becke line is a halo of light that surrounds the particle but that vanishes when the refractive index of the liquid matches that of the glass. This method allows the examiner to determine the refractive index of the glass. For the analysis of fibers, the hot stage heats the fiber while the analyst observes the behavior and records observations such as fiber swelling, curling, contraction, burning. The melting point of the fiber (if there is one that can be reached with the hot stage) is also useful in identifying the type of fiber, as is a study of changes in the optical properties of the fiber as heat is applied. Hot stages can control temperature accurately and can increase it by tenths of a degree per minute. Hot stage analyses can also be automated to some extent, simplifying tasks such as determining the refractive index of many glass fragments.

HPLC/HPLC-MS and variants See HIGH-PERFORMANCE LIQUID CHROMATOGRAPHY.

human hair See HAIRS.

human leukocyte antigen (HLA) A GENETIC MARKER SYSTEM widely used in paternity testing and organ transplantation. Although the subject of research in the 1980s, an effective and reliable method of typing HLA in BLOODSTAINS never emerged and forensic applications were limited. With the advent of DNA TYPING, the need for developing such methods was all but eliminated. The term *HLA* is still seen, however, used in conjunction with DNA TYPING in the form of the HLA-DQA1 system. This was the first system targeted by polymerase chain reaction (PCR) DNA TYPING techniques.

HLAs are human leukocyte ANTIGENS, meaning that they are found associated with white blood cells (leukocytes) and in tissues. The HLA system is part of the histocompatibility complex called the major histocompatibility complex (MHC), which consists of four different sites or loci on chromosome number 6. Since there are so many different factors that can be typed within the HLA system, it has a high DISCRIMINATION POWER, on the order of 90 percent. This means that successful typing can eliminate approximately 90 percent of all possible donors. However, the cost of typing combined with the technical challenge of typing HLA antigens in bloodstains prevented any significant application to forensic casework.

humidity cabinet An enclosed cabinet in which tests and analysis can be undertaken while preventing excessive drying. Humidity cabinets are frequently used for some types of IMMUNODIFFUSION tests to prevent gels from drying and in some methods of latent FINGERPRINT development.

hung jury A jury that is unable to reach a verdict.

hybridization In DNA TYPING, the process of "unzipping" the double helix of the DNA molecule into two separate strands (templates) and allowing the exposed bases to bind with their complement to create two new strands. The terms ANNEALING and *reannealing* are sometimes used interchangeably with the term *hybridization;* however, fundamentally all refer to complementary base pair bonding with DNA fragments.

hydrocarbon An ORGANIC compound consisting of hydrogen atoms attached to a carbon chain backbone. Many ACCELERANTS used in ARSON, such as gasoline, are mixtures of hydrocarbons such as octane (C_8H_{18}) or benzene (C_6H_6).

hydrocodone A synthetic drug classified as an OPIATE and used for treatment of pain and severe coughing, much as is codeine. Depending on what the drug is combined with, it is found on *Schedule II or III* of the CONTROLLED SUBSTANCES ACT.

hydrogen bond A type of chemical bond that can occur between a hydrogen atom covalently bonded to another atom such as oxygen, nitrogen, or fluorine and an atom that is part of another molecule. When attached to oxygen or fluorine, electrons are drawn strongly toward these atoms and away from the hydrogen, leaving it with a partial positive charge (symbolized as δ^+) that can form a bond with atoms that have unbonded electron pairs. Hydrogen bonds among water molecules are critical and lead to many of water's unique properties. Hydrogen bonds are also important in determining the shape and structure of proteins.

hydrolysis A chemical reaction in which water is added to a compound or chemical group or in which addition of water results in a breakdown of the molecule. The process of making soap from fats and a strong base (saponification) is an example of a hydrolysis reaction.

hydrophilic Literally, "water loving." This term most often refers to chemical compounds or chemical groups and their interactions with water. Small alcohols such as ethanol (four carbons or less) are hydrophilic and remain dissolved in water. *See also* LIPOPHILIC; LIPOPHOBIC.

hydrophobic Literally, "water hating" or "water fearing." This term most often refers to chemical compounds or chemical groups and their interactions with water. Oils and fats are hydrophobic and do not remain mixed in water. *See also* LIPOPHILIC; LIPOPHOBIC.

hypervariable region (mtDNA) A genetic location (locus) that has many different alleles throughout the population. As a consequence, such loci are valuable in forensic science and can greatly aid in attempts to INDIVIDUALIZE a sample. Locations called hypervariable regions 1 and 2 are used in typing of MITOCHONDR-IAL DNA (mtDNA).

hyphenated technique A process that uses instruments that has two components, most often a separation module and a detector module. One of the most widely used instruments in forensic science, a gas chromatograph coupled to a mass spectrometer, is a "hyphenated technique" since it is usually referred to by the acronym GC-MS, GC/MS, or GCMS. In this case, the GC separates the mixture into individual components that are introduced into the mass spectrometer one at a time for identification.

hypochlorite A common disinfection agent. Sodium hypochlorite (NaOCl) is the active ingredient in bleach, and 5 percent solutions of this compound are often used to disinfect surfaces such as lab benches. A solution of hypochlorite can also be employed in conjunction with PHYSICAL DEVELOPERS to visualize latent FINGERPRINTS.

hypothesis and the scientific method The procedures and framework in which science is practiced and the way in which forensic scientists approach their analyses. A hypothesis is often referred to as an educated guess, meaning that it is an idea offered to explain an observation or the result of an experiment. A hypothesis is not a theory or a natural law but rather a starting point from which scientists can work their way to a verified truth. The first step in the process is the collection of initial data either by experiments (empirical data) or by observation. These data are then analyzed, and on the basis of the analysis, a hypothesis is offered. If no reasonable hypothesis can be made, then the process returns to the beginning for the collection of additional information. A key element of any hypothesis is that it must be testable. A specialized type of hypothesis often used in forensic science is what is called the NULL HYPOTHESIS. In this approach, the forensic analyst starts with a hypothesis that can be disproved. For example, if he or she is testing a blood sample, the null hypothesis might be "This

blood is from the suspect." It is stated as a hypothesis, and by using comparative DNA TYPING, the analyst has the opportunity to disprove the null hypothesis.

During testing, a hypothesis may be modified and retested several times before it moves to the next stage, sometimes called a theory. However, many consider a hypothesis and a theory interchangeable, and the difference is not significant here. The theory is also subject to rigorous testing and attempts to disprove it. As in the earlier phases, several iterations and modifications can take place. What eventually emerges is the final answer, either a proven theory (or natural law) or, in the case of forensic analysis, a final answer that is the best that can be obtained with the evidence available. Using the previous example, the final result would be that the blood sample was from the suspect or was not. In reality, however, many times the best answer is more ambiguous and less specific, such as "This glass may be from that window" or "This fiber is consistent with the carpet in that house." The forensic scientist uses the scientific method to obtain the best answer possible, but it may not be a definitive answer as far as the case is concerned.

Some critical characteristics of the scientific method are that it is objective (versus subjective); is based on experiment, observation, and fact; includes testable ideas; and is quantitative. Science is also fluid in that current theories are constantly open to study, revision, and review. It is also governed by a process called peer review in which scientific findings are reviewed by scientists knowledgeable in the field before publication. When findings are published, other researchers usually attempt to reproduce the experiments and confirm the findings. In this way, the scientific method is designed to be self-correcting and to move in a forward direction constantly, albeit at a sometimes methodically slow pace. In the context of a forensic analysis, the scientific method is used to ensure that the best possible results are obtained.

hypothetical question A question asked of an expert witness in a trial that is designed to elicit an opinion based on evidence.

I

IABPA (International Association of Bloodstain Pattern Analysis) A professional organization devoted to the interpretation of bloodstain patterns.

IAFIS (Integrated Automatic Fingerprint Identification System) *See* AFIS.

IAFN (International Association of Forensic Nursing) A professional organization of registered nurses working in the forensic field, broadly defined. The organization was founded in 1993 and currently has approximately 1,800 members.

IBIS system An automated system used to assist in the analysis of BULLETS and CARTRIDGE CASES. IBIS was developed by the BUREAU OF ALCOHOL, TOBACCO, FIREARMS, AND EXPLOSIVES along with Forensic Technology, a Canadian company. Much as automated fingerprint database software is, IBIS is designed as a screening tool and does not replace the firearms examiner in making final identifications. The software for analysis of bullets is called Bulletproof and the software for cartridge cases is called Brasscatcher; both of these names are occasionally used to describe the entire system. IBIS has two computational components, one for data acquisition and one for analysis, and an analysis produces a list of candidate matches from the central databases. IBIS and its sister system (DRUGFIRE) have been particularly valuable in linking seemingly unrelated crimes by showing that the same weapon was used in both or all instances. IBIS and DRUGFIRE (developed by the FBI) are not currently compatible, so data are not interchangeable among them; efforts are under way to address this problem.

ice Street slang for AMPHETAMINE or for a dangerous form of methamphetamine made by slow evaporation and recrystallization of methamphetamine as a hydrochloride salt. This process yields large, clear crystals that can be smoked. Ice is considered to be both toxic and addictive.

ICP techniques Instrumental techniques used for inorganic and ELEMENTAL ANALYSIS. ICP, for *inductively coupled plasma,* refers to the method used to convert a sample to its constituent atoms or ions. In an ICP torch, gaseous argon (Ar) is ionized by a Tesla coil to form Ar^+ and free electrons. These ions are accelerated and confined in a stable magnetic field. The resulting high-energy collisions generate tremendous heat in the range of 10,300°C (~18,500°F). Under these extreme conditions, most chemical compounds are broken apart, forming free atoms and ions in this plasma "flame," although it is not fire in the same sense as combustion. Plasma is considered to be a separate state of matter in addition to solid, liquid, and gas. It consists of free electrons (electrons not associated with a specific atom), atoms, and ions and is characterized by an intense glow reminiscent of a flame. Plasmas are also found in the Sun.

There are two ICP instrumental techniques used in forensic science. ICP-AES is a form of EMISSION SPECTROSCOPY (atomic emission spectroscopy) in which the heat of the plasma is sufficient not only to atomize the sample but also to place many of the atoms into the EXCITED STATE. In such a state, an atom emits characteristic wavelengths of light in the visible and ultraviolet ranges that can be used to identify specific elements in the sample and to determine their concentrations. The second forensic application of ICP is in ICP MASS SPECTROMETRY (ICP-MS) in which the plasma is the source of elemental ions.

identification In the analysis of physical evidence, determination of the chemical substance or substances or the identification of a material. Unfortunately, this term is often misunderstood or misused in the forensic context. For example, a statement such as "The blood was identified," meaning that it was shown to be from one specific person, technically is incorrect. What is meant is that the stain was INDIVIDUALIZED or proved to be from a COMMON SOURCE. To individualize evidence, a comparison is always required, but comparison is not needed for identification; however, identification is often required before classification and individualization. It makes no sense to classify or perform DNA TYPING on a ketchup stain, so the identification of the stain as blood is part of the overall forensic analysis.

Identification News The journal of the INTERNATIONAL ASSOCIATION FOR IDENTIFICATION (AIA) published from 1951 to 1987. It was preceded by the *IAI Monthly Newsletter* and succeeded by the *Journal of Forensic Identification*.

Identi-Kit Sketching kit used by law enforcement officers to create drawings of suspects. The original kits, introduced in the 1950s, were simple and included templates and overlay tools that allowed for the creation of facial parts such as nose and chin. The individual elements were combined into a sketch. Modern systems are computer-based.

IED *See* IMPROVISED EXPLOSIVE DEVICE.

ignition point A term that can refer to the ignition temperature or to the physical location at which a fire was ignited and began to burn.

ignition source In arson and fire investigation, the source of the initial spark or other combustion-inducing event. The source of ignition may be an electrical item (such as a heated wire), a flame (such as match), a spark, or an INCENDIARY DEVICE.

ignition temperature The temperature at which a flammable material ignites. For gasoline, the most common ACCELERANT used in ARSON, the ignition temperature is in the range of approximately 500°F to 800°F, depending on type and grade.

ilium The upper portion of the pelvis, which has the familiar wing shape. *See also* Appendix V.

illumination The way an object is lighted, critical in forensic microscopy and photography. In microscopy, there are two basic types of illumination, incident and transmitted. Incident radiation is that light which is reflected off an opaque object such as a bullet or paint chip. Transmitted light, as the name implies, travels through a transparent or semitransparent object such as a hair or fiber. Four important characteristics of illumination are intensity, color, temperature, and angle. The intensity is the brightness or strength of the light; the angle is the angle at which the light intersects the surface being observed. The appearance of features, particularly three-dimensional features, can be altered as the angle of illumination (also called angle of incidence) is changed. The temperature of light does not refer directly to heat, but rather to the theoretical concept of an ideal blackbody emitter. A blackbody emits characteristic wavelengths of light that depend on the temperature to which it is heated. At a temperature of about 5700°C, a blackbody emits light equivalent to natural daylight, which is a mixture of all colors of the rainbow at equal intensities. The color of light refers to the temperature that such a blackbody emitter would have to be to emit the same mixture of colors and intensities as the light source in question. The temperature is thus considered to be a measure of the "whiteness" of light.

illumination, epi *See* EPI ILLUMINATION.

illumination, incident Lighting of a specimen being viewed under a microscope that originates between the eyepiece (objective) and the specimen itself and may originate at the sides as well. The specimen is viewed by REFLECTED LIGHT rather than by TRANSMITTED LIGHT. Although the term EPI ILLUMINATION is sometimes used interchangeably with *incident illumination,* they are not the same.

illumination source The lamp or other light source that illuminates a sample being examined under a microscope.

image enhancement/digital image enhancement Techniques that are becoming increasingly useful and important in the analysis of QUESTIONED DOCUMENTS and FINGERPRINTS and in photographic identification. Images are obtained either from a traditional camera (or video camera) that uses film or from a digital camera. These cameras may be mounted on special devices such as microscopes. The images are digitized and then manipulated by using specialized image enhancement software, usually with the goal of highlighting or making more visible the portions of the image that are of interest. This is often referred to as minimizing or removing the "noise" in the image. Care must be taken to document the steps used and to ensure that the image is merely enhanced, not materially altered to change the underlying feature. Image enhancement techniques are used in fingerprint identification, bite analysis, firearm and toolmark analysis, and facial identification, to name just a few applications.

immediate cause/immediate cause of death The medical/physiological cause of death. In criminal matters, a medical cause of death is not necessarily the legal cause. Cause of death is also distinct from the circumstances of death, which comprise the situation and conditions that led up to the fatal encounter. For example, assume a man is found with a gunshot wound to the chest and is taken to the hospital, where he dies a week later of an infection that arose from surgery per-formed to save his life. The immediate cause of death is the infection, but the legal cause is the gunshot wound. The circumstances of death are whether it was an accident (the gun fired while he was cleaning it), homicide, or suicide.

immunoassay A group of techniques used in forensic TOXICOLOGY for the detection of drugs in urine, blood, and other body fluids. The term *immunosorbent assay* is also used, and these techniques yield both QUANTITATIVE and QUALITATIVE information. Several variations based on the general technique exist. Immunoassay relies on an ANTIGEN–ANTIBODY reaction between the drug being tested and an antibody specific for it. The antibody is attached to a solid surface such as the bottom of a plastic or glass well. A complex that consists of the drug and a label is added, and the reaction occurs. As a result, the labeled drug is bound to the antibody. A sample that may contain the drug, such as urine, is added to the plastic well. If there is no drug or very little drug present, the labeled drug–antibody complex remains undisturbed. However, if there is a large concentration of the drug, it displaces the labeled drug from the antibodies, releasing the labeled drug into solution. The higher the drug concentration in the sample, the more is displaced. The amount of the displaced labeled drug is then measured. Types of immunoassay include enzyme-linked immunosorbent assay (ELISA), enzyme multiplied immunoassay technique (EMIT), radioimmunoassay (RIA), and fluorescent polarization immunoassay (FPIA).

immunodiffusion IMMUNOLOGICAL TECHNIQUES used in forensic SEROLOGY to determine the species of a blood sample. CROSSED-OVER ELECTROPHORESIS is also used in this role. Immunodiffusion is a PRECIPITIN test, meaning that a positive result is evidenced by the formation of a precipitate or solid (also called an IMMUNOPRECIPITATE) that can be seen easily. In this case, the solid forms as a result of an immunological reaction between ANTIGENS found in the blood sample and ANTIBODIES found in purified antisera applied in the test. Immunodiffusion tests rely on the process of diffusion, the natural spreading of a concentrated material or reagent into the surroundings. This phenomenon is seen when a drop of food coloring is placed into a glass of still water. The concentrated coloring diffuses throughout the glass until it reaches an equal dilute concentration. Types of immunodiffusion include single and double; the names refer to how and in how many dimensions the antigens and antibodies move through the gel. In one-dimensional techniques, the antigen or antibody moves only one way. Similarly, in single diffusion only one component moves; in double diffusion, both the antigen and antibodies are mobile.

immunoelectrophoresis Immunological techniques used in forensic SEROLOGY to determine the species of origin of a blood sample or bloodstain. These techniques combine ELECTROPHORESIS with the process of IMMUNODIFFUSION to detect ANTIGENS in a stain that can react with ANTIBODIES in an antiserum to produce a visible precipitate called a precipitin. Techniques that have been applied to forensic work include rocket immunoelectrophoresis (the Laurell Technique), two-dimensional immunoelectrophoresis, and crossed-over electrophoresis. The latter technique works by placing antigens and antibodies opposite each other in small wells punched in a gel medium. An electrical field is applied and the antigens and antibodies move toward each other and meet at some location between the two wells. If the antiserum used is specific to the antigens found in the stain, a visible precipitin band, which is a milky white color, forms in the gel. For example, if one well is filled with the extract of bloodstain and the other well is filled with antihuman antiserum, a precipitin band forms if the blood is human blood. If the stain is dog blood, for example, no precipitin line forms.

immunological techniques Tests and analyses that are based on reactions between ANTIGENS and ANTIBODIES. Such reactions are the basis of the immune response, in which an organism synthe-

sizes antibodies to react with and neutralize foreign substances. If a person contracts a cold, the immune response includes synthesis of antibodies that attack the invading cold virus and eventually overcome it. Examples of forensic uses of immunological techniques include ABO BLOOD GROUP typing, IMMUNODIFFUSION and precipitin tests to determine the species of bloodstains, and IMMUNOASSAY techniques used in forensic TOXICOLOGY.

immunoprecipitate A solid material that forms as a result of an antigen–antibody (immunological) reaction. This type of precipitate forms, for example, when human serum (from blood) reacts with antihuman antiserum. This property is exploited in immunological tests such as IMMUNOELECTROPHORESIS.

impact site The location at which a bloodstain pattern originated such as the point of impact of a blow or gunshot wound. It is the point at which blood is set in motion.

impact spatter A bloodstain pattern created when blood (or a surface that it is on) receives an impact that causes that blood to spatter. The spatter then strikes another surface such as a wall or floor, creating the impact spatter.

implied consent A legal doctrine in which a person's consent (or agreement) is inferred from actions rather than by a direct statement. This concept arises often in cases of arrests for driving while intoxicated and the breath and alcohol testing associated with it.

impression evidence Physical evidence that results from the contact between two objects or surfaces. Impression evidence is also referred to as imprint evidence or markings and as a group represents one of the largest classes of forensic evidence. Examples of impression evidence include markings made on cartridge cases and BULLETS, BITE MARKS, TOOLMARKS, FABRIC IMPRESSIONS, SHOE PRINTS, TIRE PRINTS, and, in many cases, FINGERPRINTS.

Impression evidence can be divided into groups that depend on how they were made. Scraping produces impressions called STRIATIONS, examples of which include bullet markings and many tool-marks. The other way impressions can be made is by compression of one surface under the weight of another such as when shoe impressions are made in mud or a fabric impression is transferred to a bullet passing through clothing. In general, an imprint is considered to be a flat impression (thin, as a fingerprint on glass), whereas an indentation also has depth such as a muddy shoeprint. In other words, imprints are two-dimensional and indentations are three-dimensional.

improvised explosive device (IED) A makeshift explosive device made from common or easily obtainable materials. An example is the shoe bomb Richard Reid attempted to use in late 2001 while traveling on a transatlantic American Airlines flight. He was restrained by passengers and crew while attempting to light a fuse and was convicted in 2003.

incandescence The process of emission of light by an object at high temperature. Light bulbs with filaments work on this principle; electrical current heats the wire filament, causing it to glow (emit light).

incendiary devices Devices that are used to ignite an ACCELERANT in an arson fire. They can range from simple to sophisticated and include items such as highway flares, fireworks, cigarettes, matches, black or smokeless powders, or homemade mixtures such as sugar and chlorates. Mechanical or electrical components may be part of it as well, as items such as flashbulbs are used to ignite an accelerant. The incendiary device is found at the point of origin, and thus it or the portions of it that remain are critical components of the physical evidence associated with arson.

incendiary fire A fire that is caused by purposeful human action. All arsons are incendiary fires, but the reverse is not always true.

incidental accelerant A material that is present in a fire but not purposely used as an ACCELERANT. If a fire occurred in a garage where gasoline for a lawn mower was stored, that gasoline would be an incidental accelerant.

incident illumination *See* ILLUMINATION, INCIDENT.

incised wound A wound created by cutting with an object such as a knife or scissors, as opposed to a tearing injury from BLUNT TRAUMA or a puncture wound.

inclusionary evidence Evidence, including the results of a forensic analysis, that does not exclude a given possibility or disprove a given hypothesis. Inclusionary evidence is the opposite of EXCLUSIONARY EVIDENCE.

inconclusive result A result that is not useful or fails to answer investigative questions. Inconclusive results can occur in many situations such as when there is too small a sample (a tiny spot of blood so small that it cannot be reliably typed), fragmentary evidence (a fragment of a bullet), or limitations of procedures or instrumentation. Such problems can lead to incomplete, contradictory, or inconsistent results that are impossible to interpret with any confidence. Another instance that can lead to an inconclusive result is that two analysts examine the same evidence and draw different conclusions. If the two interpretations cannot be resolved, the result is considered to be inconclusive.

inculpatory evidence Evidence that suggests or supports the idea that a person or persons were involved in the crime or action in question. Finding a suspect's fingerprint at a crime scene would be inculpatory evidence; however, taken alone is not conclusive.

indels Insertion deletions, a term used in DNA TYPING. An example is what is called a SINGLE NUCLEOTIDE POLYMORPHISM (SNP) in which a genetic locus shows variation in one nucleotide, such as a substitution of ADENINE (A) for THYMINE (T), CYTOSINE (C), GUANINE (G), or the reverse, T for A, G for C.

indented writing Writing that is transferred by the pressure of the writing instrument (pen, pencil, and so on) to the paper or other material that is beneath it. Indented writing is often undetectable to the naked eye and as such can be overlooked by a criminal. Accordingly, indented writing is a common form of QUESTIONED DOCUMENT evidence.

India ink *See* CARBON INK.

individualization The process of linking physical evidence to a COMMON SOURCE. Individualization is a process that starts with IDENTIFICATION, progresses through CLASSIFICATION, and leads, if possible, to assignment of a unique source for a given piece of physical evidence. The term *individualization* is often mistaken for *identification,* which in the forensic context does not have the same meaning. In forensic science, to identify something means exactly that—to determine that a red stain is blood or that a white powder is cocaine.

Aside from fingerprints, the other kinds of evidence that can potentially be individualized in a similar manner are blood (via DNA TYPING) and IMPRESSION EVIDENCE such as markings made on bullets and TOOLMARKS. Less obvious but extremely important in forensic analysis is individualization by way of a PHYSICAL MATCH. It is not that the evidence itself is unique but rather that the way in which it was separated and pieced back together allows for linkage to a COMMON SOURCE.

inductive reasoning (inferential reasoning) A mode or process of thinking that is part of the scientific method and complements deductive reasoning and logic. Inductive reasoning starts with a large body of evidence or data obtained by experiment or observation and extrapolates it to new conditions. By the process of induction or inference, predictions about

new situations are inferred or induced from the existing body of knowledge. In other words, an inference is a generalization, but one that is made in a logical and scientifically defensible manner.

inference A conclusion, assumption, or deduction that is based on the existence of other facts. For example, if a person leaves a fingerprint at a scene, one can infer that the person was at one time present at the scene.

infernal machine An explosive device hidden in or disguised to look like an ordinary object. A letter bomb is an example of an infernal machine.

infrared microscopy *See* INFRARED SPECTROSCOPY AND MICROSCOPY.

infrared photography A technique used in the analysis of QUESTIONED DOCUMENTS to reveal indented writing or other obscured or otherwise invisible writing. Illumination of documents by infrared (IR) light can increase contrast and reveal invisible, damaged, charred, obliterated, or otherwise obscured writing. Inks can also be studied by using IR luminescence, in which inks are exposed to visible light and then give off energy that is in the infrared portion of the electromagnetic spectrum. Infrared photography works the same way as conventional photography except that IR-sensitive film is used.

infrared spectroscopy and microscopy (IR, FTIR) A form of absorption spectroscopy in which electromagnetic radiation in the infrared (IR) range is absorbed by molecules. The pattern of absorption across the infrared range is unique for each different molecule, and as a result, an IR spectrum provides specific identification for compounds. However, definitive identification is only possible if the sample being studied is pure. IR methods are most widely used for the analysis of organic compounds such as drugs, synthetic fibers, and plastics but can also be used for many inorganic materials such as might be found in soil or paints. The only molecules that cannot be analyzed by IR spectroscopy are those such as oxygen and nitrogen (O_2 and N_2), which consist of two identical atoms. Such molecules are called homonuclear diatomics, meaning they consist of two atoms (they are diatomic) that are identical (homonuclear).

As all spectrophotometric techniques do, IR spectroscopy depends upon molecules' absorbing ELECTROMAGNETIC RADIATION. IR energy is not sufficient to break a molecule apart or to promote electrons to higher energy states. Rather, when a molecule absorbs infrared radiation, the energy is converted to vibrational motion within the molecule. To illustrate this effect: atoms within a molecule can be thought of as tiny steel marbles and the chemical bonds connecting them as springs. When IR radiation is absorbed, it causes the spring to flex and bend, but the energy is not sufficient actually to break the spring. Since the motion is three-dimensional, there are many different types of motion that can occur. To perform an IR scan, a plot is made of the degree of absorbance of each wavelength in the range by the sample. The resulting plot is the IR spectrum of that compound. The region of the spectra from 1500 to 500 WAVE NUMBERS (the inverse of the wavelength) is called the FINGERPRINT REGION and is distinctive enough to identify specific compounds.

Most IR spectrophotometers sold since 1990 use a device called a MICHELSON INTERFEROMETER coupled to a mathematical operation called a FOURIER TRANSFORM to obtain spectra. In this approach, all wavelengths of light are presented to the sample simultaneously rather than sequentially (one at a time). The resulting interference pattern (somewhat like a hologram) is then mathematically analyzed to yield the spectrum. This type of instrument and technique is referred to as *Fourier transform infrared* (FTIR). This instrumental design allows for many IR variations applicable to forensic science and forensic chemistry. These include ATTENUATED TOTAL REFLECTANCE, diffuse reflectance infrared Fourier transform spectrometry (DRIFTS), internal reflectance,

and microspectrophotometry (micro-FTIR), in which the instrument is coupled to a microscopy.

infrared spectrum *See* INFRARED SPECTROSCOPY AND MICROSCOPY; FINGERPRINT REGION.

inhalants A class of psychoactive components ingested by inhalation. Also called "glue sniffing" or "huffing," ingestion of inhalants became significant in the 1960s and remains most popular among teenagers and young adults. Inhalants are volatile (evaporate easily), and most are gases at normal room temperatures and pressures. Abusers typically fill a plastic bag or soak a rag with the substance to dose themselves. Unlike drugs of abuse, inhalants are components of common products such as paint or cleaning solvents and are not regulated under the CONTROLLED SUBSTANCES ACT. Many are propellants used in spray cans containing paint or other similar materials. Inhalants are considered to be central nervous system depressants with effects similar to alcohol, although achieved much faster. Abuse of inhalants carries high risks, particularly of kidney, liver, and brain damage.

inherent luminescence Emission of light by excitation with an ALTERNATE LIGHT SOURCE that is caused by a compound naturally present in the sample. For example, some fingerprints fluoresce naturally when exposed to Ultraviolet (UV) light even without treatment. Similarly, drugs such as LSD (LYSERGIC ACID DIETHYLAMIDE) fluoresce without any chemical alternation or treatment.

ink-jet printer A computer printer or related device such as a fax machine that creates printed documents by spraying ink onto paper in a pattern dictated by the software. The ink is delivered in fine jets and is supplied by replaceable ink cartridges.

inks The composition of inks is much like that of paints—a solvent base (water or other), coloring materials (either INORGANIC pigments or ORGANIC dyes, natural and synthetic), and other additives that control thickness and final appearance. The oldest ink, called India ink but originally used in China, consists of carbon black (ground charcoal) suspended in water and containing adhesive gums and varnishing components. Iron gallotannate inks contain inorganic colorants along with tannic acid and are used in fountain pens. Inks used in ballpoint pens contain synthetic dyes dissolved in organic solvents and additives to maintain a thick consistency. Newer gel pens contain synthetic dyes impregnated into gel.

inorganic compounds and inorganic analysis Chemical compounds can be broadly classified as organic or inorganic on the basis of their composition. Organic compounds contain carbon atoms that are bonded to other carbons or atoms that are nonmetals such as hydrogen or oxygen. All other compounds are inorganic. There are some exceptions to these rules such as the cyanide ion (CN^-) and carbonate ion (CO_3^{2-}), which are considered to be inorganic even though they fit the definition of an organic compound. The use of the term *organic* as it applies to chemicals is often confused or misrepresented in the media and in advertising, when *organic* is used to mean "natural." This definition is completely separate from the chemical one.

insects/insect activity Insects are invertebrates that are small and that have segmented bodies. They belong to the class Insecta within the phylum Arthropoda, which also includes spiders and crustaceans such as lobster. *See also* ENTOMOLOGY, FORENSIC.

instrumental analysis Methods of analyzing substances and compounds (organic and inorganic) that are based on the use of an instrument as opposed to simple equipment. Older methods of analysis are classified as traditional or "wet chemistry" and rely on simple equipment such as glassware and analytical scales (analytical balances). One such traditional method used in forensic science is THIN LAYER CHROMATOGRAPHY.

Instrumental analysis is widely used in forensic science for drug analysis, toxicology, arson, and trace evidence, to name a few applications; examples include ATOMIC ABSORPTION SPECTROPHOTOMETRY, INFRARED SPECTROPHOTOMETRY, and GAS CHROMATOGRAPHY–MASS SPECTROMETRY.

instrumental neutron activation analysis *See* NEUTRON ACTIVATION ANALYSIS.

interdigital space The space between the fingers or toes that can sometimes be used to obtain ridge characteristics and MINUTIAE that can be analyzed as latent FINGERPRINTS are.

interferences Substances that can affect or alter the results of an analysis or chemical test. PRESUMPTIVE TESTS such as those for blood have a number of interferences that may lead to FALSE POSITIVE or FALSE NEGATIVE results.

interlaboratory variation A term used to describe the small variations in results that can occur when different laboratories (forensic or other) analyze the same sample. Some analyses, such as DNA TYPING or FINGERPRINT analysis, should not produce any differences. However, areas such as drug analysis may. For example, if a single sample is submitted to 50 different laboratories to determine its purity, a range of values could result. One lab may find a value of 50.0 percent, another might find 50.5 percent, and yet another, 49.8 percent. A determination of interlaboratory variation in such cases helps to set a baseline of variation that is normal and natural, the result of small random discrepancies. Measurement of variation is an important part of QUALITY ASSURANCE/QUALITY CONTROL.

internal reflection spectroscopy A form of infrared (IR) spectroscopy (IR, FTIR) that is based on the absorbance and alteration of infrared radiation in terms of reflections of that radiation inside a sample. ATTENUATED TOTAL REFLECTANCE is an example of internal reflection spectroscopy.

internal standard A method or compound used in a chemical quantitation. Internal standard quantitation is highly accurate and provides a method for correction of problems associated with loss or suppression of an analyte during sample preparation and analysis. The most common forensic application is in chemistry (DRUG ANALYSIS and TOXICOLOGY). An internal standard is a compound or element that is different from the analyte of interest, but still closely related to it. For example, LIDOCAINE or PROCAINE would be a good choice as an internal standard for the analysis and quantitation of cocaine, as long as neither compound is found in the sample itself.

International Association of Identification (IAI) An international forensic science society initially focused on FINGERPRINTING that has expanded to include several forensic disciplines. The IAI was formed in 1914 by Harry Caldwell of the Oakland City Police Department at a time when scientific methods of identification were undergoing a transition from body measurements (ANTHROPOMETRY or BERTILLONAGE) to fingerprints. The society currently has more than 5000 members and offers professional certifications in BLOODSTAIN PATTERNS, CRIME SCENE PROCESSING, footwear and TIRE IMPRESSIONS, FORENSIC ART, forensic PHOTOGRAPHY, and latent fingerprints. The *Journal of Forensic Identification (JFI)* is a bimonthly scientific publication directed to members and other forensic science professionals.

International Journal of Environmental Forensics (Journal of Environmental Forensics, Environmental Forensics) A quarterly publication devoted to ENVIRONMENTAL FORENSIC science. The journal's first issue was published in March 2000 under the auspices of the Association for Environmental Health and Science (AEHS).

intoxication Ingestion of a substance such as alcohol or a drug to the point that physical and mental functions are impaired.

in vitro A process or study that takes place outside an organism or tissue, in a place other than where it would naturally occur. In vitro fertilization (which produces "test tube babies") is an example of an in vitro process.

in vivo A process or study that takes place within an organism or tissue.

iodine (I) The chemical element iodine, the ionic form of which is iodide (I⁻). It is widely used in forensic science in many applications, such as IODINE FUMING for FINGERPRINTS. The natural form of the element is I_2, a compound that sublimes into the gas phase, facilitating such fuming operations. Iodine forms an intensely colored blue complex in the presence of starch, a property that is also used in fingerprint development and in a presumptive test for SALIVA.

iodine fuming A process used to visualize latent fingerprints and one of the oldest in use. Crystals of IODINE (I_2) are placed in an enclosed cabinet (FUMING CABINET) and gently warmed, causing the iodine to vaporize without passing through a liquid phase, a process called SUBLIMATION. Although the mechanism of reaction is not completely understood, it appears that I_2 is physically absorbed, imparting an orange color to the prints.

ion A charged species created by a gain or loss of electrons. Cations are positively charged ions such as Na⁺, and anions are negatively charged such as Cl⁻. Polyatomics are a group of atoms that are covalently bonded (through electron sharing) but that act as an ion. Ammonium (NH_4^+) and nitrate (NO_3^-) are examples of polyatomic ions.

ion chromatography (IC) An instrumental technique that can be used to detect anions (negatively charged atoms or molecules such as Cl⁻) and cations (positively charged species such as Na⁺). IC has been applied in forensic science for the analysis of GUNSHOT RESIDUE and EXPLOSIVES. The ions of interest include ammonium (NH_4^+), nitrate (NO_3^-), and chlorate (ClO_4^-), species that are often detected by using color change or PRESUMPTIVE TESTS. The advantages of IC in these cases include specificity (presumptive tests are subject to FALSE POSITIVE and FALSE NEGATIVE results) and increased sensitivity, down to the part-per-billion (ppb) range. A part per billion is one microgram (μg) per liter of water, and a microgram is 1/1,000,000 gram.

IC instrumentation works on the basis of ION EXCHANGE. There are two modes of ion exchange used in ion chromatographs, single ion exchange and suppression ion exchange. Single ion exchange works very much as a water softener does, using resins that are specific for different cations, depending on what is of interest to the analyst. Suppression ion systems are used to reduce the ion concentration of the solution by using an additional step. An example of ion suppression would be creation of a column that replaces cations (positively charged ions) with the H^+ ion and anions (negatively charged ions) with the OH^- ion. When these two ions combine, they form water (H_2O), which is not ionic. Ion exchange still occurs, but by careful selection of resins, ions combine to form covalently bonded molecules (also called molecular compounds) that do not form ions in water. Regardless of the type of ion exchange column used, the detection system is most often a conductivity detector that relates concentrations of ions to the conductivity of the solution.

ion exchange The process of replacing one ion in solution with another while maintaining charge balance. Commercial ion exchangers are used to soften water, for example, to remove calcium and magnesium ions. To perform an ion exchange, the sample is forced through a column bed that contains an exchange resin. In a water softener, the goal of ion exchange is to remedy hard water problems by removing Ca^{2+} and Mg^{2+}. To do this, the water supply is directed over a bed charged with sodium ions (Na^+). In the column, the calcium and magnesium displace the sodium

at a one-to-two ratio to maintain charge balance. For every magnesium or calcium that is removed by the resin, two Na^+ cations are released into solution. Water hardness is removed by trapping calcium and magnesium in the bed, but as a result, soft water has a higher concentration of sodium, which can present problems to people with high blood pressure. Thus, ion exchange does not remove ions from solution; it only replaces one ion with others of the same overall charge.

ionic compound A chemical compound formed by IONS (CATIONS and ANIONS) that are held together by electrostatic attraction (like charges repel; unlike charges attract). Table salt (NaCl) is an ionic compound made up of the sodium cation (Na^+) and the chloride anion (Cl^-).

ionization energy/ionization potential The amount of energy required to remove an electron from an isolated atom. The higher the ionization energy, the more difficult the electron is to remove. The magnitude depends on the attractive force between the electron (negatively charged) and the protons (positively charged) in the nucleus. Accordingly, the ionization potential depends on the size of the atom (and thus the distance of the electron from the nucleus) and the number of protons in the nucleus.

ion mobility spectrometry (IMS) A portable instrument used in forensic science to detect drugs, explosives, tear gas, and chemical warfare agents. Originally called plasma chromatography, IMS works similarly to ELECTROPHORESIS, except that the charged species are separated in the gaseous state rather than in a gel. IMS works at atmospheric pressure, making it ideally suited for use as a portable monitoring system, and the military in several nations use IMS routinely for battlefield detection of chemical weapons. An air sample is drawn into the instrument and directed through a ionization region, where the radiation emitted by radioactive nickel-63 causes ionization and formation of ion-molecule clusters such as $H(H_2O)_3{}^+$. In this example, the ion is H^+ and the water is the molecule. An electronic pulse of a wire shutter (which looks like a screen door) admits clusters into the drift region, where they move against a flow of a drift gas such as air or nitrogen. The clusters are separated on the basis of the ratio of their size to their charge, much as separation is accomplished in gel electrophoresis, as the smaller clusters move ahead of the larger ones. The clusters arrive at the detector and the pulses are recorded as peaks, the height of which is proportional to concentration. IMS has been used to detect drugs in closed shipping containers and also is being explored for a wider role in explosives detection at airports.

ion trap A type of MASS SPECTROMETER in which ions are introduced in pulses into a circular ring. There, they move in an orbital path that is determined by their mass, charge, and the electrical fields created in the ring. At certain settings, only one mass-to-charge ratio achieves a stable trajectory. Thus, specific ions can be trapped in the ring before being introduced into the detector. Ion trap techniques are useful for biomolecules such as proteins and in tandem (dual) mass spectrometers.

IR *See* INFRARED SPECTROSCOPY AND MICROSCOPY.

isoelectric focusing (IEF) A technique used in forensic SEROLOGY to type ISOENZYME systems. In isoelectric focusing, a pH gradient is created in a gel (much as used in ELECTROPHORESIS), meaning that the pH changes with gel position. Protein molecules are charged, but at a given pH, they become neutral and cease to move. This point is called the ISOELECTRIC POINT, and proteins with different structures can be distinguished on the basis of where they stop moving in the gel. IEF proved to have a higher resolving power than traditional gel electrophoresis and thus could distinguish more types within some of the isoenzyme systems; however, the reagents are more expensive and prevented the routine use of IEF in

forensic serology. With developments in DNA technology, isoenzyme typing itself is rarely performed anymore.

isoelectric point (IEP) The pH at which a molecule, typically a protein, has no net charge. Proteins contain acidic groups such as OH and $NH_3{}^+$ that can ionize to O^- and NH_2. At low pH values (high concentrations of H^+), protonation of O^- to OH is favored, but so is protonation of NH_2 to make $NH_3{}^+$, a charged species. Conversely, high pH (low H^+ concentration) favors O^- and NH_2. Depending on the structure of the protein, there is some pH at which all acidic groups are protonated, and at that point, the protein molecule has no net charge. This property is exploited in ISOELECTRIC FOCUSING.

isoenzyme systems Before the widespread acceptance of DNA TYPING, isoenzymes were used in conjunction with ABO BLOOD GROUP SYSTEM typing to INDIVIDUALIZE blood and body fluid evidence to the extent this was possible. These isoenzymes are found on the surface of red blood cells and are thus sometimes called red cell isoenzymes. Enzymes are biological catalysts necessary to speed up reactions that would otherwise be far too slow in an organism. The term *isoenzyme* refers to the group of red cell enzymes that are POLYMORPHIC, meaning that they exist in multiple different forms and that the form a person has is determined genetically. Since heredity (genes) determines the form, the isoenzymes are considered to be GENETIC MARKER SYSTEMS. The analysis and typing of isoenzymes were accomplished by using gel ELECTROPHORESIS or closely related techniques. The six common systems typed were phosphoglucomutase (PGM), adenylate kinase (AK), acid phosphatase (ACP or EAP), glyoxalase I (GLO I), esterase D (ESD), and adenosine deaminase (ADA). The application of DNA TYPING has all but eliminated the routine use of isoenzymes in forensic biology.

isotope An atom of an element that has the same number of protons in the nucleus but a different number of neutrons and thus a different mass. Isotopes may be natural or synthetic, and contrary to a popular misconception, not all isotopes are radioactive. For example, hydrogen atoms have one proton in the nucleus, and the predominant form of hydrogen has no neutrons. Deuterium is a nonradioactive isotope of hydrogen containing one proton and one neutron in the nucleus, whereas tritium is a radioactive isotope containing one proton and two neutrons. Both isotopes are chemically identical to hydrogen.

isotope dilution A technique used in MASS SPECTROMETRY to obtain extremely accurate quantitative analysis of samples. The technique can be used for inorganic or organic analysis. Since the natural isotopic abundances are known, the addition of a known amount of a given isotope to a sample containing that element allows for ratios of isotopes to be related to concentration of the isotope in question. For example, to determine the amount of benzene (C_6H_6) in a sample, deuterated benzene (benzene-d6, C_6D_6) can be added to the sample. Since the deuterium ISOTOPE of hydrogen is chemically identical to hydrogen, it behaves exactly as benzene would during the course of the analysis; however, since its mass is different, the mass spectrometer can determine the concentrations of each type of benzene separately. By using these data and knowledge of the natural abundance of deuterium, it is possible to determine the concentration of benzene with a high degree of accuracy.

jacketed bullets Lead BULLETS that are encased in a metal jacket made of copper or a similar alloy. Jacketing of a bullet allows it to feed smoothly in automatic and semiautomatic weapons. Jacketing may be partial or complete (full metal jacket).

Jaffe test A presumptive test for urine based on the reaction of CREATININE with picric acid. A positive result is shown by a red color.

Journal of Environmental Forensics
See INTERNATIONAL JOURNAL OF ENVIRONMENTAL FORENSICS.

Journal of Forensic Identification
See ASSOCIATION OF IDENTIFICATION; IDENTIFICATION NEWS.

junk DNA Segments of DNA that do not appear to code for proteins and that consist of repeated sequences of NUCLEOTIDES. The loci typed in current DNA TYPING techniques consist of junk DNA.

K

karyotype The forty-six human chromosomes (in 23 pairs) that contain all human genetic material.

Kastel Meyer color test A PRESUMPTIVE TEST used to identify BLOOD. The chemical substance used is phenolphthalin, which is made by boiling phenolphthalin in an alkaline solution containing potassium hydroxide (KOH). The test is also referred to as the K-M or KM test, or as the phenolphthalein test, even though the starting compound is phenolphthalin. Until recently, phenolphthalein was used as the active ingredient in some laxative products, and it is still commonly used in chemical analyses (titrations) of acids and bases. The KM reagent is considered to be relatively safe and is certainly safer than BENZIDINE, which was once widely used to identify blood but abandoned as a result of its known carcinogenic properties. Phenolphthalin reacts with the hemoglobin in blood in the presence of hydrogen peroxide (H_2O_2) to cause a pink color to form. As for other presumptive tests, the KM test is not specific for blood and can produce FALSE POSITIVE results with substances such as horseradish.

keratin A tough fibrous protein found in hair and skin. The interlinking of keratin proteins provides its strength and makes it insoluble in water and resistant to chemical and biological attack.

kerosene A liquid hydrocarbon that is sometimes used as an ACCELERANT in ARSON. It has a flash point of 100°F and is commonly available as a fuel for kerosene heaters. It is also used as a jet fuel.

ketamine A drug that is increasing in popularity particularly among juveniles and young adults taking part in parties or "raves." Ketamine is used as an animal anesthetic, and the only illicit source of the drug is theft, mostly from veterinary clinics. It can also be obtained from foreign countries, particularly Mexico. As a veterinary drug, it is supplied as a liquid or a soluble powder that can be injected, sprinkled on other material and smoked, added to drinks, or snorted. The effects of ketamine have been compared to LSD and PCP, and the use of the drug is rapidly increasing. Street names for the drug include "Special K." It is listed on Schedule II of the CONTROLLED SUBSTANCES ACT.

ketimine A chemical compound formed as part of the reaction of NINHYDRIN with the amino acid residues in latent FINGERPRINTS.

ketones A class of organic compounds defined by the presence of a carbon–oxygen double bond on a nonterminal carbon. The C=O in ketones is bonded to two other carbons. Acetone is a ketone.

Kevlar A synthetic fiber that is related to nylon and is classified as an aramid, meaning it is a polymer composed of aromatic amides. Kevlar is well known for use in body armor and bulletproof vests, an amazing application for a fiber.

keyhole wound A wound created by bullet that is tumbling as it enters the flesh.

kinetics Generally, the study of motion. In forensic science, the term usually refers to the speed of chemical reactions.

Kingston model A statistical model proposed in 1964 that describes the frequency and individuality of fingerprint patterns.

Kirk, Paul (1902–1970) Considered to be the father of modern American CRIMINALISTICS, and head of the first criminalistics program in the country. He became involved in criminalistics as a result of collaboration between the Berkeley Police Department and the University of California at Berkeley encouraged by August VOLLMER. Kirk established the criminalistics program in 1937, and by 1948, it was a department under the university's School of Criminology. Kirk was active in research in many areas of evidence including TRACE, HAIR, and FIBERS as well as in teaching and casework. He authored a pioneering textbook, *Crime Investigation,* in 1953 and the second edition in 1974. The 1950s were among Kirk's most active years, and in 1955 he analyzed crime scene evidence from the Sam SHEPPARD case, in which Sheppard had been accused of murdering his wife, Marilyn.

Kirk's philosophy of criminalistics was GENERALIST in the sense that he believed forensic scientists should have a broad scientific education and knowledge of many aspects of physical evidence. He considered INDIVIDUALIZATION the primary skill and distinction of forensic science.

knots A type of physical evidence important in cases of STRANGULATION, HANGING, or restraint of victims by rope or cordage. One of the central aspects of such cases is distinguishing among murder, suicide, and accidental death, and the form and complexity of a knot can be critical elements in making that assignment.

known samples (knowns) *See* CONTROL SAMPLES (CONTROLS).

Koppanyi test *See* DILLE-KOPPANYI TEST.

Kriminologie The name of the periodical started by the forensic science pioneer HANS GROSS more than a century ago. It is still published.

Kumho Tire **case (***Kumho Tire Company, Ltd. v. Carmichael* **[526 U.S. 137 1999])** A decision by the United States Supreme Court that extended the "gatekeeper" role of trial judges in determining whether the testimony of an expert should be admissible. In the *DAUBERT* decision of 1993, the court assigned to judges the responsibility for deciding whether the testimony of a scientific expert were admissible in the case in question. The *Carmichael* decision extended the gatekeeper role to include the testimony of any expert in any field, not just in a scientific or technical discipline.

L

laboratory accreditation *See* CERTI-
FICATION.

laceration A slice or cut that is created
by the impact of blunt force rather than
by actual cutting or stabbing by a sharp
instrument. The edges or margins of the
wound tend to be irregular, and strands of
connected tissue (bridging) may be seen
spanning the wound.

lactose Also known as milk sugar, a
compound that is sometimes encountered
as a CUTTING AGENT (diluent) for drugs. It
is a disaccharide, meaning that it is composed of two simple sugars, here galactose
and glucose. Lactose is the primary sugar
in mother's milk. Some people, specifically
those whose body does not produce the
enzyme lactase used to break down this
sugar, are lactose-intolerant.

laminated glass A type of GLASS found
in windshields that is produced by sandwiching a layer of plastic polymer between
two sheets of safety glass. Because of the
presence of this plastic, laminated glass
tends to hold together and retain the shape
created by the impact.

land The elevated portion of rifling that
is cut into the barrel of FIREARMS. *See also*
LANDS AND GROOVES.

lands and grooves Structures that are
cut into the barrel of a firearm. The lands
and grooves are cut in a twisting pattern,
and, as a result, the bullet emerges from
the barrel spinning. A barrel that has
lands and grooves is said to be rifled. A
spinning bullet does not tumble or wobble
and therefore is much more accurate than

a projectile fired from a smooth-bore
weapon such as a musket.

Landsteiner, Karl (1868–1943) An
Austrian physician and Nobel Prize–win-
ning researcher in the field of IMMUNOL-
OGY. Landsteiner discovered the ABO
BLOOD GROUP SYSTEM in 1900–01. Typing
of this group would be the mainstay of
forensic SEROLOGY until DNA TYPING supplanted it in the early 1990s. In the late
1800s the first attempts at blood transfusions had led to many deaths when the
blood of the donor caused the red blood
cells of the recipient to clump together
(agglutinate). Landsteiner noted that this
reaction was not universal in that the blood
of some individuals was compatible. These
observations coupled with his research led
to the identification of the ABO system, the
first blood group system identified. Landsteiner also recognized that a person's
blood group was inherited and thus would
be useful in paternity cases. Because the
ABO system is inherited, it also served as a
GENETIC MARKER system in forensic analysis of blood and body fluids for nearly a
century. Landsteiner's discovery led to systematic typing for blood transfusions and
saved untold thousands of lives. As a result,
he was awarded the Nobel Prize in medicine in 1930. Landsteiner eventually moved
to New York and continued to work in the
field of immunology, participating in the
discovery of several more blood group systems including the Rh system.

lanthanides A group of elements listed
in the Periodic Table of Elements (*see*
APPENDIX III) referred to as the rare earth
elements. It encompasses cesium (atomic
number 58) onward, and not all of the
theoretically predicted elements in the

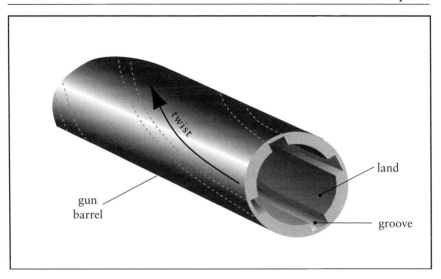

View of a rifled barrel of a firearm. The lands are the protruding structures; the grooves are cut into the metal, and the pattern twists down the barrel.

series have been observed or created. The known lanthanides currently end at lutetium, atomic number 71. The lanthanides are always shown below the main body of table for clarity.

larva/larvae The wormlike entity that hatches from the eggs of insects such as flies. The larva of the common blowfly, an insect useful in forensic ENTOMOLOGY, is a maggot.

laser printer A computer or FAX printing device that works by imparting an electrical charge to paper as directed by software in the computer. Initially, a drum surface is charged electrostatically in a fine grid pattern. A laser selectively scans the grid, discharging any grid that it strikes to create the pattern contained on the document to be printed. Toner is applied to the drum and adheres wherever the charge remains, and paper is then fed through using a roller or similar system. Toner adheres to the paper and is set in place by heat.

lasers Light sources that emit very intense radiation at a specific wavelength.

Lasers can be used to induce fluorescence, which in turn can be exploited in several ways. Fluorescent detectors are among the most sensitive available for analytical instruments as in HIGH PRESSURE LIQUID CHROMATOGRAPHY and can be exploited as part of an IMMUNOASSAY used in forensic TOXICOLOGY. Lasers can also be components of ALTERNATIVE LIGHT SOURCES used in fingerprint visualization. Finally, lasers can be used to vaporize a surface for introduction into instruments such as those used in MASS SPECTROMETRY. This technique, called laser ablation, is suited for the examination of paint layers and any other samples in which a characterization of the surface is needed.

latent prints (latent fingerprints) FINGERPRINTS (or other prints such as palm prints) left at a scene on an object that are not visible, barely visible, and/or potentially visible; fingerprints that are not visible under normal lighting conditions; prints that are invisible or nearly so. This definition is somewhat loose, and the term *latent prints* is often used to refer to any type of fingerprint regardless of visibility. Latent prints can be visualized with powders, chemical developers, and physical developer, and by many other types of apparatus.

Lattes, Leone (1887–1954) A professor at the Institute of Forensic Medicine in Turin, Italy, who was instrumental in applying KARL LANDSTEINER's discovery of the ABO BLOOD GROUP SYSTEM to forensic casework. In 1915, he developed a test that came to be known as the Lattes procedure or the Lattes crust test, in which red blood cells were added to dried bloodstains to determine the ABO blood type of the stain. For example, a stain from a person with type A blood contains anti-B antibodies. When type A cells are added to the stain, nothing happens, but when type B cells are added, they clump together (agglutinate). Although novel at the time, the Lattes procedure did not work well with old stains, and interpretation of results was difficult. However, it laid the groundwork for later more sensitive and robust testing methods such as ABSORPTION-ELUTION that were widely used until DNA TYPING supplanted ABO typing in forensic SEROLOGY.

LCMS (liquid chromatography mass spectrometry) An instrumental analysis technique used for identification of drugs (particularly in TOXICOLOGY) and other large molecular species. The instrument consists of a high-pressure liquid chromatograph coupled to a MASS SPECTROMETER. The most common interface between the liquid chromatograph and the mass spectrometer detector is based on the ELECTROSPRAY technique.

LD_{50} See LETHAL DOSE.

lead A heavy metal, symbol Pb, of forensic interest primarily as a poison and as the metal used to make bullets. Lead is easily smelted and extracted and thus became one of the first metals to be exploited by humans. Most lead poisoning is inadvertent or accidental, versus homicidal or suicidal, probably because there are many other materials such as ARSENIC and CYANIDE that are more deadly and

thus better suited to the purpose. Modern exposures to lead primarily result from old paints (leaded paints were banned in 1978), lead in gasoline (although use of tetraethyl lead as an additive ceased in the 1990s), and mining and pollution. Lead has also been used in pipes and is an ingredient in solder.

leading The fouling of the barrel or other parts of a firearm by the lead contained in the bullets. This problem has been greatly reduced by the introduction of jacketed ammunition, in which the lead bullet is encased in copper or a copper alloy.

lens/lenses Devices that are used to focus light and that can be designed to create a magnified image. As shown in the figure, a simple convex lens (wider at the center than at the edges) serves to converge light to a single point, where an image is created. The PRINCIPAL FOCUS (X in the figure) is the point at which an image is formed from an object placed at an infinite distance away from the opposite side of the lens. In the case of a simple lens, the distance of X from the lens is also the FOCAL LENGTH (f). A lens can be characterized by the equation

$$\frac{1}{f} = \frac{1}{p} + \frac{1}{q}$$

Any pair of values of p and q that satisfy this equation are referred to as conjugate foci.

Additionally, the magnification (m) is equal to q/p.

lethal dose The amount of a substance such as a poison or toxin that is necessary to cause death. Factors that influence the size of a lethal dose include health of the person, age, family history and genetics, and weight. For example, two extra-strength aspirin tablets is a normal dose for an adult, but that same dose could be fatal to an infant. To account for this size dependence, dosages of drugs are normally defined in units of milligrams of the substance per kilogram of body weight. Toxi-

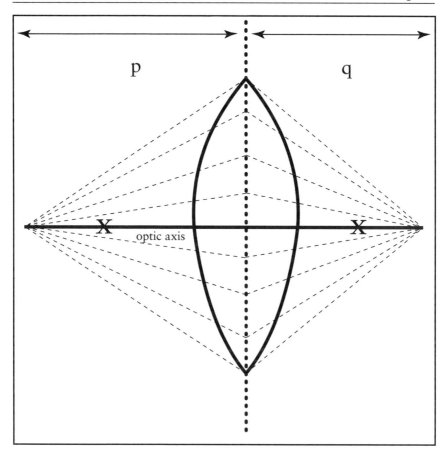

p

q

optic axis

X   X

A simple convex (converging) lens. The optic axis runs down the center of the lens, *p* is the distance from the object being studied to the lens, and *q* is the distance from the lens to the image that is created by it. The points *X* on the optic axis represent the principal focus for each side of the lens.

cologists often refer to a value such as the LD_{50}, or "lethal dose 50," a value most often derived from animal studies. In such studies, a large population of experimental animals such as mice or rats is given increasing doses of the substance being tested. The dosage that is fatal to half the population is the LD_{50}. Although a useful estimate, an LD_{50} is specific only to the animal of interest and is not always directly transferable to human beings. Some LD_{50} values exist for humans, but these data are normally derived from studies of accidental or unintentional expo-

sures. LD_{50} is reported in terms of body (milligrams per kilogram [mg/kg]) or as a concentration in blood or other body fluids. This is the case with CARBON MONOXIDE, for which dose is reported as %COHb or percentage of the hemoglobin that is bound to CO instead of to oxygen. Fatalities can occur in the range of 50 percent or above, but the exact value of %COHb that is fatal varies among individuals.

leucomalachite green A PRESUMPTIVE TEST for blood that is less frequently used than the KASTEL MEYER or LUMINOL test.

It is prepared by combining leucomalachite green and sodium perborate ($NaBO_3$) in water and acetic acid.

levels I, II, and III Terms that can be used to describe features in fingerprints. Level I consists of the pattern such as a loop; level II consists of ridge patterns such as deltas or bifurcations (MINUTIAE); and level III consists of patterns and features of the pores.

lidocaine Also known by the trade name *Xylocaine,* a white powder sometimes used as a CUTTING AGENT for COCAINE because it is, like cocaine, a local anesthetic that produces localized numbness. It also reacts with COBALT THIOCYANATE, a reagent commonly used as a presumptive test for cocaine. However, the color change is different from that observed with cocaine and would not produce a FALSE POSITIVE result if interpreted by an experienced forensic chemist.

Lieberman test (Liebermann reagent) A PRESUMPTIVE TEST used in DRUG ANALYSIS. The reagent is prepared by dissolving potassium nitrite (KNO_2) in concentrated sulfuric acid and is used to test for COCAINE (a yellowish color indicates the possible presence of the drug) and MORPHINE (a black test result is positive).

lie detector *See* POLYGRAPH.

ligature The object used to cause STRANGULATION in a murder, suicide, or accidental death. Ligatures are often ropes or other cords but can be anything that will encompass the neck and to which pressure can be applied. Towels, scarves, belts, sheets, and phone cord have all been used as ligatures, and the KNOTS used to tie or secure them can become critical evidence. Ligatures can also produce distinctive impressions on the neck that can be physically matched to the ligature. For example, if a belt with a woven pattern is used to strangle someone, the weave pattern may be visible in the skin in much the same way a bite mark is. A ligature is also any cording that is used to bind a victim.

likelihood ratio A quantity useful in the statistical analysis, presentation, and interpretation of forensic analyses that starts with at least two different theories or hypotheses concerning given scenarios ("competing hypotheses"). Assume, for example, that a single hair is found on the clothing of a homicide victim and that a suspect has been identified. Investigators, and ultimately a court, will want to know whether the hair is the suspect's. However, since hairs cannot be linked to a unique COMMON SOURCE in the way fingerprints can, the answer to that question becomes more complex than a simple yes or no. Instead, the results of the analysis can be reported on the basis of probabilities and a likelihood ratio. One hypothesis is that the hair is that of the suspect; one competing hypothesis might be that the hair is very similar to the suspect's but was deposited on the victim as a result of some random process. The relative likelihood of these two hypotheses can be expressed numerically by using a likelihood ratio. The likelihood ratio could take into account several separate probabilities, such as the probability that the victim and the suspect had contact, the probability that a transfer occurred, or the probability that the hair remained in place long enough to be detected. Similar probabilities could be estimated for the competing hypothesis and the summation of these probabilities used to create the likelihood ratio, that would be expressed as

LR = probability that the suspect is the source of the hair/probability that random person is the source

The type of evidence involved and the amount of hard data and databases available are among the many factors that determine the reliability and usefulness of a likelihood ratio. For example, fingerprints or DNA TYPING results are supported by much larger databases and knowledge of frequencies than are HAIRS, fibers, GLASS, and other kinds of physical evidence. Despite these limitations, likelihood ratios can be useful for examining alternative explanations and encouraging the generation of alternative explanations for forensic findings. *See also* BAYESIAN STATISTICS.

limit of detection (LOD) In an analytical procedure, the smallest amount or concentration of an analyte that can be detected by that protocol. The LOD may or may not be the same as the LIMIT OF QUANTITATION for the procedure. An instrument may be capable of detecting quantities too low to be reliably quantitated, and in such cases, the LOD is a lower concentration that the LOQ.

limit of quantitation (LOQ) In an analytical procedure, the smallest amount or concentration of an analyte that can be reliably quantitated by that protocol. The LOQ may be the same as the LIMIT OF DETECTION, or it may be a higher concentration.

limnology, forensic Limnology is a subdiscipline of biology devoted to the study of freshwater bodies and ecosystems. Thus, forensic limnology is an extension of this study into civil and criminal investigations and cases. Freshwater systems includes lakes, streams, ponds, and marshes; a case that could involve limnology, for example, would be one in which a body was dumped in a lake. The principles of limnology could be used to determine the time of the year when the dumping occurred. Typical types of organisms that are used are freshwater DIATOMS, other types of algae, larger aquatic plants, and microscopic and small animals.

LIMS (Laboratory Information Management System) Computer software and hardware used to create a database system integrating laboratory data, record keeping, reporting, and other tasks.

Lindbergh kidnapping A 1932 crime that until the O. J. SIMPSON trial was considered to be the "crime of the century." As in the Simpson trial, scientific evidence played a critical role, and the atmosphere and media frenzy surrounding both trials were similarly described as circuslike.

Charles A. Lindbergh became a national and international hero in 1927 after flying the Atlantic alone in the *Spirit of Saint Louis*. He later married Anne Morrow, and their first child, a son named Charles Jr., was 20 months old when he was kidnapped around 9:30 P.M. on March 1, 1932. The child's body was later found in woods close to the home. The arrest of the primary suspect, Bruno Hauptmann, did not occur until 1934.

During the time between the kidnapping and murder of the child and Hauptmann's arrest, forensic investigations of the physical evidence, including the ransom notes (QUESTIONED DOCUMENTS); TRACE EVIDENCE; psychological and psychiatric studies; and, perhaps most damning for Hauptmann, analysis of the ladder, were undertaken. ALBERT S. OSBORN, a pioneer in the field of questioned documents, performed analysis of the handwriting found in the ransom notes. Arthur Koehler, a wood expert employed by the U.S. Forest Service, undertook a meticulous evaluation of the ladder, including the wood, construction techniques, and TOOLMARKS found on the wood. Eventually, he was able to trace the lumber used to a lumberyard and mill located in the Bronx. Marks made by planers on the wood in the ladder matched a planer at the yard. A search of the attic above Hauptmann's apartment revealed a missing floorboard. Nail holes and tree ring patterns from a rail of the ladder lined up perfectly where the floorboard had been. Furthermore, Koehler was able to demonstrate this at the trial as well as to show how the planer marks from the ladder matched Hauptmann's planer. Hauptmann was convicted, and after a series of appeals and reviews, including one by the Supreme Court, he was executed on April 3, 1936.

line quality In handwriting, the term used to describe the appearance of a stroke. The criteria used to describe line quality include the smoothness, darkness, and angle of the writing instrument to the paper or other surface.

LINES *See* REPETITIVE DNA.

lingual Related to the tongue or referring to it in some way. Some drugs are administered sublingually, meaning that they are dissolved under the tongue

linguistics, forensic The evaluation of the use of language in forensic applications. Techniques of forensic linguistics are used in such areas as attempting to identify a person's region of origin, determining the author of a document or determining whether two documents were written by the same person, and attempting to clarify the meaning of statements made in court or to law enforcement officials. Linguists can also work with QUESTIONED DOCUMENT examiners and VOICE RECOGNITION experts. Threatening letters or phone calls, ransom notes, email communications, and disputed wills are examples of the types of evidence forensic document examiners analyze. One of their most common tasks is in the area of speaker or author identification, comparison, authentication, and analysis. Some authors consider forensic stylistics to be a separate specialty focusing on the style of speech (oral or written) characteristic of a group or individual.

Analysis of documents or speech involves study of the types of words used, word choice (for example *pop, soda,* or *Coke* to describe a carbonated beverage), grammar, accent or dialect, spelling, error patterns, and sentence structure. Statistical analysis and comparison to general population usage patterns are sometimes employed in an attempt to determine where a speaker might have been born or currently live. Similarly, if a will is suddenly changed on the eve of a person's death, linguists can compare known writings of the person with the will to see whether the questioned writing follows the same pattern as that of the new will. Many of these types of analyses are used as investigative tools more often than in direct courtroom testimony. For example, linguistic analysis alone would not be able to prove that the author of a threatening letter lived in a certain region, but knowing that it was likely could provide a great deal of help to investigators.

linkage In genetics, genetic marker loci that are physically close to each other on the same chromosome and are inherited together. This process is referred to as linkage or genetic linkage.

linkage disequilibrium/linkage equilibrium In genetics, in linkage equilibrium the frequency of a combination of different types can be calculated by multiplying the frequency of the individual types together. For example, if a person is typed for two different genetic marker systems and the first type is found in one in a thousand people and the second is found in one in 100, the frequency for the combined types is one in 100,000 (100 × 1,000). Such combinations are said to be in linkage equilibrium, meaning that the two types are inherited independently of each other—in other words, they are not linked. If in this example the combined types were found to exist not in one in 100,000 but in one in 5,000, then the two genetic loci are not inherited independently of each other and the system is said to exhibit linkage disequilibrium. In DNA TYPING, there are 13 loci that are determined for the CODIS system. The frequency of a particular combination of types can be calculated by multiplying the frequencies for all 13 together; thus for these 13 loci, there is linkage equilibrium, and they are inherited independently of each other.

lipids Informally, "fats." Lipids are organic compounds that are grouped together on the basis of their solubility in nonpolar solvents. Water is a polar solvent and as such lipids are water insoluble. The category of lipids includes fatty acids (such as found in butter and margarine), GLYCERIDES, steroids, waxes, and LIPOPROTEINS. Lipids are a component of FINGERPRINT residues.

lip impressions *See* LIP PRINTS/LIPSTICK.

lipophilic Characteristic of substances that dissolve in or associate with LIPIDS, which are nonpolar molecules. The general rule regarding solubility is that "like dissolves like," meaning that nonpolar materials such as fats dissolve in other nonpolar materials such as oils, but not in polar solvents such as water. Substances that are lipophobic are "fat loving," meaning that they associate with or dissolve in nonpolar lipids. *See also* HYDROPHILIC; HYDROPHOBIC.

lipophobic Characteristic of substances that do not dissolve in or associate with LIPIDS, which are nonpolar molecules. The general rule regarding solubility is that "like dissolves like," meaning that nonpolar materials such as fats dissolve in other nonpolar materials such as oils, but not in polar solvents such as water. Substances that are lipophilic are "fat fearing," meaning that they do not interact with nonpolar lipids. Water is lipophobic. *See also* HYDROPHILIC; HYDROPHOBIC.

lipoproteins Proteins in the blood that transport LIPIDS. Examples of lipoproteins include low-density lipoproteins (LDLs), which transport cholesterol.

lip prints/lipstick A form of impression evidence that can be left on glasses, cigarettes, and other surfaces. There are two aspects of lip prints that can be of forensic interest: the pattern itself and the composition of the material such as lipstick or lip balm that might have been the medium by which the impression was made. The study of lip patterns, also called celioscopy, would be expected to be similar to dactyloscopy, the study of fingerprints. However, research to date indicates that unlike fingerprints, lip prints are not unchanging and that there appear to be some inherited characteristics such that relatives can have similar lip prints. In contrast, even identical twins have different fingerprints. Thus, the patterns of lip prints may have some investigative use, but this type of forensic evidence has not proved to be of much value.

liquid accelerant A combustible liquid used to start a fire. The most commonly used is gasoline.

liquid chromatography *See* HIGH-PRESSURE LIQUID CHROMATOGRAPHY.

liquid chromatography mass spectrometry *See* LCMS.

live scan A group of optical techniques used to acquire FINGERPRINTS. Rather than the finger's being inked and placed on paper, the finger is placed on a surface and scanned optically, creating a digital image.

lividity Also referred to as LIVOR MORTIS, the settling of blood that occurs in a body after the heart stops beating. At the time of death, circulation ceases and blood is no longer being pumped under pressure to all areas of the body. In addition, since blood is not reaching the lungs and is not being oxygenated, it takes on a bluish purple tint. Exceptions can occur if the victim has been poisoned with substances that alter the color of the blood such as CARBON MONOXIDE (CO). CO imparts a distinctive cherry red color that also alters the appearance of the lividity stain. Lividity occurs in the areas of the body in which gravity naturally causes settling, and lividity stains can appear similar to bruises. The only places blood will not settle are locations where pressure is being applied. For example, if a person dies while seated, pressure is being applied to the buttocks and prevents blood from pooling there.

The first signs of lividity begin to appear about an hour after death and reach a maximum after three to four hours. After about 12 hours, no additional lividity occurs. Thus, lividity stains are useful in determining the time since death or the POSTMORTEM INTERVAL (PMI). The stain pattern can also be used to determine whether a body was moved during the period in which lividity was developing. If a stain has already formed and the position of the body is altered within the 12-hour window after death, a new pattern can form while the old stains partially fade. Similarly if the body is moved after 12 hours, the lividity pattern may not match the position in which the body is finally discovered. Such inconsistencies are usually uncovered at AUTOPSY, when all of the clothing of the victim is removed.

livor mortis *See* LIVIDITY.

Locard, Edmund (1877–1966) A pioneering French forensic scientist who was instrumental in taking new theoretical ideas of what was then called police science and applying them to casework. Locard was trained in both law and medicine and was influenced by the writings of HANS GROSS as well as the fiction of SIR ARTHUR CONAN DOYLE. In 1910, Locard established a forensic laboratory in Lyons, France. The lab was primitively equipped, but even so, Locard was able to establish a reputation and to increase the visibility of forensic science in Europe. Locard was interested in microscopic evidence, particularly DUST, and believed that such TRACE EVIDENCE was crucial in linking people to places. Although he never used the exact phrase himself, Locard is most famous for LOCARD'S EXCHANGE PRINCIPLE, which evolved from his studies and writings. The principle is stated simply as "Every contact leaves a trace," and reflects his belief that every contact between a person and another person or a person and a place results in the transfer of materials between the entities involved. Most of this TRANSFER EVIDENCE, such as dust, is microscopic, and it may not last long, but the transfer does occur and it is the task of the forensic scientist to find those traces and use them to establish the link. An example is shown in the figure, in which a victim and a suspect have contact at a crime scene, resulting in transfers of trace evidence among all elements present. When the contact is finished, the scene contains evidence of both people, perhaps in the form of blood, HAIRS, FIBERS, or FINGERPRINTS. Likewise, each person carries away traces of the scene (dust, carpet fibers, and so on) and of the other person. Not all of these may be detectable, but they do, according to Locard's exchange principle, occur.

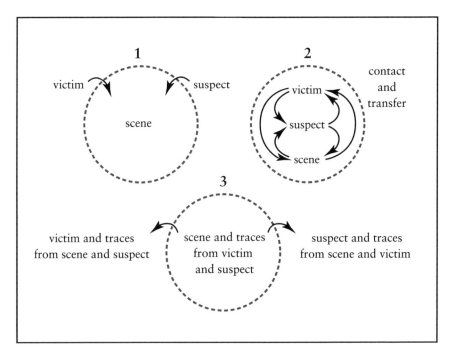

Schematic of Locard's exchange principle, which states that any contact between people or between a person and a place results in the exchange of material. The characteristics of the transferred material are physical evidence of the contact.

The success of his laboratory and methods encouraged other European nations to form forensic science laboratories after the conclusion of the First World War. In Lyons, he founded and directed the Institute of Criminalistics at the University of Lyons, and he remained a dominant presence in forensic science into the 1940s.

Locard's exchange principle *See* LOCARD, EDMUND.

locus/loci Location on a chromosome where a gene or other base pair sequence is found.

LOD *See* LIMIT OF DETECTION.

loop A type of fingerprint ridge pattern.

LOQ *See* LIMIT OF QUANTITATION.

lorazepam *See* ATIVAN.

Love Canal A milestone in the history of the environmental movement and subsequent environmental legislation and one of the earliest implementations of the principles of ENVIRONMENTAL FORENSICS. The canal had been started in the 1890s but was never completed and never used as a transportation waterway. Instead, the large ditch was used by Hooker Chemical to bury waste materials during the 1940s and early 1950s, before laws regulated these practices. Wastes were placed in barrels or poured into the soil and the ditch was eventually covered. In 1953, the property was given to the local school district, which built a school on it. After wet years in the 1970s waste began to seep into the school and into residences built alongside the converted canal. After national publicity and a prolonged dispute, residents were eventually bought out and resettled. The incident was crucial in the passage of subsequent legislation such as "Superfund."

lower explosive limit (LEL) The minimal concentration of vapors in air at which explosion or burning can occur when a combustion source such as a flame

is present. Below the LEL, no combustion can occur. *See also* UPPER EXPLOSIVE LIMIT.

low explosive Rather than literally exploding as a HIGH EXPLOSIVE does, a low explosive burns very quickly and must be in a confided space to explode. Accordingly, low explosives are occasionally referred to as burning explosives. Examples of low explosives include BLACK POWDER and SMOKELESS POWDER (used as PROPELLANTS in AMMUNITION), which are frequently used to make homemade explosives and pipe bombs. Another low explosive, made infamous by the Oklahoma City bombing on April 19, 1995, is composed of ammonium nitrate and 6 percent fuel oil (ANFO). A similar mixture of urea nitrate and other materials was used in the first attack on the World Trade Center in 1993. The maximum burning speed of low explosives is about 1,000 meters/second (m/s). Low explosives are sensitive to heat, friction, and sparks and are thus not very stable. The detonation of a low explosive generates what is referred to as pushing power, in which large objects are moved rather than shattered. Fragments of the container in which the explosive was placed are relatively large, and there are often significant amounts of residues found after the explosion.

low-velocity impact spatter A bloodstain pattern that results when blood moving slowly (approximately five feet per second or less) strikes a surface. Blood dripping from a nosebleed onto the floor creates a low-velocity impact spatter. These patterns show larger drops and less secondary spatter than medium- and high-velocity impact spatters.

LPI Lines per inch; refers to typesetting or printing by a device such as a typewriter or computer printer. LPI depends on factors such as FONT size and spacing.

Luma-Lite A commercial ALTERNATIVE LIGHT SOURCE used to detect enhanced latent FINGERPRINTS.

luminescence The emission of ELECTROMAGNETIC RADIATION by an atom or molecule as the result of some form of excitation. The emitted electromagnetic energy can be of any kind; the most familiar is that in the visible range of the electromagnetic spectrum. There are three forms of luminescence, which are based on the way excitation occurs: chemical (chemoluminescence or chemiluminescence), heat-generated (thermal luminescence), and photoluminescence (caused by absorption of light). An example of thermal luminescence can be seen (literally) when table salt (sodium chloride, NaCl) is placed in a flame. The heat is absorbed by the sodium atoms and causes them to emit a distinctive yellow light. This light is the same as produced in sodium vapor lamps commonly used as streetlights. Glow-in-the-dark watches are based on photoluminescence—the material in the watch face absorbs visible light, which is later emitted. In these cases, the emitted light is of a longer wavelength (lower energy) than the light that was originally absorbed. Thus, if a material absorbs energy in the ultraviolet region of the electromagnetic spectrum, it may emit in the visible region, which is lower in energy and has longer wavelengths. In forensic science, chemoluminescence is exploited to detect traces of blood using the LUMINOL reagent. Luminol reacts chemically in the presence of HEME (in hemoglobin), and in the process of the reaction, some of the chemical energy is converted to light that can be seen as long as the surroundings are dark. FLUORESCENCE, another form of luminescence, is exploited in many areas of forensic chemistry, toxicology, biology, and microscopy.

luminol A PRESUMPTIVE TEST for blood that is favored because of its extreme sensitivity, particularly to old blood or traces that have been left after an attempted cleanup of a crime scene. Luminol, unlike other tests for blood, does not work by producing a color change in the presence of hemoglobin. Rather, the reaction produces light in a process called chemoluminescence—a "glow in the dark" effect. Since the glow is faint, this test is not appropriate for areas that cannot be made dark such as

large outdoor scenes. However indoors, luminol is one of the most useful tests for blood. In addition, the chemicals used in the luminol reagent (3-aminophthalhydrazide, sodium carbonate ([Na_2CO_3], and sodium perborate [$NaBO_3$]) do not interfere with any subsequent DNA TYPING that is performed on bloodstains that are discovered. Finally, since no color change has to be observed, the color of the material on which the blood has been deposited is not a limitation, as it can be for other reagents. Luminol is sometimes referred to as an "enhancement chemical," and it can be sprayed over large areas.

Lycra A commercial elastic fiber used in swimwear and sportswear.

lysergic acid diethylamide (LSD) A potent hallucinogen in the same family as MESCALINE (PEYOTE) and phenylcyclohexyl piperidine (PCP). Doses as low as 25 μg (one microgram [μg] is equal to one one-millionth of a gram) are sufficient to induce the drug's effects, which include alteration of sensory perceptions, hallucinations, a feeling of floating or being "out of body," and extreme mood swings. Physical symptoms such as dilatation of the pupils, sweating, and increased heart rate also occur. LSD is an ergot ALKALOID that is produced from lysergic acid. Lysergic acid is a compound produced by a fungus that attacks grasses, a family that includes grains such as wheat and rye. Another PRECURSOR chemical, lysergic acid amide, is found in the seeds of the morning glory flower. There has been speculation that historical incidents of mass hysteria or otherwise unexplained behavior, such as the Salem witch trials, might have been partially attributable to grains that were infected with the fungus and subsequently used in food such as bread in which the preparation and baking process produced LSD.

LSD was first made synthetically in 1938 and its hallucinogenic properties unearthed by accident when the chemist who made it accidentally ingested a small amount, resulting in the first confirmed LSD "trip." For a short period, it was used in conjunction with psychotherapy, but that use was soon abandoned. LSD is now listed on Schedule I of the CONTROLLED SUBSTANCES ACT. Although not as popular as in the 1960s and 1970s, the drug is still encountered, sold most often on tiny blotter papers with patterns on them or soaked into sugar cubes or tiny tablets in many different colors. LSD does not produce physiological dependence but psychological dependence does appear to occur. A unique aspect of LSD use are "flashbacks" that can occur years after the drug was used.

lysis A rupturing such as the rupturing of cells. For example, red blood cells lyse in distilled water as a result of ion imbalance. The interior of the cell has a high concentration of ions and water flows into the cell to equalize the concentration. As a result, the cell quickly bursts. Before DNA TYPING, cells must lyse to release the DNA.

M

machine pistol A type of handgun that fires in the same way as a machine gun: fully automatically.

macrofibril Protein fiber that is part of the substructure of hair.

macromolecules Large molecules or polymers such as proteins that have or can have very complex three-dimensional structure. HEMOGLOBIN is a macromolecule.

macroscopic crime scene The context or larger setting in which a scene is found; an overall view of the larger scene. For example, if a murder occurs in a bedroom, the macroscopic scene may be the house or even the neighborhood.

macroscopic examination Assessment of evidence or other materials on a scale compatible with human vision or under low magnification such as with a magnifying glass; most often refers to simple examination without the aid of any tools or instruments. Macroscopic examination always precedes microscopic examination.

macrotransfer The transfer of readily visible material evidence between people and/or places as described in LOCARD'S EXCHANGE PRINCIPLE. Paint, blood, and glass are examples of materials that can be part of a macrotransfer.

magazine In an automatic or semiautomatic firearm, the device that holds cartridge cases so that they can be fed into the barrel before firing.

maggots Larvae of flies that hatch from laid eggs.

magnaflux A method used in SERIAL NUMBER RESTORATION that is also referred to as the magnetic particle method. In this approach, metal from which a serial number was obliterated is magnetized and magnetic powder applied. Because the stamping procedure often used to create serial numbers can distort the magnetic properties of the metal beneath the stamp, the magnetic particles may arrange themselves such that the obliterated serial number is visible.

magnetic flake powders Fingerprint powders that contain ferrous (magnetic) components. These powders can be delivered or applied to a fingerprint by using a magnetic brush, in a procedure that affords greater control than many other methods.

magnetic particle method *See* MAGNAFLUX.

magnetometer A metal detector; a device that detects changes in magnetic fields. Magnetometers can be useful in finding CLANDESTINE GRAVES.

MALDI Matrix-assisted laser desorption/ionization. A sample introduction technique used in MASS SPECTROMETRY (MS). Analytes are usually large molecules such as DNA, which are placed into a special solution or matrix that is subjected to a laser pulse. The result is the creation of gas phase molecules that can be introduced into the mass spectrometer.

malingering A purposeful attempt to fake a disease or condition such as amnesia or paralysis. Physicians, forensic psychiatrists, and psychologists may be called on to detect malingering.

Malpighi, Marcello (1628–1694) An Italian botanist who studied the structure of plants (plant morphology) as a professor at the University of Bologna. In forensic science, he is remembered as the first person to use magnification techniques to study the ridge detail and pore structure of human skin. As a result, he is considered one of the pioneers of the study of FINGERPRINTS even though he focused on neither their potential to identify individuals nor their value in criminal investigations. In honor of his work, a portion of the skin was named after him. The Malpighian layer (sometimes called the stratum malpighii) is found in the epidermis (the outer layer of skin) and refers to the combined basal and prickle cell layers (stratum germinativum and stratum spinosum).

maltose Malt sugar, a disaccharide composed of two linked glucose molecules.

Mandelin test A versatile PRESUMPTIVE TEST used in drug analysis. The Mandelin reagent is prepared by dissolving ammonia meta-vanadate in cold concentrated sulfuric acid (H_2SO_4). Addition of this reagent to cocaine produces an orange color, whereas in codeine it yields an olive green, in heroin a brown, and in amphetamine a bluish green. As with all presumptive tests, the results of a color test are not sufficient to identify a drug, but it is useful for screening purposes and for direction of further analysis.

mandible The bone structure that makes up the lower jaw. *See also* APPENDIX V.

manner of death The classification of a death as accidental, homicidal, indeterminate, natural, or suicidal, for which the abbreviation *NASH* is sometimes used. The determination of the manner of death when it is suspicious, questioned, or unattended is determined by the MEDICAL EXAMINER or CORONER.

marginal abrasion *See* CONTUSION RING.

marijuana (cannabis) A general term for drugs derived from the plant *Cannabis sativa.* and specifically for the leaves and flowering tops of those plants. Marijuana is the most widely abused illegal substance in the United States, and it is classified as a hallucinogen and is listed on Schedule I of the CONTROLLED SUBSTANCES ACT. The marijuana plant can grow to more than five feet in height and is also used as a source for hemp, the fibers of which can be made into rope or clothing products. The active ingredient in marijuana and its derivatives is Δ^9-tetrahydrocannibinol ("delta-9"), usually abbreviated simply as *THC.* Through careful cultivation, the THC content of the plant has steadily increased and now is in the range of 3–5 percent for leaves and flowers. HASHISH or hash oil is the oily resin excreted by the flowering tops and has a higher concentration of THC, in the range of 10–20 percent. Sinsemilla, a variety of marijuana plant developed in the late 1970s, has THC content in the range of 10 percent. Within any marijuana plant, the THC content is highest in the resins and lowest in the seeds.

marine forensics A type of wildlife forensics that specializes in marine animals. The National Oceanographic and Atmospheric Administration (NOAA) maintains a forensic marine service as part of the Center for Coastal Environmental Health and Biomolecular Research (CCEHBR).

Marquis test A PRESUMPTIVE TEST used in drug analysis. The Marquis reagent is prepared by adding a 40 percent solution of formaldehyde to concentrated sulfuric acid (H_2SO_4). When it is added to HEROIN or related opiate ALKALOIDS such as MORPHINE, the solution turns a purplish color. LSD (LYSERGIC ACID DIETHYLAMIDE), when present in sufficient quantities, creates a blackish color; AMPHETAMINES and methamphetamine, an orange; METHADONE, a yellow-tinted pink; and MESCALINE, a reddish orange. As with all presumptive tests, the results of a color test are not sufficient to identify a drug, but it is useful for screening purposes and for direction of further analysis.

Marsh, James (1794–1846) A distinguished English chemist best known for the development of a reliable test for the presence of ARSENIC in the body. The test variants were used by forensic TOXICOLOGISTS well into the 20th century, when methods of instrumental analysis such as ATOMIC ABSORPTION eventually replaced it. Marsh was also the first to present the results of an analytical toxicology analysis in court, in the year 1836, and his test was used by M. B. ORFILA, who is considered to be the father of forensic toxicology. The Marsh test became a powerful and reliable test for arsenic at a time when arsenic poisoning (accidental, suicidal, and especially homicidal) was rampant.

The test works on the basis of chemical reactions called OXIDATION/REDUCTION or

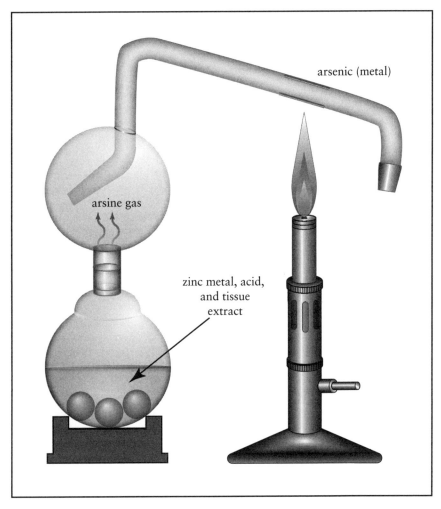

The Marsh test, developed by the English chemist James Marsh. Arsenic in tissues is vaporized and then reduced in the heated zone to metallic arsenic, which is deposited in visible plating on the interior of the glass tubing.

redox reaction. First, solid zinc metal is added to a glass vessel containing a powerful acid such as hydrochloric (HCl) or sulfuric (H_2SO_4) acid. The tissue or body fluid in question is prepared and added to the vessel, where the reaction of the zinc and the acid has created hydrogen gas (H_2). The hydrogen reacts with the arsenic compound (usually in the form of arsenic trioxide, AsO_3) to produce arsine gas (AsH_3). Heating of the gas causes arsenic metal to plate out on the glass or ceramic container into which the gas has flowed. The test, although reliable, required that the chemist performing it be skilled and practiced in the procedure. Tests based on similar chemical reactions are still used for screening (PRESUMPTIVE TESTS) for the analysis of other heavy metals such as MERCURY.

Marsh test   *See* MARSH, JAMES.

Mars red   A fluorescent powder that has been used to visualize latent FINGERPRINTS.

mass disasters   Also referred to as mass fatality incidents, events characterized by a large death toll that exceeds the capacity of local resources. Mass disasters used to be primarily associated with transportation disasters such as airplane or railroad accidents, but attacks such as occurred on September 11, 2001, have added terrorist acts to the list of potential causes. From the forensic perspective, mass disasters have two critical aspects: first, recovery and identification of remains, and, second, documentation, recovery, and preservation of evidence necessary for the subsequent investigation.

mass graves   Burial sites holding large numbers of bodies. Massacres often lead to the creation of mass graves and COMMINGLED REMAINS.

mass spectrometry (MS)   A versatile instrument system that separates ions on the basis of their size-to-charge ratio. Mass spectrometry can be utilized for the analysis of organic compounds such as

drugs as well as for the analysis of elements such as lead or arsenic. Mass spectrometry as used in forensic science is not a "stand-alone" technique; rather, a mass spectrometer is coupled to different sample introduction instruments or devices, yielding what is called a HYPHENATED TECHNIQUE. Examples include GAS CHROMATOGRAPHY–MASS SPECTROMETRY (GC-MS), high-pressure liquid chromatography/mass spectrometry (HPLC-MS), inductively coupled plasma–mass spectrometry (ICP-MS), and PYROLYSIS mass spectrometry. A slash (/) is often used in place of a hyphen, but the meaning is the same, and the notations are used interchangeably.

The design of a generic mass spectrometer is shown in the figure. The sample is introduced through an inlet into a chamber that is kept at a very low pressure, in the range of 1×10^{-5} torr (about one one-billionth of normal atmospheric pressure). This low pressure is essential to prevent collisions between ions (charged particles) created by the instrument and atmospheric components. The sample molecule (if it is an ORGANIC compound) is ionized to form charged fragments (F_1^+, and so on) of the original molecule. The M^+ ion is called the molecular ion in that it is created by stripping a single electron away from the original compound. In organic mass spectrometry, the type used for the analysis of drugs, for example, this molecular ion is important because it has the same molecular weight as the parent molecule. Smaller fragments, which can be as small as individual atoms, also form. The process of ionization is different for inorganic materials such as MERCURY or ARSENIC, as the inductively coupled plasma creates the ions before they are introduced into the mass spectrometer. Thus, in a sample containing gold (Au), the plasma creates gold ions (Au^+) that have the characteristic atom weight of gold (197 atomic mass units). However, these atoms do not fragment and are detected as is.

A number of different types of ionization schemes are available for mass spectrometers used for organic molecules. The

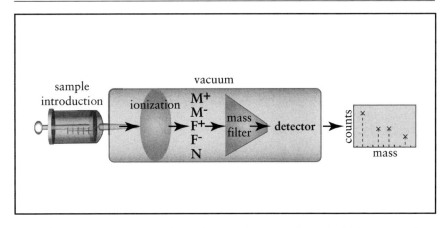

A generic mass spectrometer. Samples are introduced into a region under high vacuum, where the sample is ionized. Neutrals (Ns), along with molecular ions (M) and smaller fragments (F) are created. Ions are separated on the basis of their size-to-charge ratio.

most common is electron impact (EI), in which ionization results from collision with a beam of electrons. Other ionization modes include chemical ionization (CI), particularly useful when a strong M⁺ signal is needed; electrospray, common with HIGH-PERFORMANCE LIQUID CHROMATOGRAPHY (HPLC) systems; fast atom bombardment (FAB) and matrix-assisted laser desorption/ionization (MALDI), used for large molecules such as proteins; and laser ablation, for surface analysis of organic and inorganic materials.

Once ions are produced, they are directed into a region called the mass filter, where they are separated out into individual masses and the abundances recorded. The separation is based on the size of the ion as well as the charge. The results are plotted in a MASS SPECTRUM that shows the relative abundance of each mass. There are a number of different types of mass filters used in mass spectrometers including time-of-flight (TOF), quadrupoles, ion traps, and magnetic sector. The most common type found in forensic labs for both organic and inorganic applications is the quadrupole, in which manipulation of the electrical fields allows the ions to be filtered such that only one mass is detected at any given time. The quadrupole is nearly standard for GC; ion traps and

TOF designs are found in many HPLC systems. There are also MS/MS systems in which fragment ions that emerge from the mass filter are directed into another mass spectrometer, which repeats the process and produces another series of smaller fragments. This type of information is extremely useful when attempting to identify a material that has not been found in the mass spectral library.

mass spectrum A graphical output from a mass spectrometer showing ion mass versus intensity of that mass. The output is usually normalized, meaning that all intensities are scaled relative to the largest intensity, which is assigned a value of 100. A mass spectrum represents the fragmentation pattern of a molecule or the mass of an atom in INDUCTIVELY COUPLED PLASMA–MASS SPECTROMETRY (ICP-MS). Mass spectra are used to identify compounds or elements and provide a starting point for obtaining quantitative data.

mass-to-charge ratio In MASS SPECTROMETRY and related techniques, separation of charged species based on the size-to-charge ratio of the ion. For example, the mass-to-charge ratio of a sodium cation (Na⁺) would be the same as its mass, since the charge is +1. However, the size-to-charge ratio of a calcium cation (Ca^{2+}) would be half the mass since the charge is +2.

material fatigue Faults or damage incurred in a material such as a steel girder after repeated application of dynamic loads. Fatigue can be visible or microscopic and typically is seen in the form of cracks or fractures. Such damage can cause a structure or part to fail under a load that it could normally withstand if undamaged.

Material Safety Data Sheet (MSDS) Data sheet that is supplied by chemical manufacturers and vendors that describes in detail the safe handling, storage, and disposal procedures for a given material. The MSDS also provides known toxicological data and dosage limits.

maternal lineage Genetic information such as found in MITOCHONDRIAL DNA that is passed directly from mother to child with no contribution from the father.

mathematics, forensic A term that usually refers to statistical methods applied to DNA evidence or to statistics and probability considerations in the evaluation of evidence.

McCrone, Walter (1916–2002) An American considered to have been the preeminent microscopist of his generation, who founded the McCrone Research Institute (www.mcri.com) in Chicago, Illinois, in 1960. McCrone earned his Ph.D. from Cornell University in New York in 1942 and worked for a time at the Armour Research Institute. His interests in microchemical analysis led to the formation in 1956 of McCrone Associates, who emphasized this field as well as the analysis of crystals and general microscopy. The McCrone Research Institute followed, and it remains a premier training facility for industrial and forensic microscopists worldwide. Of his many publications, one of the best known is *The Particle Atlas,* which is used by forensic microscopists to identify unknown materials. He became

known outside the scientific community during the analysis of the Shroud of Turin in 1980. The shroud was purported to be the burial cloth of Jesus. McCrone, along with a group of other scientists, conducted numerous tests and concluded that the shroud was a clever forgery dating not to the first century, but to the 14th. He was awarded the American Chemical Society's Award in Analytical Chemistry in 2000 for his numerous contributions to MICROSCOPY and microchemical analysis.

McDonald, Jeffrey An army Green Beret captain and physician who was convicted of the murder of his pregnant wife, Colette McDonald, and two young daughters in 1979. The case was the subject of a popular book and miniseries entitled *Fatal Vision* (written by Joe McGinniss), and discussions and legal maneuvering continued more than 20 years after the fact. From the forensic perspective, the case was notable because of extensive physical evidence such as HAIRS, FIBERS, BLOODSTAIN PATTERNS, and murder weapons (an ice pick and wooden club). At the trial, fiber evidence, blood spatters, and a bloody footprint were crucial in reconstructing the crime scene and suggesting the course of events during the killings.

MDB (7-(p-methyoxybenzylamineo)-4-nitrobenz-2-oxa-1,3-diazole) A type of dye that is used to visualize latent FINGERPRINTS. It is also a component of a dye mixture used for the same purpose called RAM.

MDMA *See* ECSTASY.

mechanism of death The specific medical, biochemical, and/or physiological process or failure that causes death. For example, in a stabbing, blood loss can lead to shock, and often this shock is the mechanism of death even though it was precipitated by a stab wound.

Mecke's test A PRESUMPTIVE TEST used in drug analysis, often in conjunction with other tests such as the MARQUIS TEST. The Mecke reagent is prepared by dissolving selenious acid in concentrated sulfuric acid (H_2SO_4). When added to HEROIN, the reagent turns yellow, which fades into a greenish color. When present in sufficient quantities, LSD produces an olive green color that turns blackish, and PSILOCYBIN produces a yellow-green that turns brownish. As with all presumptive tests, the results of a color test are not sufficient to identify a drug, but it is useful for screening purposes and for direction of further analysis.

medical examiner (ME) A forensic pathologist appointed by a jurisdiction such as a state, county, or city to oversee death investigations. Medical examiners did not appear in the United States until 1877, when Massachusetts became the first state to abolish the CORONER system in favor of an ME system. New York City followed in 1915, and many jurisdictions have since converted to the newer system. However, both systems still coexist in the United States. An ME is a physician with specialized training in forensic pathology who is qualified to conduct an AUTOPSY and tissue analysis. As is a coroner, an ME is charged with determining the cause, manner, and circumstances of a death. At a crime scene involving a death, normally it is the ME's office that has jurisdiction over the body; law enforcement agencies are responsible for the rest of the scene. Similarly, the ME is responsible for collecting any physical evidence discovered during the autopsy and for delivering it to the appropriate forensic laboratory.

medicine, forensic See PATHOLOGY, FORENSIC.

medicolegal investigation An older term generally referring to the investigation of a suspicious death or the application of forensic medicine and forensic pathology.

medium-velocity impact spatter A bloodstain pattern created by a medium-velocity blow such as from a fist or club. It is characterized by droplets that are smaller than a simple drip (as in a nosebleed) but larger than the misting produced by a high-velocity impact.

medulla In a HAIR, the central portion that runs lengthwise; it is often likened to a canal running down the middle of a hair. Characteristics of the medulla vary among species and among individuals.

medullary index The diameter of the MEDULLA relative to the diameter of the HAIR. This ratio is sometimes employed during microscopic analysis and comparison of hair.

megabyte (Mb) A unit of storage in computers corresponding to 1,048,576 bytes.

megahertz (MHz) A unit of frequency corresponding to 1 million cycles per second. These units are used to describe speed of computer chips or frequency of electromagnetic radiation such as encountered in NUCLEAR MAGNETIC RESONANCE.

meiosis The process of cell division that results in the halving of the chromosomes by separating the pairs. Meiosis produces the sex cells (sperm and egg), which each carry 23 chromosomes instead of the usual complement of 46. See also MITOSIS.

melanin An organic polymer that forms pigment granules found in hair; these granules impart the color to hair. Melanin is created by melanocyte cells found in the hair follicle. Melanin is also responsible for skin color.

melting point The temperature at which a substance changes from a solid to a liquid state. In forensic science, melting points of chemical compounds are occasionally used to assist in the identification of unknowns. However, the most common use is the determination of the melting points of synthetic materials such as fibers as an aid to their identification.

menstrual blood Blood that is excreted from the womb during a woman's menses. This type of blood contains high amounts

metal fouling

of fibrinogen, and this characteristic can sometimes be used to differentiate a stain of menstrual blood from stains deposited in other ways.

meperidine See DEMEROL.

Merck Index A handbook of chemical compounds and drugs used as a reference by forensic CHEMISTS, TOXICOLOGISTS, and PATHOLOGISTS. The *Index,* currently in its 13th edition, is published by Merck Sharpe and Dohme Research Laboratories and contains data on more than 10,000 compounds as well as tables, lists, and extensive cross-referencing information.

mercury (Hg) A toxic heavy metal unique in that it is the only metal that is a liquid at room temperature. As any metal does, it conducts electric current and can be used as a component in explosive devices that use mercury switches. However, mercury is best known as an environmental pollutant and poison. Despite its toxicity, mercury is rarely encountered in homicide cases, in which ARSENIC or CYANIDE is more commonly employed. Mercury exists in many different forms besides the familiar elemental form (also called quicksilver), which is actually one of the least toxic forms unless the vapor is inhaled. Mercury can exist as solid salts such as mercuric chloride ($HgCl_2$), a corrosive salt that can be fatal in doses of one to two grams, and as mercuric sulfide (HgS), a coloring agent. One of the organic forms, dimethyl mercury, $(CH_3)_2Hg$, is also extremely toxic. People are most often exposed to mercury by consuming contaminated food such as fish, in which mercury from an aquatic environment can concentrate.

mescaline A hallucinogenic compound that is contained in peyote, which is the "button" on top of cactus found in Mexico and the southwestern United States. This button, also called a mescal button, consists of the flowering head of the *Lophora williamsii* cactus. The compound was first isolated in 1896 and made synthetically about 20 years later. Chemically,

mescaline is related to AMPHETAMINE. Although mescaline is listed on Schedule I of the CONTROLLED SUBSTANCES ACT, peyote is legal for use in rituals in the Native American Church, primarily in New Mexico and Arizona. Mescaline can also be sold illicitly in tablets or gel capsules. The hallucinogenic effect can be pronounced and lasts about 12 hours.

metabolite/metabolism A breakdown product of a substance such as a drug or toxin produced by a metabolic process in the body. The substance ingested is called the parent compound and is classified as a xenobiotic, meaning it is foreign to the body. For example, if someone were to ingest COCAINE, because that compound is not normally found in the body, it is considered to be foreign. In other cases, the substance may be found naturally in the body, but in trace concentrations. In forensic TOXICOLOGY, it is often the metabolites that are the target of analysis in fluids such as blood and urine.

As illustrated in the figure on page 158, the process of metabolism involves stepwise changes in the molecule using different kinds of chemical reactions. Each stage of the process is catalyzed by the actions of enzymes, which are biological catalysts. The first-stage metabolites may undergo additional processes, leading to a large number of by-products. Final metabolic products have three principal fates. Volatile products such as carbon dioxide (CO_2) can be exhaled, as in the case of the metabolism of ALCOHOL (ethanol). Water-soluble products (called HYDROPHILIC, "water loving") are excreted in the urine or in other body fluids. Products that are fat-soluble (HYDROPHOBIC) build up in fatty tissues. Heavy metal poisons such as MERCURY can produce fat-soluble products. Most metabolic processes are concentrated in the liver, but not exclusively. Some metabolism can take place in the stomach and intestines (gastrointestinal [GI] tract) and in other organs such as the kidneys.

metal fouling Accumulation of metals in components of FIREARMS. Lead

157

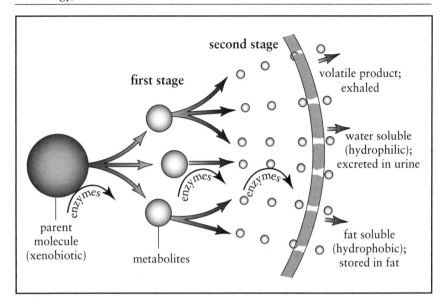

General flow chart for metabolism. The parent molecule is metabolized in one or more stages, each producing transformed products called metabolites. The ultimate fate of a given metabolite can be exhalation as a gas, excretion in urine or other aqueous body fluids, or deposition in fat.

fouling of the barrel is an example of metal fouling.

metallurgy, forensic An area of expertise that is often part of or closely associated with forensic ENGINEERING and revolves around the analysis of metal components of objects. Steel is often the subject of forensic metallurgical analysis because it is commonly used as a structural component in everything from planes to buildings. Forensic metallurgy was used in the investigation of the crash of TWA flight 800 in 1996, when there was considerable doubt early in the investigation as to what caused the center fuel tank to explode. Potential causes included a bomb or missile, both of which would have left characteristic evidence. For example, if a bomb exploded from within the plane, the metal skin of the aircraft would have been peeled backward and outward, whereas a missile striking the fuselage and exploding would have left a much different signature on the fuselage.

In this case, forensic metallurgy pointed to an explosion from within. Other applications of forensic metallurgy include the analysis of bombs such as pipe bombs and more traditional forensic engineering analysis such as was used to unravel the sequence of events that led to the loss of the space shuttle *Columbia* in February 2003. Metal fatigue, joint failures (welded, bolted, or other), are often a central concern in many incidents.

metal salts An ionic chemical compound consisting of a metal cation and a nonmetal anion. Silver chloride ($AgCl$) is an example of a metal salt. This term is most often associated with techniques used to visualize latent FINGERPRINTS.

metameric pairs In microscopy, two objects such as fibers that appear to be the same color under one kind of illumination and a different color under another type of lighting. The term also applies when two fibers appear to have the same color when viewed in visible light but actually have different spectral characteristics. Microspectrophotometry using visible, infrared, or fluorescent techniques can be used to differentiate them.

methadone (Dolophine) A synthetic OPIATE that is used to treat addiction to HEROIN and other narcotics. Methadone helps to relieve withdrawal symptoms and eases cravings and is often used as part of a detoxification ("detox") treatment.

methamphetamine *See* AMPHETAMINE.

methaqualone (Quaalude) A drug classified as a sedative-hypnotic that was developed in the 1950s and the subject of abuse in the 1960s and 1970s. The drug was removed from the U.S. domestic market in the 1980s and is no longer commonly encountered in forensic drug analysis.

method bias A bias in results that is caused by the analytical method, as opposed to other factors such as the analyst or sample. For example, if a white powder sample containing 50 percent cocaine is subjected to an inefficient extraction procedure, the results will be biased low since some of the cocaine will not be extracted. The bias is the fault of the method, not the analyst or any intrinsic problems with the sample itself. Method bias is addressed through METHOD VALIDATION and QUALITY ASSURANCE/QUALITY CONTROL procedures.

method validation A procedure or procedures used to ensure that an analytical method is reliable and that any results obtained are reproducible. Method validation is an important part of QUALITY ASSURANCE/QUALITY CONTROL and is applicable in many forensic areas such as drug analysis and DNA TYPING. Typically method validation requires that an analyst or analysts perform the method on a sample of known composition or type to prove that the results are acceptably accurate and reproducible (precise). Once a method has been validated, it is reasonable to expect that any trained analysis

using that method properly will obtain the same results as any other analyst working in any other laboratory.

methylphenidate (Ritalin) *See* STIMULANTS.

micelle A microscopic structure formed when SURFACTANTS such as soap molecules are placed into water. As shown in the figure on page 160, surfactants are long-chain molecules that consist of a nonpolar (water-insoluble) portion or tail topped by a polar water-soluble group. When sufficient surfactant is placed into water (exceeding the critical micelle concentration [CMC]) micelles will form. Micelles are exploited in electrophoresis as explained in the next entry.

micellar capillary electrophoresis/ micellar electrokinetic electrophoresis A type of capillary electrophoresis (CE) that combines elements of chromatography with CE to allow for separation and analysis of neutral (uncharged) species. To facilitate this, a surfactant (soaplike molecule) is introduced into the sample solution at a concentration at which MICELLES form. The micelles encapsulate the neutral molecules of interest with a charged "shell," allowing the micelles to move in the electrical field applied along the capillary column. As in other forms of chromatography, the analytes partition between the solvent phase and the interior of the micelle (sometimes called the pseudostationary phase), resulting in high-resolution separations.

Michel-Levy chart A colored chart used in polarizing light microscopy to determine the BIREFRINGENCE of samples.

Michelson interferometer A device used in modern INFRARED SPECTROPHOTOMETERS that allows many wavelengths of ELECTROMAGNETIC RADIATION to be directed at a sample simultaneously rather than sequentially. The advantage of this approach is that a sample can be scanned much faster and can be scanned multiple times to improve the quality and sensitivity

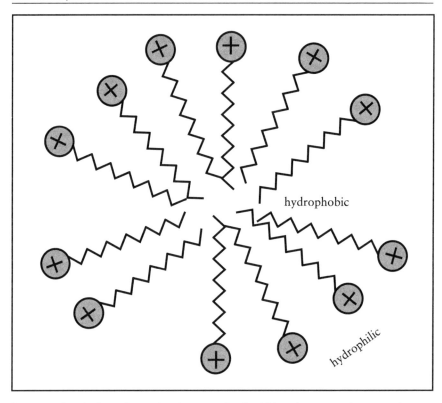

hydrophobic

hydrophilic

Structure of a micelle made up of surfactant molecules. This is the process that occurs in soaps and detergents, which contain surfactants. In water, micelles can carry away materials trapped inside and can break the surface tension of water.

of the resulting spectrum. An interferometer works on the principle of interference of coincident light waves. If two beams are in phase, the interference is constructive, and the resulting combination has the same maxima and minima at twice the amplitude (height). On the other hand, if the beams are completely out of phase, the resulting interference is destructive, and the net result is a canceling out of both. Between these two extremes, different combinations of constructive and destructive interference occur. The Michelson interferometer takes advantage of this condition, coupled to a mathematical operation called a FOURIER TRANSFORM that enables a computer to break down a composite pattern into individual components, in effect undoing the interference and

reconstructing the original components. The Michelson interferometer creates an output called an interferogram that is composed of a complex signal generated by multiple wavelengths and multiple cycles moving through the extremes of constructive and destructive interference. The Fourier transform is then applied to the signal to recover the intensity of the original radiation at each wavelength used.

microanalysis The analysis of evidence using magnification ranging from a simple magnifying glass through MICROSCOPY to SCANNING ELECTRON MICROSCOPY.

microbial degradation Processes of decomposition and breakdown of materials by the action of microorganisms, principally bacteria and fungi. Although the public usually associates this process with the decomposition of human bodies (PUTREFACTION), microbial degradation is an issue for any kind of biological evidence including blood and body fluid stains. Any organic material that a microbe can use as food, from remains to ropes and cloth, is subject to degradation. The rate of degradation depends on the suitability of the environment for the microbes, which generally favor warm, moist conditions. Blood and other body fluid are also subject to putrefaction, and for this reason, bloodstained materials should be dried and stored refrigerated or frozen to slow the process.

microchemical analysis Chemical tests performed on tiny amounts of samples, the results of which are observed under a microscope. CRYSTAL TESTS used in drug analysis and identification of blood are types of microchemical analysis, as are several tests that are performed on fibers.

microcrystalline tests *See* CRYSTAL TESTS.

microfibril Part of the substructure of a HAIR.

micrometer (micrometry) The instrument and process used for taking measurements of objects that are being magnified. This is usually accomplished by using a calibrated scale in a reticle that attaches to the objective lens.

microorganism A collective term referring to organisms that can only be seen under magnification. Bacteria, viruses, fungi, and yeasts are microorganisms, and all can play an important role in forensic science, particularly in decomposition.

microsatellite DNA In DNA, a segment that consists of SHORT TANDEM REPEATS (STRs) with two, three, four, or five nucleotides in the repeated segment. *See also* MINISATELLITE.

microscopic examination The visual evaluation of evidence or other materials

that requires the use of an instrument for magnification such as a microscope or magnifying glass.

microscopy A fundamental tool of forensic science that encompasses visual techniques based on light and electromagnetic radiation as well as others based on electrons (SCANNING ELECTRON MICROSCOPY). The simplest form of microscopy, still widely used, employs a simple magnifying glass. Other types of forensic microscopy use COMPOUND MICROSCOPES (optical or biological microscopes), STEREOMICROSCOPES, POLARIZING LIGHT MICROSCOPES, COMPARISON MICROSCOPES, PHASE CONTRAST microscopes, and FLUORESCENCE microscopes. MICROSPECTROPHOTOMETRY, the coupling of microscopes with SPECTROPHOTOMETRY, is gaining more widespread application in forensic science as well.

All optical microscopes are based on magnification that results from light's passing through a LENS. A lens can produce two different kinds of images. A real image is one that can be projected onto a screen (as in a movie) or onto the retina of the person looking at it. A virtual image—what a magnifying glass produces—is an image that is not physically real and can only be seen when looking through the lens. A compound microscope, also called a biological microscope, builds on this design by using two lenses in series, creating two stages of magnification (see figure on page 162). Although high magnification (800×) is possible with compound microscopes, increasing magnification has a cost. The higher the magnification, the smaller the field of view, as in the case of the magnifying glass. Second, as magnification increases, the depth of focus (how "deep" into a sample the focus remains sharp) decreases. Other variants of microscopy build on this foundation of magnification by lenses.

microspectrophotometer/microspectrophotometry The combination of a spectrophotometric technique with a microscope, allowing for analysis of very small samples or areas. Spectrometry is

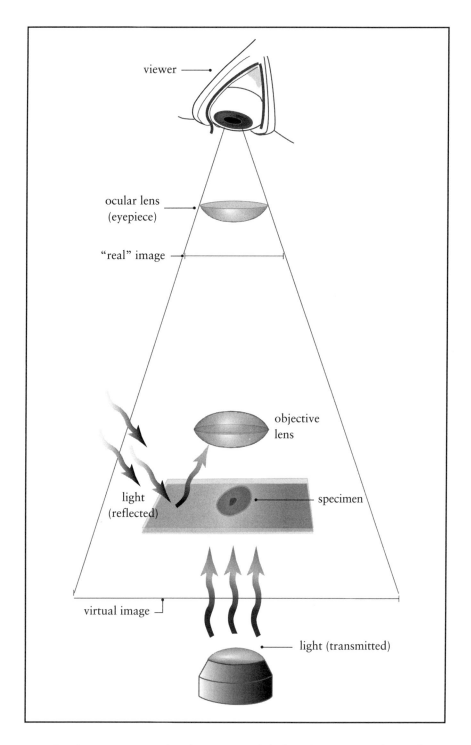

viewer

ocular lens
(eyepiece)

"real" image

objective
lens

light
(reflected)

specimen

virtual image

light (transmitted)

Principles of microscopy as employed in a compound (biological) microscope. The sample is placed on the stage and illuminated by transmitted light, or in some cases reflected light. The two lenses (ocular and objective) combine to project a virtual image into the eye of the viewer with a magnification factor that is equal to the product of the individual magnifications.

based on the study of the interaction of matter with ELECTROMAGNETIC RADIATION or light. Since microscopes exploit many portions of the electromagnetic spectrum, linking the two has proved relatively simple. In forensic work, the most widespread type of microspectrophotometry uses the INFRARED (IR) portion of the spectrum. IR microspectrophotometers are used for the analysis of drugs, paints, dyes and coatings on fibers, and inks. Subtle differences in surface composition can be studied on the microscopic level and can be used to create surface composition maps. Variations of micro-IR include ATTENUATED TOTAL REFLECTANCE, diffuse reflectance (DRIFTS), and IR-polarizing microscopy. Microspectrophotometers are also available for the visible region of the spectrum and for ultraviolet (UV) fluorescence and for IR scattering, a technique called RAMAN microspectrophotometry.

microtransfer The transfer of microscopic evidence between people and/or places, as described in LOCARD'S EXCHANGE PRINCIPLE. Dust and fibers are examples of materials that can be part of a microtransfer.

microwave assisted digestion A sample preparation technique used to digest or dissolve matrices such as glass or soil before analysis, typically for identification of elemental composition using techniques such as inductively coupled plasma–atomic emission (ICP-AES) or mass spectrometry (ICP-MS).

military explosives See EXPLOSIVES.

mineral fibers Fibers that are composed of inorganic materials such as asbestos and fiber glass.

minisatellite Genetic locus that is classified as having a VARIABLE NUMBER OF TANDEM REPEATS. Variations between individuals are based on different numbers of repeats of the same BASE PAIR sequence. The sequence in a minisatellite is limited to 10–100 base pair repetitions. See also MICROSATELLITE.

minutiae The small features that are critical in matching two FINGERPRINTS. Ridge endings and BIFURCATIONS are examples of minutiae.

misdemeanor A less serious crime than a FELONY and one that carries a less severe penalty, such as a fine.

misting A bloodstain pattern that is created by a high-velocity impact such as a gunshot wound. *Misting* is also a term used in the context of ATOMIC ABSORPTION SPECTROPHOTOMETRY and refers to dispersal of the liquid sample into a very fine mist before introduction into the flame.

mitochondria Cell organelles that are located outside the nucleus and are active in cellular respiration and energy production. The mitochondria have their own DNA (mtDNA), which is separate from the nuclear DNA in typical DNA TYPING procedures. (See figure on page 164.)

mitochondrial DNA See mtDNA.

mitosis The process of cell division in which chromosomes are duplicated completely. In humans, mitosis starts with one cell containing 23 pairs of chromosomes (46 total) and ends with two cells with 46. See also MEIOSIS.

MNSs blood group system See BLOOD GROUP SYSTEMS.

mobile phase In CHROMATOGRAPHY, the phase of material that is moving. In GAS CHROMATOGRAPHY (GC), the mobile phase is the carrier gas, whereas in HIGH-PRESSURE LIQUID CHROMATOGRAPHY (HPLC), the mobile phase is a solvent or solvent mixture. The mobile phase may be inert, as in GC, or it may be active in the separation process, as in HPLC.

modus operandi (MO) The methods, techniques, and approaches that a criminal uses to commit a crime. A person's MO can change or evolve over time as the criminal gains more experience; however, the pattern of the MO can be very useful in linking crimes to the same perpetrator.

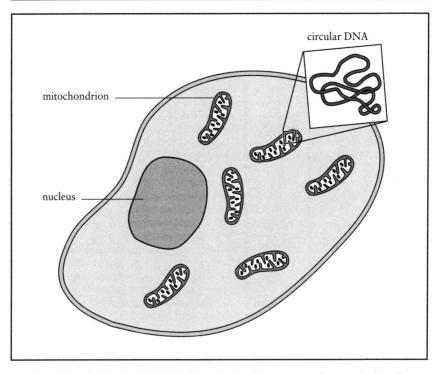

Mitochondria and mitochondrial DNA. The mitochondria are organelles outside the cell nucleus that have their own mitochondrial DNA (mtDNA). This DNA is inherited from the mother and can be typed.

molar A term that has two meanings in forensic science. It is a type of TOOTH and thus considered in ODONTOLOGY (forensic dentistry). The word also refers to a unit of concentration used in chemistry. A one-molar (1M) solution of a given substance is one MOLE of that substance that is dissolved in one liter of water.

molar absorptivity A value used in BEER'S LAW, a relationship widely employed in spectrophotometry. This quantity, symbolized by ε, is constant for a given wavelength and expresses how strongly a given substance will absorb that wavelength of light.

molar extinction coefficient (ε) *See* MOLAR ABSORPTIVITY.

molarity A unit of chemical concentration in moles per liter of solvent (moles/l or M).

molecular compound *See* MOLECULE.

molecular ion In MASS SPECTROMETRY, the ion that is created by the loss of a single electron. Because this loss results in a negligible change in the mass of the molecule, the molecular ion (often denoted as M⁺) retains the mass of the parent molecule.

molecular weight The combined atomic weights of the atoms that a MOLECULE comprises. The units of molecular weight are usually expressed in grams per mole. For example, the molecular weight of water (H_2O) is 18.01 g/mole, the sum of 2 × 1.008 g/mole for hydrogen plus 15.9994 g/mole for oxygen.

molecule A group of atoms held together by a covalent chemical bond, in which outer shell electrons are shared among the atoms. This contrasts with ionic compounds such as table salt (NaCl), in which atoms are linked through electrostatic attraction of unlike charges (Na$^+$ and Cl$^-$). Water (H$_2$O) is a covalently bonded molecule.

Molotov cocktail An incendiary device, often referred to as a firebomb, that is made by placing a flammable liquid (usually gasoline) into a glass container. A wick such as a cloth sticks out of the neck of the container. The device is used by lighting the wick and throwing the container, which breaks on impact, spreading flames.

morphine A derivative of OPIUM that constitutes anywhere from approximately 5 to 20 percent of the extract. Morphine is also a METABOLITE of some of the OPIATE ALKALOIDS such as heroin. The name *morphine* is from the name of the Greek god of dreams, Morpheus. Morphine was first extracted from opium in the early 1800s, and by the time of the Civil War, it was widely used for its potent pain-relieving ability, desperately needed by the wounded in that conflict. Morphine is classified as a narcotic analgesic (pain reliever) that in large doses can cause respiratory depression and death. It is still used for the relief of severe pain such as that of terminal cancer. Morphine has many forms including salts such as morphine hydrochloride and morphine sulfate.

morphology A term used to describe structure. In forensic science, the term *morphology* may refer to microscopic evidence (the morphology or structure of a hair, for example) or biological structures and features.

mouth swabs *See* BUCCAL SWABS.

MR microscopy Microscopic examination using medical magnetic resonance imaging (MRI) techniques. It has been employed to study details of injury patterns and damage.

MSDS *See* MATERIAL DATA SAFETY SHEET.

MS (MSD) Mass spectrometer or mass selective detector. These abbreviations are often used to describe mass spectrometers used as detectors for chromatographic instruments as in GAS CHROMATOGRAPHY.

mtDNA (mitochondrial DNA) DNA that is found outside the cell nucleus in structures called the MITOCHONDRIA. The mitochondria are key in cell energy production and have their own DNA, which is inherited solely from the mother. Thus, even if a person is not available to provide comparison samples (deceased, missing, and so on), comparison standards can be obtained from anyone in the maternal line. Because it is found in even greater abundance than nuclear DNA, mitochondrial DNA can be used on very small samples, and problems of degradation are reduced. Mitochondrial DNA is found in a loop structure and there are three regions where variations occur, called HV1, HV2, and HV3 for hypervariable regions 1–3, which can be amplified by using polymerase chain reaction (PCR) techniques.

multimetal deposition (MMD) A technique used to visualize latent fingerprints involving the application of gold followed by silver. The general procedure can be used on porous and nonporous surfaces.

multiplex system/multiplexing Analytical techniques in DNA TYPING and in older ISOENZYME SYSTEMS typing in which more than one locus or genetic marker system is typed at one time by one combined procedure.

murder A crime that involves the killing of another human being. In U.S. law, there are six categories in which a killing can be classified, and often forensic evidence is critical in deciding the category in which a person will be charged. In general, first-degree murder is a willful, deliberate, and premeditated killing undertaken with "malice aforethought," or an intent to cause harm present before the crime is committed. Second-degree murder has all of the same elements as first-degree except premeditation. Thus, in murder cases it is often important to use physical evidence to demonstrate premeditation, for example, if a murderer purchased a gun a few days before killing someone or placed a sheet in the trunk of his or her car the day before it was used to transport a body to a remote dumping ground. In both incidences, the killer has demonstrated planning and thus premeditation. Manslaughter is the second broad category of killing, also divided into two degrees. Manslaughter in the first degree (more commonly called voluntary manslaughter) is an intentional killing but without malice aforethought. Many "crimes of passion," in which one person kills another while in the grips of a sudden rage, fall into this category. Involuntary manslaughter is generally a killing that results from negligence, such as a vehicular homicide in which one person's negligent driving results in the death of another. The other two categories of murder are justifiable homicide (such as legitimate self-defense) and accidental death (such as a hunter accidentally killing another during a hunting trip, although evidence might be developed to elevate the charge to involuntary manslaughter if the killer had been drinking alcohol).

mutation rate A term related to DNA TYPING that refers to the number of mutations that occur in a gene per unit time.

muzzle energy/muzzle velocity The speed or energy of a bullet when it emerges from the barrel of a firearm after the cartridge is fired. *See also* SILENCERS.

muzzle flash and muzzle blast Muzzle blast is a shock wave that is produced when a firearm is discharged. In a gun, when the trigger is pulled, a firing pin strikes the PRIMER, causing it to explode. This tiny explosion ignites the PROPELLANT, producing a large volume of hot gas that rapidly expands, pushing the BULLET down the barrel of the weapon. When this wave of hot expanding gas contacts the atmosphere, the muzzle blast results. The muzzle flash is also produced by the ignition of the propellant and appears as a flash of fiery tendrils exiting any openings in the gun. Any gun that accelerates the bullet faster than the speed of sound (approximately 330 meters per second or 1,080 feet per second) will produce a cracking sound, which is a small sonic boom similar to those produced by supersonic aircraft. SILENCERS are designed to reduce or eliminate the muzzle blast and the cracking sound.

myotomy The cutting of the muscles in the jaw to release rigor mortis.

NA (numerical aperture) A number that measures the size of the "cone" of light that can be produced by the objective or condenser in a microscope. It is calculated as $n \times \sin \times u$ where n is the refractive index of the material between the objective lens and the condenser and u is equal to half the angle between the two edges of this cone of light. The NA is calculated when the diaphragm on the microscope, located below the stage and controlling the amount of light reaching the specimen, is completely open. The NA number is related to the resolution; for maximal resolution, the NA of the objective lens should be matched to that of the condenser.

NAA *See* NEUTRON ACTIVATION ANALYSIS.

NAFE National Academy of Forensic Engineers, a professional organization that was formed in 1982. It is affiliated with the National Society of Professional Engineers (NSPE).

narcoanalysis The use of drugs, informally referred to as "truth serums," to elicit information or a confession from a suspect. BARBITURATES such as sodium pentothal or scopolamine are most commonly used. Courts have refused to admit the results of narcoanalysis and the technique is rarely used in criminal investigations.

narcotics *See* DRUG CLASSIFICATION.

NASH An informal acronym for four legal causes of death—natural, accidental, suicidal, and homicidal.

National Crime Information Center (NCIC) Now a part of the FBI, the NCIC was established in 1967 as a national repository for data needed and used by law enforcement agencies throughout the United States and Canada. The center has resources for fingerprint searching, files of mug shots, and a convicted sex offender registry. NCIC is housed under the FBI's Criminal Justice Information System (CJIS).

natural fibers Fibers such as cotton, silk, and wool that are not synthetic but rather derived from plants or animals.

natural size A photograph that is at one-to-one (1:1) scale, such that size of an object in a photo is exactly the same size as the object itself.

NBD (NBD-Cl or NBD-F) Reagent (7-chloro- or 7-fluoro-4-nitrobenzo-2-oxa-1,3-diazole) used to visualize latent FINGERPRINTS. Both forms react with the amino acids present in fingerprint residues to yield products that fluoresce.

near IR (NIR) A range in the electromagnetic spectrum with wavelengths of .78 µm–2.5 µm (780 nm–2500 nm). Spectrophotometry in this region can be combined with the visible (VIS) and ultraviolet (UV) range and is referred to as UV-VIS-NIR spectrophotometry. Unlike in IR SPECTROPHOTOMETRY, an NIR spectrum is not characteristic enough to identify a compound, but the technique has been used to supplement other approaches in the analysis of fibers and in other forensic examinations.

nebulizer The component in an atomic absorption spectrophotometer that takes the liquid sample and converts it to a fine mist that is then introduced into the flame.

A typical design uses a glass bead that is placed into the flow stream that causes misting, much as a thumb placed on a garden hose nozzle does.

necropsy A dissection of a dead body, which may be animal or human; for human dissection, the term AUTOPSY is used.

neonatal line A line distinguishable in the DENTIN of a tooth that arises from the stresses of birth. This stress disrupts the normal metabolic processes, including those in the cells that form teeth. The neonatal line can be used to estimate the age of remains of young person whose baby teeth have not been completely replaced with adult teeth.

Nernst equation An equation that describes the response of an electrode. In its most common form, it is expressed as

$$E = E^0 - \frac{0.05916V}{n}\log\frac{[A]^a}{[B]^b}$$

For an OXIDATION/REDUCTION half-reaction $aA = bB + ne^-$. $E°$ is the standard half-cell potential (obtained from a table), n is the number of electrons, and the values in

brackets are the respective concentrations in MOLARITY of the two species raised to the power of their coefficients a and b. The constant .05916 is in units of volts (V).

Nessler's reagent A PRESUMPTIVE test for urine. It is a solution of mercuric iodide, which detects ammonia that is created by the degradation of urea.

neuropsychology, forensic See PSYCHIATRY, FORENSIC.

neutron activation analysis (NAA) Also called gamma ray spectroscopy, this technique was used in a few cases that required ELEMENTAL ANALYSIS. The most notable application was to the analysis of bullets and fragments recovered during the investigation of the KENNEDY ASSASSINATION. In NAA, a sample is placed into a nuclear reactor (research-type, not those used in nuclear power plants), where it is bombarded by neutrons. These neutrons are called thermal neutrons since they are moving relatively slowly and thus can be absorbed by atoms in the sample. This step, called activation, produces a radioactive ISOTOPE of the original atom, which is unstable and decays, giving off gamma (γ)

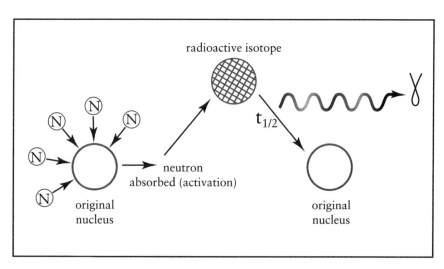

Steps in neutron activation analysis (NAA). A nucleus is bombarded by neutrons, creating radioactive isotopes that decay and give off gamma rays of characteristic energies.

ray radiation. Other types of radiation (alpha and beta) may also be given off but are not the radiation measured in NAA. The rate of decay of the activated species depends on the isotope and is expressed as the half-life ($t_{1/2}$), which is the time required for half of the radioactive atoms to decay. Because the wavelength of gamma rays emitted is characteristic of an element, the identity of an element can be determined by using gamma ray spectroscopy.

NFN (nonflammable ninhydrin) A spray solution of NINHYDRIN that uses a propellant that does not burn, such as a hydrochlorofluorocarbon (CFHC). These propellants also cause less harm to the ozone layer than the freon and chlorofluorocarbon (CFC) propellants that were once commonly used in spray cans.

nicotine An additive component of tobacco that is a drug that can be fatal if a large quantity is ingested at one time.

NIDA National Institute on Drug Abuse, a branch of the National Institute of Health (NIH) under the United States Department of Health and Human Services (HHS). NIDA serves as a clearing house of information on drugs and drug abuse and maintains an extensive website at http://www.nida.nih.gov.

NIJ The National Institute of Justice, which is the research and development arm of the United States Department of Justice (DOJ). It is a primary source of research funds for forensic science and related research. The NIJ maintains an extensive website at http://www.ojp.usdoj. gov/nij.

ninhydrin A versatile compound used to visualize latent FINGERPRINTS on porous surfaces such as paper and cardboard. It can also serve as a developing agent for use in THIN LAYER CHROMATOGRAPHY in drug analysis and TOXICOLOGY. Ninhydrin is also sometimes referred to as triketohydrin-hene hydrate. Because of its versatility and sensitivity, it has become one of the primary tools in latent fingerprint visualiza-

tion and has displaced iodine fuming as the method of choice for analysis of porous materials. The first synthesis of ninhydrin is attributed to Ruhemann in 1910, and the characteristic purple produced by ninhydrin reacting with latent prints is called Ruhemann purple. Ninhydrin reacts with the AMINO ACIDS and their degradation products that are part of any latent print. Ninhydrin can be swabbed or sprayed onto a surface, or an entire article can be dipped into a solution. Once ninhydrin is applied to an article, development of the prints may take hours or even days, but increasing heat and moisture in the development environment can accelerate it.

ninhydrin analogs Chemicals that exhibit characteristics similar to those of NINHYDRIN that are used for the development of latent FINGERPRINTS. As ninhydrin does, they react with the amino acid component of fingerprints. The most commonly used analog is 1,8-diazafluorenone (DFO).

NIR See NEAR IR.

NIST (National Institute of Standards and Technology) A federal agency within the Department of Commerce that is responsible for maintaining collections of reference standards and assisting in standardization of measurements across science, industry, and commerce. It was formed in 1901, houses a number of standards used in forensic science, and assists in some investigations. NIST maintains an extensive website at http://www.nist.gov.

nitrated cotton Cotton lint treated with nitric acid and sulfuric acid, a precursor to some types of firearms PROPELLANTS.

nitric acid (HNO_3) A versatile strong acid used in presumptive tests and in sample preparation.

nitric acid test An older chemical test (PRESUMPTIVE TEST) used in DRUG ANALYSIS. When the acid is added to morphine, it changes from a red to a yellowish color. Heroin changes from yellow to green.

nitrocellulose (NC) A low explosive in SMOKELESS POWDER, the PROPELLANT used in modern AMMUNITION. If nitrocellulose is the only ingredient, the powder is called single base; if it is mixed with NITROGLYCERIN, the powder is called double base. Nitrocellulose is also an ingredient in varnishes and lacquers and was at one time used in automotive paints. There are also formulations of dynamite that use nitrocellulose as an ingredient. When NC is mixed with nitroglycerin in the proper proportions, the resulting mix is a gel that is water-resistant and can be used in wet environments. The plastic celluloid is derived from nitrocellulose and camphor. The forensic analysis of nitrocellulose is accomplished principally by THIN LAYER CHROMATOGRAPHY.

nitrogen phosphorus detector (NPD) A specialized detector used with GAS CHROMATOGRAPHY (GC). This detector, also called a thermionic or alkaline flame detector, has a design similar to that of a FLAME IONIZATION DETECTOR. Sample exiting the GC column enters the detector, where hydrogen gas (H_2) and air (containing oxygen, O_2) are mixed in, creating a flame that burns while the detector is operating. An element coated with an alkaline salt such as rubidium sulfate (Rb_2SO_4) is inserted into the reaction area and allows for efficient ionization of molecules containing nitrogen or phosphorus contained in the flame. Once the molecules are ionized, they are attracted to a collector that registers an electrical current that is based on the number of ions that strike it. Since many pharmaceuticals, illegal drugs, and their metabolites contain nitrogen, the NPD is useful in many forensic applications.

nitroglycerin (NG) A shock-sensitive material used as an explosive and also as a drug to combat a heart condition called angina pectoris. Nitroglycerin, informally called "nitro," was synthesized in 1847 and its potential for mining and other operations was immediately recognized. However, the instability and sensitivity of the compound caused many accidents and prevented its widespread use until 1867, when Alfred Nobel found that it could be stabilized by mixing it with a diatomaceous earth formulation. Nobel went on to become a wealthy man, and a portion of his estate was used to establish the Nobel Prize, administered from his native Sweden.

As shown in the figure, nitroglycerin is a relatively small molecule with nitrogen groups substituted for the –OH groups on a glycerin molecule. Glycerin is a common ingredient in consumer products and is used as a moisturizing agent in lotions, an ingredient in soaps, and a sweetener. It is a thick oily liquid, as is nitroglycerin, and old dynamite is often found with an oily

The structure of nitroglycerin compared to that of glycerin.

seepage on the outer coating. Nitroglycerin can also penetrate the skin in much the same way that glycerin in cosmetic and lotions does, and for treatment of angina pectoris, NG can be administered as an ointment that penetrates the skin and acts as a vasodilator, relieving the chest pain associated with angina.

nitrous oxide (N$_2$O) Also known as dinitrogen oxide or "laughing gas," a gas used by dentists as a relaxant and anesthetic and also abused as an INHALANT drug.

NMR *See* NUCLEAR MAGNETIC RESONANCE.

noise Spurious signals found in the outputs of instruments that arise from sources other than the sample itself. Electronic noise from nearby appliances and equipment as well as from components of the instrument can be sources. The ratio of analytical signal to the noise (signal-to-noise ratio [SNR]) is critical in determining the LIMIT OF DETECTION of an instrument.

noncoding region A region of DNA that does not code for a protein.

nondestructive testing An examination or test that does not alter or damage the sample. Nondestructive tests are valued in forensic science because they preserve the evidence in the same condition in which it was recovered. Truly nondestructive tests are rare, but many types of forensic testing come close to this goal. For example, consider the analysis of FIBER evidence. Visual and simple microscopic examination of the fiber does not damage it; nor does simple analysis such as some forms of MICROSPECTROPHOTOMETRY. However, if the analyst wishes to determine the melting point of the fiber, the sample is altered and potentially destroyed during heating.

noninvasive analysis An analysis that can be completed without opening a container or body. An X ray or computed tomography (CT) scan is an example of a noninvasive technique.

nonpolar solvents Solvents that are comprised of molecules that do not have a dipole moment. Such molecules have even distributions of electrons and do not dissolve in polar solvents such as water. In the familiar oil and vinegar example, the oil is nonpolar and does not dissolve in vinegar, which is a dilute solution of acid in water, which is a polar molecule.

nonrequest standard Handwriting samples collected as part of a QUESTIONED DOCUMENT examination. Nonrequest standards are not collected from the person after the writing in question has taken place; rather, these are samples of previous writing such as notes, check signatures, diary entries, and so on, that are not created purposely for the investigation.

normal hand forgery In handwriting and QUESTIONED DOCUMENT analysis, the forging of writing or a signature using the same style as normal writing. The forger does not attempt to disguise his or her personal writing style.

normal phase In HIGH-PRESSURE LIQUID CHROMATOGRAPHY, a combination of a polar stationary phase (column) and a nonpolar solvent such as hexane. This contrasts with REVERSED PHASE, which is the more widely used phase in forensic applications.

NTSB (National Transportation Safety Board) A federal agency formed in 1967 to investigate aviation, railroad, pipeline, marine, and other types of transportation accidents. The board maintains a "Go Team," who can rapidly deploy to the site of accidents and call in any needed expertise, including that of manufacturers and other companies whose equipment or personnel were involved in the incident. The five board members who oversee the NTSB are presidential appointees who serve for five years. The NTSB investigates accidents, announces findings, and makes recommendations to prevent future incidents, but the board has no regulatory power. NTSB investigations usually fall under the category of forensic engineering, and increasingly, the NTSB works alongside law enforcement agencies when it is possible that a criminal or terrorist act was responsible for an accident. The NTSB maintains an extensive website at http://www.ntsb.gov.

nuclear fast red   A stain used to identify sperm cells. This stain imparts a red color to the nucleus of the sperm, which is particularly valuable if the tails of the sperm cells are missing. When the stain is combined with a dye called picroindigocarmine, a CHRISTMAS TREE staining procedure stains the tails green.

nuclear magnetic resonance (NMR)
An instrumental technique used in organic chemistry and biochemistry to study molecular structure. NMR has also been adapted for medical use in magnetic resonance imaging (MRI). The technique is based on the absorption of ELECTROMAGNETIC ENERGY in the radio frequency range by spinning atomic nuclei. This contrasts with other types of SPECTROSCOPY such as infrared (IR) spectroscopy, in which the electrons (outside the nucleus) absorb energy. Any atom that has an odd mass such as hydrogen (mass of 1, indicated as ^{1}H) is amenable to NMR, as is any atom with an even mass number but an odd number of protons. Nitrogen is an example of this type of atom; it has seven protons in the nucleus and a mass of 14. NMR is most often applied to hydrogen and a naturally occurring isotope of carbon, ^{13}C. Most (~99 percent) naturally occurring carbon is ^{12}C, which has six protons and six neutrons in the nucleus. About 1 percent is ^{13}C, which has six protons (an even mass number) and seven neutrons, leading to an odd mass overall. When NMR targets the hydrogen, the process is often referred to as "proton NMR," since a hydrogen atom consists of one proton. By coupling proton NMR with ^{13}C NMR, the backbone of carbon and hydrogen of an organic compound can be deduced.

nucleoside   A unit consisting of a sugar (deoxyribose) linked to a PURINE or

PYRIMIDINE base such as adenine (A), thymine (T), cytosine (C), or guanine (G). The phosphate esters of nucleosides are NUCLEOTIDES, the building blocks of DNA.

nucleotide   A unit consisting of a sugar (deoxyribose), a phosphate group, and a PURINE or PYRIMIDINE base such as adenine (A), thymine (T), cytosine (C), or guanine (G). Nucleotides are the building blocks of DNA. (See figure on page 174.)

nucleus   The structure that controls the function of all organelles in the cell and contains the chromosomes and nuclear DNA that is targeted in current SHORT TANDEM REPEAT (STR) DNA TYPING techniques.

null hypothesis   A fundamental component of statistical testing and of the scientific method. The null hypothesis provides a starting point for the analysis of data and for testing of a given idea that arises from a forensic analysis. Simply put, the null hypothesis assumes that two quantities or entities are the same or are from the same source (a COMMON SOURCE). Any differences between them can be attributed to small random errors and not to any statistically significant problems with the hypothesis itself. In drug analysis, for example, a chemist might be asked to determine the percentage of COCAINE in two samples of a white powder. If one sample is analyzed and found to contain 51.0 percent cocaine and the second contains 51.2 percent, the null hypothesis would be that the percentage of cocaine in each is the same and the difference is due only to small random errors and not to any significant differences in the cocaine concentrations. One of the key aspects of a null hypothesis is that it must be stated in such a way that disproving it is possible. If a null hypothesis is falsified, then a new hypothesis must be developed and similarly tested until all attempts to falsify it fail. When this situation occurs, the null hypothesis is accepted as the correct explanation.

nursing, forensic   The application of nursing skills to legal matters and law

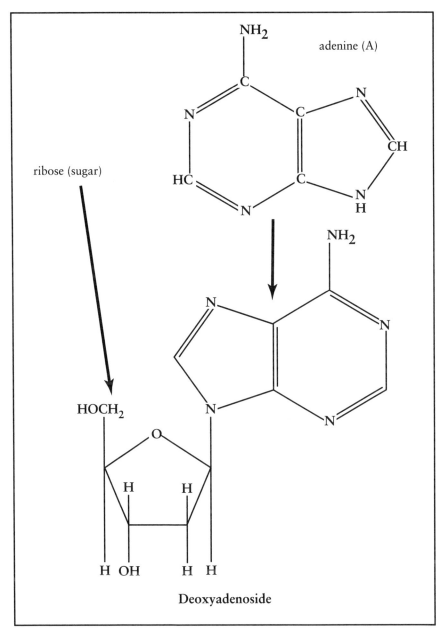

NH₂

adenine (A)

ribose (sugar)

HOCH₂

Deoxyadenoside

An example of a nucleoside, a combination of a base (adenine [A]) and the sugar deoxyribose.

enforcement. Nurses are often the first to treat or see people who have been the victims of crime and violence, but they also have contact with those who are suspects in criminal activity. Relatives and friends of both victims and suspects may also be involved in such scenarios and thus within the scope of forensic nursing. Nurses in

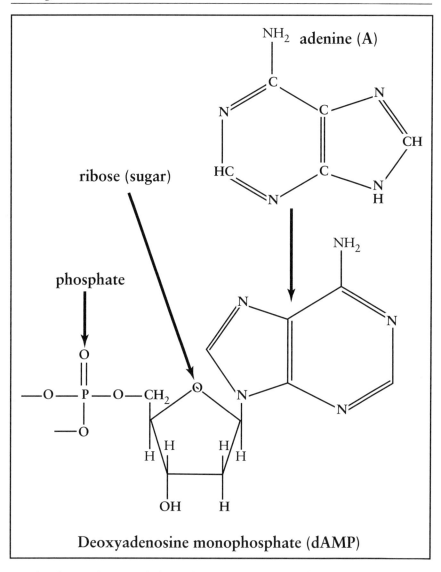

Deoxyadenosine monophosphate (dAMP)

A nucleotide, a combination of a base, ribose (sugar), and a phosphate group (PO_4^{3-}). This is the nucleotide formed by adenine (A), the complement of thymine (T).

these situations need to be familiar with collection, of physical, and behavioral evidence, as well as evidence related to communication, written or other. Clinical forensic nursing (CFN) focuses on nursing in settings such as emergency rooms and other treatment facilities such as prison clinics and forensic (psychiatric) hospitals. Forensic nurses may also be associated with the office of the MEDICAL EXAMINER or CORONER and participate in death investigations. The types of investigations that forensic nurses can become involved in include civil and criminal cases and span the gamut from crimes of violence to car accidents, workplace accidents, substance abuse, product tampering, neglect, and medical malpractice. A sexual assault nurse examiner (SANE) is a nurse trained to collect evidence in and to counsel victims of sex crimes.

nylon A common synthetic fiber that is a polyamide. The first completely synthetic fiber, introduced in 1938. Used in manufacturing women's stockings, which were soon referred to generically as nylons, nylon now includes many polyamide fibers.

Oak Ridge National Laboratory (ORNL) A national laboratory that is part of the United States Department of Energy. It is located in Oak Ridge, Tennessee.

objective (objective lens) The first component of the imaging system in a microscope. It consists of the lens itself and the mounting. *See also* MICROSCOPY.

oblique lighting In microscopy or photography, lighting of the subject from an angle as opposed to light that is shining straight down or up through the object being studied. In microscopy, oblique lighting is lighting situated at an angle to the optical axis of the system.

obliteration A term that has two meanings in forensic science. Obliterations are encountered during the analysis of QUESTIONED DOCUMENTS, in which someone has erased or otherwise removed or hidden some kind of writing or printing. Any type of writing, be it handwriting or printing by commercial presses, computer printers, typewriters, or copiers, can be obliterated or altered. The obliteration can be achieved by erasing and overwriting, crossing out or scribbling over, cutting or scraping the paper surface away, using erasing substances such as a "whiteout" compound or correction ribbon on a typewriter, or using chemical means such as a bleaching agent to remove the writing from the paper. The other type of obliteration is the purposeful removal of serial numbers on evidence such as stolen property or guns. This type of obliteration is addressed in SERIAL NUMBER RESTORATION.

Occam's razor A proposition relating to science expressed by WILLIAM OF ORANGE. Paraphrased, it holds that when there is more than one viable explanation for an observation, the simplest should be the first choice.

occlusal Characteristic of the grinding, chewing, or biting surface of a TOOTH; the part of the tooth's surface that has contact with the teeth in the other jaw (upper or lower).

ocular The eyepiece of a microscope.

odontology, forensic The application of dentistry to legal matters in the areas of personal identification, age determination, BITE MARKS, analysis of wounds and trauma to the jaw and teeth (particularly in potential child abuse cases), and evaluation of alleged dental malpractice or negligence. Since teeth are physically and chemically resilient, they endure and are likely to survive the severe trauma associated with MASS DISASTERS. The American Board of Forensic Odontologists (ABFO), along with the Odontology section of the AMERICAN ACADEMY OF FORENSIC SCIENCES, are the principal professional organizations in the field. Dental identification tasks can fall into one of two categories, identification by comparison to antemortem records and identification where no comparison is available. Odontologists can contribute by estimating age (AGE-AT-DEATH ESTIMATION) on the basis of the condition of the teeth and identification of the type of teeth present. *See also* ORDER OF ERUPTION.

oil immersion A technique in MICROSCOPY in which a specimen is covered in an immersion oil that has the same refractive index as glass. Oil immersion

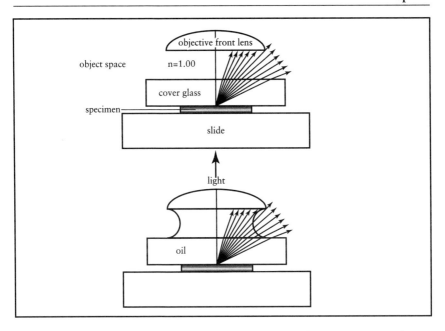

Oil immersion microscopy. In the top frame, light rays (indicated by arrows) emerging from the sample (located between the microscope slide and the coverslip) are refracted as they leave the glass medium and enter air. When the oil is used to fill this space, refraction is minimized, allowing the lens to collect more light and improving the image quality.

allows for the use of a 100X objective lens, the strongest available in conventional light microscopy, which is able to maintain resolution and brightness. As shown in the figure, oil immersion works by eliminating refraction of light passing through a sample, then through the coverslip (glass), then through air, and finally through the glass of the objective lens. In effect, there is no change in refractive index from the glass of the coverslip to the objective lens. The result is an increase in the effective NUMERICAL APERTURE of the lens system.

oligonucleotide   In DNA, a short chain of bases on a sugar-phosphate backbone. The primers used in SHORT TANDEM REPEAT (STR) DNA TYPING are oligonucleotides. (See figure on page 178.)

oligospermia   A condition in males characterized by a low sperm count.

opacity   The imperviousness of a sample or object to the passage of light. Although the term is most often applied to visible light, it is sometimes used to refer to any form of electromagnetic energy. Certain types of quartz are opaque in appearance but are transparent to ultraviolet radiation.

opiate alkaloids   *See* ALKALOIDS.

opiates   A class of drugs that occur naturally in the OPIUM poppy and can be synthesized from materials extracted from it. Examples include MORPHINE and HEROIN.

opium   A mixture of compounds obtained from the unripened seedpods of the plant *Papaver somniferum*. This poppy plant grows in large areas of Asia, southwest Asia, and Mexico. To obtain opium, the seedpod is cut and a brownish

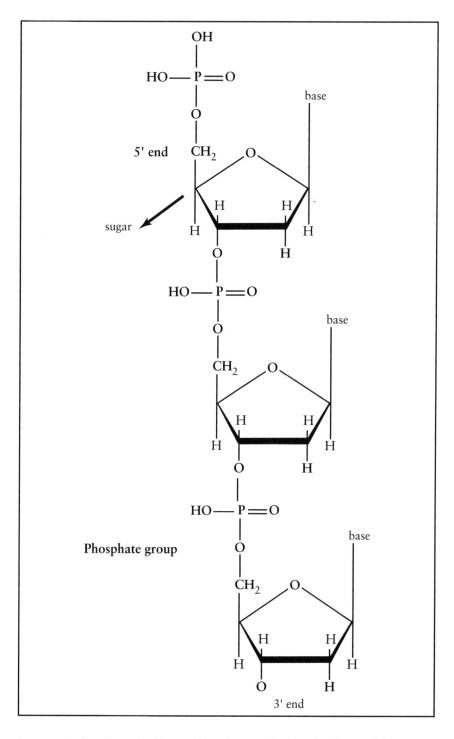

An example of an oligonucleotide, consisting of a sugar-phosphate backbone and the accompanying bases (adenine, thymine, guanine, or cytosine), three in this example. The two different ends are distinguished by the 3' and 5' notation.

milky substance is extracted. This extract is then dried and can be crushed into a powder that gradually lightens in color. Opium consists of a mixture of ALKALOIDS, including MORPHINE and CODEINE, and can be consumed directly by ingestion or smoking. Opium was first seen in large quantities in the United States in the 1800s when immigrants carried it from China and the Orient, and the substance is now listed on Schedule II of the CONTROLLED SUBSTANCES ACT.

optic axis In a microscope, an imaginary straight line that traverses the center of curvature of the lenses in the optical path.

optimization The process of adjustment of instruments or procedures to obtain the best results possible.

order of eruption In forensic ODONTOLOGY, the order in which (or age when) a tooth typically erupts from the gums, which can be used in estimating the age of the deceased. (See figure on page 180.) *See also* AGE-AT-DEATH ESTIMATION.

order of magnitude A power of 10. The difference between 10 and 100 is one order of magnitude; similarly the difference between 0.1 and 100 is four orders of magnitude.

Orfila, Mathieu B. (1787–1853) Orfila, considered the founder of forensic TOXICOLOGY, was born in Spain and moved to France, where he worked and became professor of forensic chemistry and dean of the medical faculty at the University of Paris. He began publishing early, with his first paper on poisons in 1814 when he was 26 years old. As a toxicologist, he concentrated on methods of analysis of poisons in blood and other body fluids and tissues. He became involved in a famous arsenic poisoning case in 1839, when a young woman, Marie Lafarge, was accused of using arsenic to murder her much older husband. Initial results of the analysis of the husband's exhumed remains were nega-

tive, but Orfila was able to detect arsenic in the exhumed remains. Marie was eventually convicted of the crime. His testimony in the case was one of the earliest examples of sound scientific testimony by a recognized scientific expert in a court of law.

organic compounds and organic analysis Organic compounds are defined as those that primarily contain carbon (C) and hydrogen (H), a class of compounds centrally important in forensic science. Examples of organic compounds are sugars, petroleum products, drugs, plastics, and proteins. Carbonate (the ion CO_3^{2-}) and carbon dioxide (CO_2) are not considered organic compounds since they do not contain hydrogen. Organic chemistry, the study of organic compounds, is the foundation of biochemistry, which revolves around organic chemistry that occurs in living organisms. Likewise, organic analysis involves detection, quantitation, and study of organic compounds. Organic compounds may be natural or synthetic. For example, MORPHINE and CODEINE occur naturally as part of OPIUM, whereas HEROIN is synthesized from morphine. Organic analysis can be conducted by using myriad techniques. In forensic science, the most common include simple chemical tests such as PRESUMPTIVE TESTS for different drug classes, instrumental techniques such as INFRARED SPECTROSCOPY, and many different types of CHROMATOGRAPHY.

organized offender A criminal whose acts are planned and executed carefully.

origin determination A term that applies to many areas of forensic science. For example, in arson the point of origin is the site where the fire started; in the analysis of blood and body fluid or anthropology, *origin* can be used in the context of species of origin. In a broad sense, *origin* refers to the location of where an object, sample, or evidence is found.

orthotolidine test (tolidine test) A PRESUMPTIVE TEST for blood that is also referred to by the shorthand notation of *o*-tolidine. This test works similarly to most

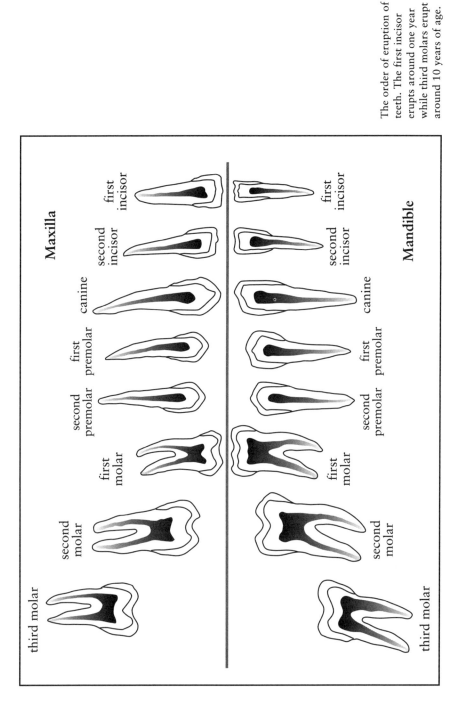

The order of eruption of teeth. The first incisor erupts around one year while third molars erupt around 10 years of age.

other tests for blood in that the *o*-tolidine reacts with hemoglobin in blood in the presence of hydrogen peroxide (H_2O_2) to cause a bluish green color to form. As in other presumptive tests, the tolidine test is not specific for blood and can produce FALSE POSITIVE findings with substances such as horseradish. Tolidine is also a carcinogenic material, and thus this test is not as widely used as the KASTEL MEYER TEST.

Osborn, Albert (1858–1946) A pioneer of forensic document examination (analysis of QUESTIONED DOCUMENTS) and author of a 1910 text that is still considered a foundational work and reference in the field. He also was a founding member of the American Society of Questioned Document Examiners (ASQDE) and served as its first president from 1942 until 1946, the year of his death. Osborn's sons continued in the field, and both Albert and his son were involved in investigation of the LINDBERGH KIDNAPPING CASE.

Osborn grid method A method for comparison of FINGERPRINTS that overlays a grid on enlarged images of the prints and compares features found within the grid square.

ossification/ossification centers The process of bone formation from tissue such as cartilage. This process begins a few weeks after conception.

osteoarthritis A disease characterized by deterioration of joint that leaves distinctive signatures on bone.

osteology The study of the structure and function of bone; closely related to forensic ANTHROPOLOGY. It is also referred to as skeletal biology.

osteometry The taking of skeletal measurements using special tools and specific procedures. In forensic science, measurements of skeletal remains allow estimation of stature, age at death, and sex.

osteons (Haversian system) Subunits of bone structure that possess a layered, concentric structure around a canal containing a blood vessel. The number of osteons in a given bone sample is sometimes useful for estimating AGE AT DEATH. (See figure on page 182.)

Osterburg model A statistical model proposed in the late 1970s that attempted to describe the individuality of FINGERPRINTS. It utilized a grid pattern and description of the types of features found within each grid.

OTC Over the counter; drugs and medicines that can be purchased without a prescription.

otoscopy *See* EAR IDENTIFICATION.

Ouchterlony test An IMMUNODIFFUSION test used to determine the species from which blood or other body fluid originated. To perform the test, AGAROSE is poured into a Petri dish and allowed to gel. A central hole is punched in the gel and sample extract containing antigens is placed into it. A series of other wells are punched into the gel around the central hole a few millimeters away. Antibodies from different species such as human, dog, and cat, are placed into the wells. Over a few hours, the antigens in the sample diffuse outward toward the antibodies that are diffusing inward. If the antigens and antibodies are from the same species, an IMMUNOPRECIPITATE forms in the gel, indicating the species of the sample.

overkill Injuries inflicted during a crime that exceed what is necessary to kill or incapacitate a victim.

overstamping A method used to obliterate or obscure a serial number stamped into a metal object such as a gun. Through overstamping a false serial number is stamped over the existing one.

ovoid bodies Microscopic voids or pockets that may be found in the cortex of HAIR.

oxidation/reduction (redox) A coupled chemical reaction that is characterized by exchange of electrons or electron flow.

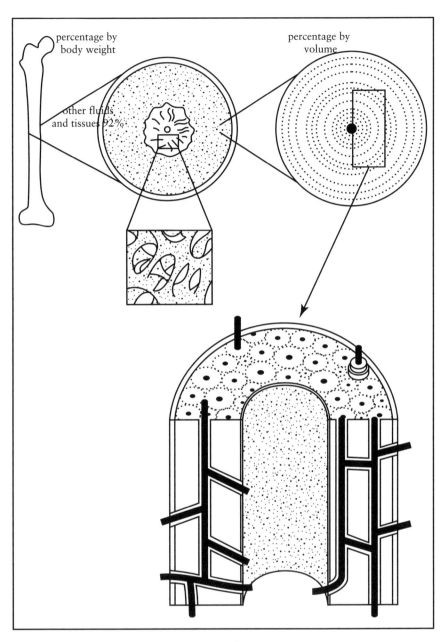

percentage by
body weight

percentage by
volume

other fluids
and tissues 92%

Structure of an osteon, the microscopic substructure of bone. It is organized around a central blood vessel and has a layered structure often compared to that of an onion. The progression of figures shows increasingly small cross sections.

Redox reactions are the basis of batteries, for example. Although electron exchange is at the heart of redox reactions, there are many different ways to describe such reactions. There are three ways to define oxidation: (1) a loss of electrons, (2) a gain of oxygen (O), or (3) a loss of hydrogen. Similarly, reduction can be defined as a gain of electrons, a loss of oxygen (O), or a gain of hydrogen. In the combustion of natural gas (CH_4, methane) in air, carbon dioxide (CO_2) is produced. Methane is the reduced form of carbon; carbon dioxide is the oxidized form. Redox is the chemical basis of PRESUMPTIVE TESTS for blood such as the KASTEL MAYER TEST, in which HEMOGLOBIN acts as a catalyst.

oxidizing agent A compound or material that causes oxidation of another species; the species that is reduced in an OXIDATION/REDUCTION reaction. Oxygen (O_2) is an oxidizing agent, and it causes the oxidation of iron (Fe) to form Fe_2O_3, or rust. Bleach is another common oxidizing agent.

oxycodone (OxyContin) A narcotic pain reliever made from thebaine, an alkaloid derived from OPIUM. Oxycodone acts in the same manner as morphine and codeine, and, as with those substances, abuse can lead to physical and psychological addiction. OxyContin, introduced in 1995, contains a time-released form of oxycodone. Since the drug is time-released, the amount of oxycodone in each tablet is much higher, and this property has led to increasing abuse and diversion of OxyContin for illegal use. Abusers crush the tablets and by doing so destroy the time-release properties of the drug. This allows them to get a large dose immediately by ingestion, smoking, or injection, with effects that mimic those of heroin.

P

p30 A protein that is specific to seminal fluid and detectable in the vaginal tract for about eight hours after sexual intercourse has occurred. Also known as prostate-specific antigen (PSA), p30, when detected in a stain, shows that it is or contains semen.

paint In general terms, a coating of coloring agents suspended or dissolved in a solvent containing other additives such as binders and drying agents. Pigments were at one time primarily INORGANIC compounds such as titanium dioxide (TiO_2, white), Fe_2O_3 (rust), or other compounds made up of cadmium, lead, and other metals. Because of their toxicity, many of these compounds have been replaced with ORGANIC pigments. However, lead paints still present a POISON hazard in many older and inner city buildings, where children eat it. Solvents employed in paint formulations can be organic such as toluene or water-based such as those used in latex paints. Water-based paints can be cleaned up with soap and water and are popular for interiors of homes. The solvent, be it water or organic, is referred to as the *vehicle*. Finally, polymers, or materials capable of polymerization (called binders), are included; when dry, they form a protective coating over the pigments.

palindrome A location or site on a DNA strand in which the base sequence on one side of the strand is the exact reverse of the sequence in the complementary strand.

Palmer method A method used to teach cursive handwriting style.

palynology A branch of botany that deals with the study of pollen and spores.

Forensic palynology is a field of forensic botany.

paper A complex mixture made from slurry containing wood and cotton fibers, binders, glues, bleaching agents and dyes, other colorants, preservatives, and coatings. In general, the finer the paper, the higher the cotton content, and such papers are sometimes referred to as containing cotton rag.

papillae/papillary dermis The boundary layer in skin that separates the DERMIS from the EPIDERMIS. The hill-and-valley pattern of the papillae determine the ridge characteristics of the finger and thus the fingerprint.

paraffin test *See* DERMAL NITRATE TEST.

parent drop In bloodstain patterns, a large central drop that, at impact, creates a series of smaller drops associated with it. The smaller drops are sometimes referred to as SATELLITES of the parent.

parent ion In some forms of MASS SPECTROMETRY, an ion, such as a MOLECULAR ION, that further fragments into smaller daughter ions. The structure of the parent ion can often be deduced from the pattern of daughter ions that it produces.

parent/metabolite ratio *See* P/M RATIO.

pars papillaris Along with the PARS RETICULARIS, one of two regions in the DERMIS portion of the skin. The papillaris is the outermost layer.

pars reticularis Along with the PARS PAPILLARIS, one of two regions in the DER-

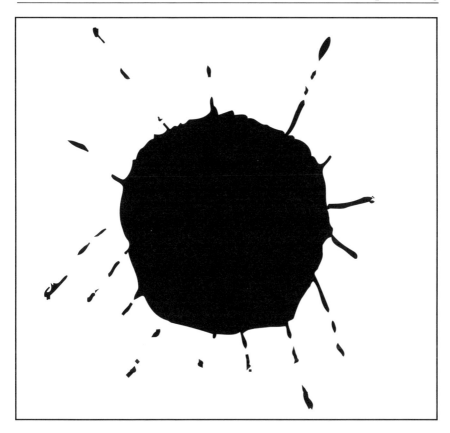

Parent drop of blood with smaller satellite droplets. The tails on the satellites point back toward the parent, and the type of surface onto which the blood falls can influence the pattern.

MIS portion of the skin. The reticular is the innermost and thicker layer and is composed of connective tissue and collagen.

partial individualization The result of a forensic analysis that does not truly INDIVIDUALIZE a piece of evidence but that narrows down the potential sources significantly. For example, a blue fiber that is determined to be from the carpet in a truck made at one factory during one period would be considered partial individualization.

partition chromatography A broad category of CHROMATOGRAPHY in which selective PARTITIONING between two dis-

tinct physical phases is exploited to separate components of a mixture. COLUMN CHROMATOGRAPHY, THIN LAYER CHROMATOGRAPHY, GAS CHROMATOGRAPHY, and HIGH-PRESSURE LIQUID CHROMATOGRAPHY are all examples of partition chromatography.

partitioning The process of separation based on different affinities for different chemical or physical components. Solvent extractions, widely used in areas such as toxicology and drug analysis, exploit partitioning. A simple example would be extraction of salt from salad oil by adding water. The mixture would separate into two phases since water is not soluble in oil. Salt, an ionic compound, dissolves in water. Thus, if the mixture is shaken vigorously and allowed to separate, the salt selectively partitions into the aqueous phase. Most forms of chromatography rely on physical or chemical partitioning to separate complex mixtures.

patent fingerprints Fingerprints that are visible, as opposed to a LATENT FINGERPRINT, which is invisible or barely visible.

paternity testing Determination of parentage of a child that often involves or revolves around a legal issue such as life insurance, inheritance, or child support. Although the term *paternity* in a strict sense applies to determination of the father, *paternity testing* generally refers to either parent. The same techniques that are used in forensic SEROLOGY and forensic biology are utilized in paternity testing, principally DNA TYPING. MITOCHONDRIAL DNA has proved useful in paternity cases when good samples are not available from one or more people involved, if one parent is dead or missing, for example.

pathology The medical specialty that involves the study of disease and its effects on the structure and function of the body.

pathology, forensic The study of the cause, manner, and mechanism of death undertaken by a medical doctor who has completed a residency in pathology, followed by additional training in forensic pathology and death investigation. A forensic pathologist may work as a MEDICAL EXAMINER, as a CORONER, or in the office of one of these officials. The primary job of the forensic pathologist is to perform AUTOPSIES of questioned or suspicious deaths to determine the cause, manner, and circumstances of death. This determination is based on the autopsy results and also on information gathered as part of the death investigation. The pathologist is also tasked with identification of the body and with estimation of the time of death (POSTMORTEM INTERVAL).

pattern matching and pattern recognition A technique used in forensic comparisons in areas such as FINGERPRINTS, FIREARMS, BITE MARKS, and TOOLMARKS. Some of these procedures can be automated, such as those for fingerprints and firearms evidence; however, many pattern recognition techniques cannot be easily encoded. For example, when an impression of a shoeprint is made at a scene and a possible source shoe located, an analyst compares them and uses pattern matching techniques to determine whether that shoe created that impression. In most pattern recognition applications, the analyst employs some type of point comparison technique with which he or she locates interesting, significant, or unique features on one of the pieces of evidence and then attempts to locate that same feature on the other piece. However, even if many points are compared, there is some element of subjectivity in any pattern matching analysis. In some cases, this subjectivity has been challenged in court.

PCP *See* PHENCYCLIDINE.

PCR Polymerase chain reaction, a technique used in current methods of DNA TYPING using SHORT TANDEM REPEATS (STRs). The process begins by heating the extracted DNA to cause the double helix to "unzip" into two complementary strands, a procedure called DENATURATION. A PCR PRIMER is then added to identify the region or locus on the strand to be typed. The primer binds to the base sequences on each side of this region, marking the boundary of the locus of interest. This process is called hybridization and is carried out as the solution is allowed to cool. The final step is to add bases—adenine (A), thymine (T), cytosine (C), and guanine (G)—that pair with the complementary bases on the isolated regions on the strands. DNA polymerase enzyme, which promotes the rezipping of the strands, is also added. The result of this cycle is the production of an exact copy of the region of interest. By the process of this thermal cycling, the origin sample is amplified thousands of times.

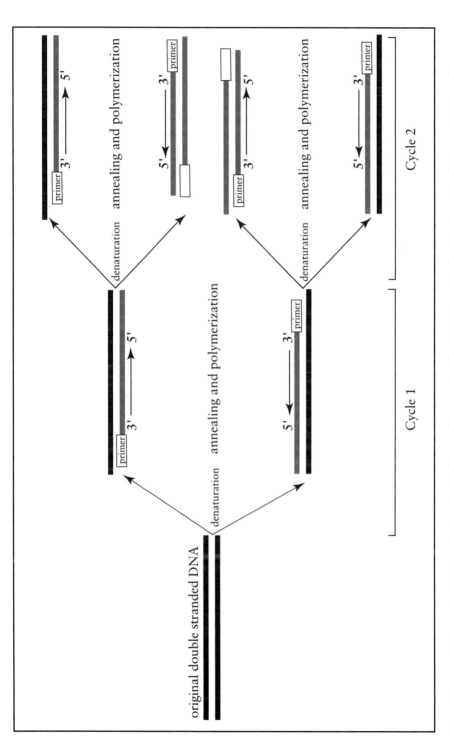

The polymerase chain reaction (PCR) cycle, described in the text. The notation 3' and 5' refer to the directionality of the chain; amplification always occurs from the 3' end of the strand to the 5'. The 3 and 5 refer to carbon atoms in the ribose sugars that form the nucleotides in the DNA chain.

PCR primer In DNA TYPING using PCR techniques, the primer is an OLIGONU-CLEOTIDE that binds specifically to the boundary region of the genetic locus that is to be typed.

PD *See* PHYSICAL DEVELOPER.

peer review A process of evaluation of research results that is used in science. Peer review of journal articles, laboratory reports, and proposals is commonly used in forensic science and in the case of review of laboratory work represents a form of QUALITY ASSURANCE/QUALITY CONTROL. The goal of peer review is to ensure the reliability, validity, and quality of scientific work.

pellets In forensic context, a term that may refer to the projectiles fired by a shotgun and defined by the GAUGE or to thin wafers prepared for some types of INFRARED SPECTROPHOTOMETRY. These pellets are prepared by adding a small amount of purified sample to a dry solid such as potassium bromide (KBr) that is transparent to infrared (IR) radiation. The powders are ground together and pressed into a thin pellet for insertion into the instrument.

pencil lead The graphite-based material that is applied to paper by a pencil. Although pencil lead is mass produced, chemical and spectrochemical analysis of these components may be useful in dating pencil notations or in eliminating possible sources. There is no lead (Pb) in pencil lead.

pen lift A characteristic seen in handwriting using a pen, pencil, or other writing instrument. It is demonstrated by a break in a line that indicates that the writing instrument has been lifted off the paper. Pen lift patterns can be useful in QUESTIONED DOCUMENT examination.

penmanship The style and pattern of handwriting used by an individual.

pen pressure The pressure applied to a writing instrument when it is in contact with paper or another surface. Pen pres-

sure patterns can be useful in QUESTIONED DOCUMENT examination.

peptidase A (PEP A) An ISOENZYME system with three common phenotypes. Before the advent of DNA TYPING, it was occasionally typed in forensic work using gel ELECTROPHORESIS.

peptide bond A COVALENT chemical bond that links two AMINO ACIDS. Specifically, it is an amide bond that forms between the carboxyl group of one amino acid and the amino group of the second. Such a bonding of two peptides is sometimes referred to as a condensation since a water molecule is produced.

perchlorate (ClO_4^-) An anion that may be encountered in explosives analysis.

perimortem Occurring at the time of death or very near to it such as a perimortem injury.

periodic acid–Schiff test (PAS) A PRESUMPTIVE TEST used to detect the presence of vaginal secretions and material. Vaginal material contains a large amount of exfoliated (shed) epithelial cells, which in turn have large amounts of glycogen in their cytoplasm. In this test, the stain used imparts a magenta color to the cells; the more intense the color, the more cells present.

periodic table A table developed by the Russian Dmitri Mendeleev (1834–1907), which organizes all known elements in an orderly row and column arrangement. The current version of the Periodic Table of the Elements, still the foundational tool of chemistry, is found in APPENDIX III. Without knowledge of nuclear structure or electron arrangements, Mendeleev was able to assign elements to families (the columns of the table) of elements that have the same type of chemical behavior. Across the table (rows), elements are organized by increasing atomic number. Within a given family, chemical behavior is similar, as in group 8, the noble gases, all of which are nearly inert. Similarly, in group 1, each element tends to react violently with water and to

Formation of a peptide bond to form a dipeptide. The carboxyl group links to the amino group, releases a water molecule, and forms the covalent bond.

form salts with elements of group 7 (halogens), such as NaCl and LiBr. Modern versions of the periodic table can include much more information such as crystal form, ionization potential, and density.

PERK kit A kit used to collect physical evidence from victims of sexual assaults. The acronym stands for *Physical Evidence Rape Kit* and typically includes a whole blood sample, swabs of any dried secretions, and swabs of the vagina, genital area, thighs, anus, and mouth. In addition, smears on slides are made from these swabs and all must be air-dried. A vaginal rinse may also be collected. Hair, fiber, and occasionally toxicological samples are also collected.

permanent dentition The 32 teeth of an adult arranged as shown in the illustration on page 190.

peroxidase An enzyme that catalyzes an oxidation reaction of an organic compound by using a peroxide. A well-known peroxide is hydrogen peroxide (H_2O_2), which is used to sterilize wounds. Most PRESUMPTIVE TESTS for blood such as the KASTEL MEYER TEST are based on the peroxidaselike activity of hemoglobin, which catalyzes a reaction that results in a color change.

personal protective equipment (PPE) Equipment used in the lab and during evidence collection to prevent transmission of blood-borne diseases, toxins, and other biohazards. Gloves and eye protection are examples of PPE.

perspiration *See* ECCRINE SWEAT.

petechial hemorrhage/spots/Tardieu's spots In most strangulation deaths, pinpoint hemorrhages called petechiae, petechial hemorrhages, or Tardieu's spots are found in the face and particularly in the eyes and on the eyelids. These hemorrhages are caused by capillary rupture

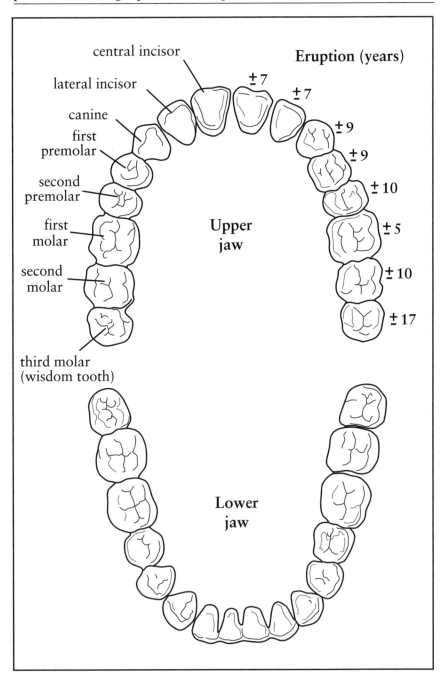

Permanent adult dentition. The arrangement on the upper and lower jaws is shown, along with names of the teeth and age at which the teeth typically erupt.

brought on by pressure. However, presence of petechiae in itself is not sufficient to determine that the death was the result of strangulation.

PETN A high EXPLOSIVE used commercially, by the military, and in mixtures employed by terrorists. PETN (pentaerythritol tetranitrate) was synthesized in the 1940s and is used by the military as a sheet explosive (a common form is Detasheet), in grenades, and as a propellant in small-caliber weapons. Detonation cord, often called "det cord" or Primacord, contains a core of PETN or RDX covered in cotton and enclosed in a weatherproof casing. PETN must be detonated by a shock wave from some type of initiator such as a blasting cap before it will explode.

petroleum distillates Organic compounds that can be obtained from the distillation of oil (petroleum). Gasoline, diesel, and compounds such as benzene are classified as petroleum distillates.

petroleum ether A mixture of volatile ethers used as a solvent. Also called "pet ether," it is used in drug and arson analysis as well as other laboratory procedures.

petroleum hydrocarbon A hydrocarbon (organic chemical compound containing hydrogen and carbon), such as ethane, octane, or benzene, that can be derived or extracted from crude oil.

peyote *See* MESCALINE.

PFTBA Perfluorotributylamine, a compound that is used to check and calibrate settings of MASS SPECTROMETERS to ensure that results are comparable from instrument to instrument.

PGM (phosphoglucomutase) An ISOENZYME that catalyzes the conversion of glucose-6-phosphate to glucose-1-phosphate during glucose metabolism. PGM is a polymorphic enzyme and thus a genetic marker system. Before the advent of DNA TYPING, PGM was routinely typed in blood wherever the sample allowed.

Phadebas reagent and tests Commercial test kits and reagents that are used to detect AMYLASE. The tests are based on a starch–iodine complex and the ability of amylase to break down starch. These tests are PRESUMPTIVE TESTS for SALIVA.

pharmacodynamics The study of the effect of drugs and their metabolites on the body.

pharmacokinetics The study of the movement of drugs through the body. After a drug or other XENOBIOTIC SUBSTANCE is introduced, it is absorbed and distributed to tissues in the body by the bloodstream. Portions of the substance are METABOLIZED and in effect act as new drugs that enter into the same cycle. As an example, heroin is metabolized to morphine, which then reenters the cycle as a "new" drug that may be further metabolized and/or eliminated. Elimination of the drug and metabolites follows, although some portions of some compounds may accumulate in fatty tissues or other locations. Elimination can be by several routes, principally in urine, but also in sweat, feces, or breath. Different drugs have different rates of absorption, metabolites, and speed of elimination. One measure of the time a drug remains in the system is the half-life, symbolized by $t_{1/2}$. This information, combined with toxicological findings, can be useful in determining how much of a drug was taken and roughly when. Heroin is metabolized to morphine and has a half-life of one to 1.5 hours; accordingly, evidence of heroin ingestion can be detected in urine for two to four days after ingestion.

pharmacology, forensic The study of drugs and their effects on organisms (primarily humans) in legal matters. Forensic pharmacology addresses such questions as when a drug was taken, how much was taken, and what effects it produced in an individual. It is intimately related to forensic TOXICOLOGY, PATHOLOGY, and drug analysis. Forensic pharmacology involves two broad topic areas, PHARMACOKINETICS and PHARMACODYNAMICS.

phase Generally, the physical state of a material, such as the liquid phase. The other two phases are solid and gas. *Phase* can also refer to a temporary state.

phase contrast microscopy A specialized microscopic technique that increases the contrast between items that would otherwise be very difficult to see, such as transparent or nearly transparent materials. Phase contrast techniques are particularly useful in forensic science to visualize sperm cells and fine structure in hair and soil. A phase contrast microscope is similar to a COMPOUND MICROSCOPE in design but includes specialized objectives, lighting, plates, and light condensers. When a specimen and oil (or other mounting medium) have similar refractive indices, seeing surface relief and distinguishing edges is difficult. The purpose of phase contrast techniques is to increase the contrast between the object and the mounting medium, making the object much easier to see.

pH buffer *See* BUFFER.

phencyclidine (PCP) An abused drug that is easily synthesized and was at one time used as a veterinary anesthetic. The compound was synthesized in 1926 and used as a cat anesthetic in the 1950s. In the early 1960s, it was introduced as a human anesthetic but soon withdrawn from the market after bizarre behavior and other side effects were noted. Production of the compound ended in 1979. As an illicit hallucinogen, PCP first appeared in the late 1960s, and it was placed on Schedule II of the CONTROLLED SUBSTANCES ACT in 1970. Because it is relatively easy to make, its PRECURSORS are also controlled.

phendimetrazine/phenmetrazine Prescription stimulants used for weight control that are abused in ways similar to those used with AMPHETAMINE and methamphetamine.

phenol A solvent (C_6H_5OH) used to extract DNA from cells and in DNA TYPING.

phenolphthalein *See* KASTEL MEYER TEST.

phenotype In genetics, the outward appearance of a trait, as opposed to the genotype, which is the actual combination of genes. For example, in the ABO BLOOD GROUP SYSTEM, a person inherits one gene from the mother and one from the father. If both genes are A, then the person's genotype is AA and phenotype is A, meaning that the person's blood is type A. If a person inherits an A gene from one parent and an O gene from another, the genotype is AO, but the phenotype is A and the person has type A blood.

phenyl-2-propanone (P2P) A PRECURSOR in a CLANDESTINE LAB synthesis of methamphetamine. This material was added to the CONTROLLED SUBSTANCES ACT (Schedule II) and now is rarely encountered.

Philips physical developer A form of PHYSICAL DEVELOPER used to develop and visualize latent fingerprints. It is used for prints on paper.

phonetics, forensic The study of the sound components of speech, usually with the intent of identifying the speaker or learning about the speaker. In general, the type of evidence of interest in forensic phonetics is called a disputed utterance or questioned utterance in the same way disputed documents are the subject of QUESTIONED DOCUMENT analysis. For example, the speech of a native of the southern United States is usually quite different than that of someone from Boston or New Jersey, even though all are speaking English. These are extreme examples, which illustrate that the sound of speech can provide useful information. Other aspects of speech of interest include pitch, speed, intonation, emphasis, and accent on syllables. Aside from helping to uncover information, phonetics can be helpful in interpretation of meaning from the sound of speech. Again, a simple example would be the difference in meaning between the following statements: "*I* like it" versus "I like *it*" versus "I *like* it." All three are written the same but spoken quite differently and because of different emphasis carry a different meaning. Although controversial, VOICEPRINTS are sometimes considered part of forensic phonetics.

phosphorescence The emission of ELECTROMAGNETIC RADIATION that occurs after absorption. Phosphorescence is a form of luminescence, which is characterized by emission that continues after the excitation energy has been removed. Glow-in-the-dark watch faces rely on phosphorescence.

photocopiers Copy machines that use a photographic process to produce a copy of a document. The term *copier* is generic and usually refers to copiers that work in much the same way as laser computer printers (xerographic process) and as photocopiers. Informally the three terms (*copy, photocopy,* and *Xerox*) are used interchangeably, although there is a distinct difference between the way a photocopier works and the way a xerographic copier works. In a photocopy process, an electrostatic charge is placed on a surface and is selectively discharged by exposure to light, creating a negative image. Projecting light through the negative creates a print, much as printing a negative creates a photograph.

photoelectric effect The ejection of electrons from a surface as a result of interaction with ELECTROMAGNETIC RADIATION. Some types of detectors for spectrophotometers exploit this property, converting light to an electrical current that can be detected and manipulated.

photograph/photography The word *photography* means "light writing," and the process is based on exposure of light-sensitive compounds such as silver chloride (AgCl) to light. The light causes the salt to darken; when it is developed, the pattern forms a negative of the original image. Color photography employs layers of these light-sensitive compounds, each of which responds to different colors. Projection of light through the negative is used to create the prints. One of the earliest types of photography was the daguerreotype, named after its French

inventor. It was also known as the tintype because of the backing. By the 1870s daguerreotype portraits were common, and by the mid-19th century, photos were being used for identification of cadavers and of criminals. Application of photography to forensic work became commonplace in the latter part of the 19th century and Alphonse BERTILLON was one its well-known advocates, both for crime scene documentation and for identification of individuals. His early work with identification cards (*portrait parle*) could be considered one of the precursors of today's "mug shots."

Today, photography includes digital imaging, which does not require film. Digital cameras are based on an array of light-sensitive diodes called CHARGE-COUPLED DEVICES (CCDs). The number of megapixels in a camera corresponds to the number of light-sensitive diodes in the array, so a 2.1 megapixel camera has 2,100,000 photosensitive sites. When light strikes a diode, a charge builds up, and more light creates a bigger charge. A scanning process discharges the diodes, allowing the intensity of light at each diode to be recorded. To create color images, filters that separate out the red, blue, and green lights are used to control the color that reaches a given diode. A processor in the camera creates the images from the CCD and filters.

photoluminescence *See* LUMINESCENCE.

photomicrography Taking of photographs (digital or film) through a microscope.

photon A discrete packet of electromagnetic energy; informally a "particle" of electromagnetic energy (light). The energy of a photon is calculated as $E = hv$, where h is Planck's constant (6.626×10^{-34} Js) and v is the frequency of the ELECTROMAGNETIC ENERGY per second (s^{-1}) or Hertz (Hz).

Photoshop A computer program sold by Adobe that is widely used to process digital and video images.

pH/pH scale The scale used to describe the acidity or alkalinity (basicity) of a water-based solution. Water can dissociate into two ions, H^+ and OH^- (hydroxide); the pH of a solution refers to the balance of those two species. The pH is defined mathematically $pH = -\log[H^+]$, where the $[H^+]$ denotes the concentration of hydrogen ions (also called protons) measured in units of molarity (number of moles H^+ per liter, M). The pH scale centers on 7, which is a neutral solution, in which the concentration of hydrogen ions equals the concentration of hydroxide at 1×10^{-7}M. A pH less than 7 is acidic (H^+ in excess); the lower the number, the more acidic the solution. A pH greater than 7 is basic (alkaline, excess OH^-); higher numbers correspond to increased alkalinity. Lemon juice, which is acidic, has a pH of about 4, skim milk about 6.6, blood about 7.4, milk of magnesia about 10.5, and lye (NaOH, sodium hydroxide) about 13.

physical anthropology A broad subdivision of anthropology that studies (among other topics), primate and human evolution and behavior. Forensic ANTHROPOLOGY is generally considered to be the subdiscipline of physical anthropology that deals with recent skeletal remains as opposed to historical or ancient remains.

physical developer (PD) Contrary to what the name implies, a chemical method for developing and visualizing LATENT FINGERPRINTS. Although it can be used alone, it is often used as an addition to other approaches such as NINHYDRIN. The process is similar to that used in black-and-white photography and uses silver chloride (AgCl) or similar salts that react with fats and other insoluble constituents of the fingerprint residue.

physical evidence Broadly speaking, any type of tangible evidence, as opposed, for example, to the testimony of an eyewitness. Physical evidence can be anything from a microscopic trace of dust to a car, but there are some generalizations that can be made. Physical evidence must be documented, collected, marked, trans-

ported, and stored in a manner consist with the type of evidence. It is subject to a CHAIN OF CUSTODY to ensure its safety and integrity from time of collection to use in court. Analysis of physical evidence involves IDENTIFICATION, comparison, determination of CLASS CHARACTERISTICS, and INDIVIDUALIZATION, when possible; occasionally, it is used in reconstructions.

physical matching The act of linking pieces of evidence that exist in pieces but once were part of the same item. Reconstructing a broken window is an example of physical matching, as is linking a specific match to the matchbook from which it was torn on the basis of the unique tearing pattern.

physical properties Properties of matter that can be measured by physical means rather than chemical means. Measuring a physical property can be accomplished without changing the chemical composition of the material being studied. On the other hand, CHEMICAL PROPERTIES can only be determined by processes that alter the chemical composition of the sample. For example, the freezing point of water is a physical property since it can be determined by freezing the water and measuring the temperature at which that occurs. The chemical formula of liquid water (H_2O) is the same as the chemical formula of ice, so no chemical change has occurred. The flammability of butane, a chemical property, can only be determined by an experiment that involves the COMBUSTION of butane, resulting in the conversion of butane to carbon dioxide and water.

Physician's Desk Reference (PDR) A family of reference books used by medical doctors. The most commonly used volume in forensic applications lists PRESCRIPTION DRUGS, their composition, action, indications, and other important information. In forensic science, the *PDR* is an indispensable reference manual in forensic CHEMISTRY (drug analysis), PATHOLOGY, and TOXICOLOGY. The first edition of the *PDR* was published in 1946; it has been published annually since by Medical Economics Company. For identification of prescription drugs seized as physical evidence, the large collection of actual size photographs is invaluable, as are the extensive cross-listing and referencing of the drugs and active ingredients.

picking When a drug tablet is mechanically punched out of a matrix, a residual of that drug that adheres to the punch surface is referred to as picking.

picroindigocarmine (PIC) A stain used to visualize sperm cells. It is a component of the CHRISTMAS TREE STAIN procedure.

pigments *See* DYES AND PIGMENTS.

pincushion method Also called the polygon method, a technique for comparing LATENT FINGERPRINTS. In this approach, points of comparison are marked with a point, and a line is drawn from one point to the next until a closed shape is created. If the shape is the same for two prints, a match is indicated.

pipe bomb An explosive device typically constructed of a closed section of metal pipe containing black powder or gunpowder and a fuse that extends out-

side the device. Projectiles such as tacks or nails are sometimes included to increase the shrapnel produced by a detonation.

pistol A generic term referring to a handgun, usually a semiautomatic one; however, revolvers are sometimes referred to as pistols.

pixel Literally, "picture element," a small dot that forms the smallest component of a digital image such as a computer screen or a digital camera photo. Increasing the number of pixels for a given image increases the resolution and clarity of the image; however, since more pixels are used, greater allocations of memory are required to store it.

plain arch One of the common fingerprint patterns, along with patterns such as loops and whorls. The other type of arch is a tented arch, which has a more pronounced point at the top than a plain arch pattern.

plain film radiography A term referring to a standard X-ray techniques such as those used at AUTOPSY.

plaintiff In a civil case, the person or entity that has taken the case to court; the party opposite the defendant(s).

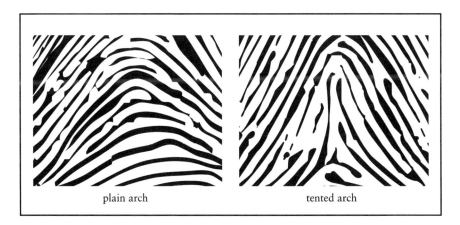

plain arch   tented arch

Arched fingerprint patterns. The plain arch on the right is smooth at the top; the tented arch has a more pronounced and pointed arch pattern.

plane polarized light *See* POLARIZED LIGHT.

plasma A term that has two forensic applications. In blood, the plasma is the liquid portion that is separated from the cellular component by centrifuging. If clotting elements such as fibrinogen are also removed, what remains is SERUM. Alternatively, a plasma is an extremely hot state of matter that is used as a sample introduction system for elemental analysis using INDUCTIVELY COUPLED PLASMA–ATOMIC EMISSION SPECTROSCOPY and INDUCTIVELY COUPLED PLASMA–MASS SPECTROMETRY. A plasma is a highly energetic state of matter (sometimes referred to as the fourth state of matter) that contains ions and free electrons.

plastic deformation/plastic indentation An alteration in the shape of a surface that, once created, does not disappear. Such deformations are called nonelastic; an example would be a fingerprint created by pressing into soft putty. The resulting pattern has three-dimensional quality, as opposed to a fingerprint on paper, which does not have depth.

plastic explosive *See* PLASTIQUE.

plastics A form of POLYMER that can be encountered as a form of physical, trace, or transfer evidence. Plastics and related compounds are found in storage bags, food containers, bottles, paints, fibers, and automobile parts, to name just a few applications. The term *plastic* generally is used to refer to synthetic polymers derived from petroleum products. Polyethylene and polypropylene are examples.

plastique The term *plastic explosive* or *plastique* usually refers to RDX (high-EXPLOSIVE) mixtures that are moldable; C4 is a complex that is 90 percent RDX.

platelets Part of the cellular component of blood, along with leukocytes (white blood cells) and erythrocytes (red blood cells). Platelets are critical for clot formation.

platinic bromide and chloride *See* GOLD BROMIDE AND GOLD CHLORIDE.

platters A key storage part of a hard disk drive of a computer. Hard drives comprise several platters that spin and align; data are typically written to both surfaces.

PNNL The Pacific Northwest National Laboratory, part of the United States Department of Energy (DOE) laboratory system. It is located in Hanford, Washington.

P/M ratio The ratio of a body fluid such as blood to the original ingested substance (parent, P) to a metabolite (M). Given that METABOLISM takes place in stages, this ratio can be used to estimate the time that has passed since the substance was ingested.

poaching The illegal killing or taking of fish and animals; the crime targeted by WILDLIFE FORENSICS.

Poincaré index A criterion used as part of automatic fingerprint classification programs such as AFIS.

point comparison *See* RIDGE CHARACTERISTICS.

point of convergence In BLOODSTAIN PATTERNS, a point (or more correctly, a region in space) where measurements of angles and elevation converge. This area in space is where the blood originated. Elastic strings or computer programs are used to locate this area.

point of origin A term that can apply to gunshots or to arson, in which it is the point as which a fire started. It is also occasionally used in bloodstain pattern analysis, referring to the POINT OF CONVERGENCE.

poison Any substance capable of causing a toxic (harmful) response in an organism. An oft-repeated phrase in TOXICOLOGY is "The dose makes the poison,"

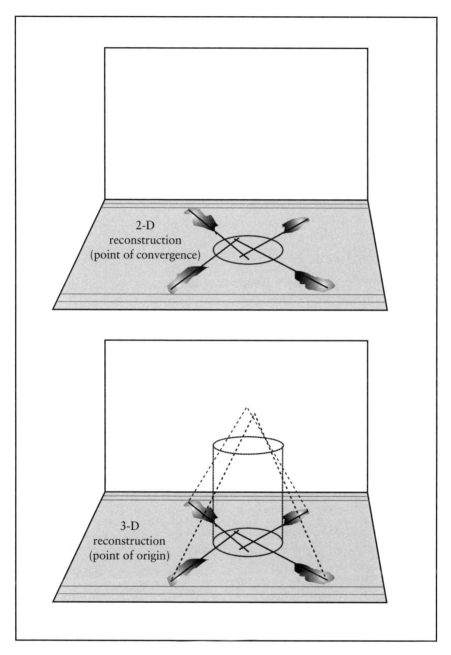

Determination of the point of origin based on bloodstain patterns. The term *point* is somewhat misleading; what actually results is an area in space from which the blood originated.

which means that any substance can cause harm, depending on how much is ingested. Even such familiar substances as water and table salt can cause death if sufficient quantities are ingested over a short period. The toxicity of a poison is measured by a

polar coordinates

quantity called the LD_{50}, *lethal dose 50*. The LD_{50} of a substance is that dose (based on body weight) that results in death for half of an experimental population such as laboratory rats or mice. For any given individual, the fatal dose of a poison depends on general health, age, weight, and a number of other variables.

Biological poisons, broadly defined, are those obtained from living organisms such as bacteria or plants. Hemlock, strychnine, venom from snakes, and botulinum toxin are examples. Inorganic poisons include hydrochloric acid (HCl), CYANIDE (both salts and gaseous HCN), lye (sodium hydroxide, NaOH), ammonia, MERCURY, lead, thallium, and ARSENIC, which is perhaps the most notorious poison of all. The term *organic poisons* usually refers to ORGANIC COMPOUNDS such as organic insecticides and pesticides such as DDT, toluene and other petroleum distillates, and many drugs. Methanol (wood alcohol) poisoning can occur with homemade or adulterated liquors in which the methanol is substituted for ethanol. Of the gaseous poisons, CARBON MONOXIDE, HCN, and forms of arsenic are the most familiar.

polar coordinates An alternative set of coordinates that can be used to document crime scenes in addition to traditional *x-y-z* coordinates.

polarimetry *See* POLARIZED LIGHT.

polarized light (plane polarized light) Light that has been filtered such that the waves of ELECTROMAGNETIC RADIATION are oscillating in only one direction. For this reason, the term *plane polarized light* is often used. One familiar application is in sunglasses; if the lenses have been polarized (have a Polaroid coating), the glare from reflections is eliminated. Polarization of light is exploited in forensic science in POLARIZED LIGHT MICROSCOPY and in DRUG ANALYSIS, principally for the analysis of COCAINE. Cocaine that is obtained from extraction of the coca plant rotates plane polarized light to the left (levorotatory), whereas synthetic cocaine rotates it to the right (dextrorotatory).

The two forms of cocaine are referred to as *l*-cocaine and *d*-cocaine to distinguish them. The instrument used to determine this is called a polarimeter and the technique is referred to as polarimetry. Essentially, a polarimeter consists of a light source, a polarizing filter, the sample chamber, and a viewer that measures the direction and angle of rotation of the plane polarized light that has passed through the sample.

polarized light microscopy (PLM) A type of microscopy originally used in geology for the analysis and identification of minerals. Polarizing light microscopes are sometimes still referred to as petrographic microscopes for this reason. The design of a polarizing light microscope is similar to that of a COMPOUND MICROSCOPE with the addition of a few key components. The light from below the sample first passes through a filter called the polarizer, which converts the beam to plane POLARIZED LIGHT. This passes through the sample and then through a second polarizing filter called the analyzer. Initially, these two filters are set such that the light passing through the analyzer is polarized in a plane that is perpendicular to the light that can pass through the analyzer. As a result, with no sample in the light path, the view appears completely dark, since the analyzer blocks all the polarized light from the polarizer. However, when a sample is placed on the stage, the oscillations of light that emerge from it are no longer polarized in one direction. Thus, some of this emerging light is able to pass through the analyzer, where the beams can interfere with each other. The result is a viewable image (magnified) with distinctive contrast, which may be highly colored.

PLM is particularly valuable for studying the optical properties of materials that are ANISOTROPIC. A material that is *isotropic* for a given optical characteristic has the same value of that characteristic regardless of the direction of the light. Solid materials that are made up of molecules that are randomly placed or molecules that are not symmetric are isotropic. Thus, a material that is isotropic has the same

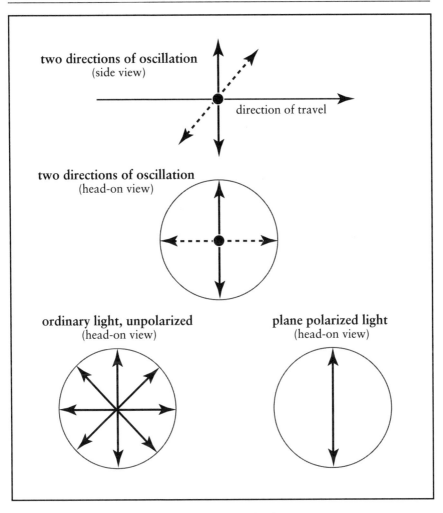

two directions of oscillation
(side view)

direction of travel

two directions of oscillation
(head-on view)

ordinary light, unpolarized
(head-on view)

plane polarized light
(head-on view)

How plane polarized light is created. See the text for details.

refractive index regardless of the direction of the light. In contrast, *antistropic* materials display different properties depending on the direction of propagation of the light. The difference in the refractive index of anisotropic materials is the BIREFRINGENCE, which can be determined quantitatively and used to identify materials.

polar solvents Solvents such as water or ethanol in which the molecules have an uneven distribution of electrons (dipole moment). Polar solvents can be used to dissolve some ionic compounds and other polar solvents, but not nonpolar solvents. This rule of "like dissolves like" explains why oil, a nonpolar material, does not dissolve in water.

Police Scientific Development Branch (PSDB) A division within the Home Office of the United Kingdom that sponsors research in forensic science within the United Kingdom.

pollen A very fine dust produced by plants containing microspores; the male component of reproduction. The study of pollen is called palynology.

polyacrylamide A gellike polymer medium used in electrophoresis.

polyatomic *See* ION.

polyester A generic name for a large class of synthetic fibers in which the monomer units are chemically bonded by ester bonds involving oxygen atoms. Polyesters are versatile and thus widespread; they are also used for insulation. Trade names for polyesters include Ceylon, Dacron, Kodel, and Mylar.

polygon method *See* PINCUSHION METHOD.

polygraph Informally called a lie detector, the polygraph is an instrument that consists of several sensors that respond to selected physiological functions. A polygraph does not detect truthfulness or lying per se; the examiner studies the chart of data and makes a determination of deceptiveness. The functions monitored are the heart rate, blood pressure, respiration rate, respiration volume, and galvanic skin response. The last is a measure of how well the skin conducts electric current. The controversy that always has swirled around the device is not related to the reliability of the sensors—there is no question that these functions can be accurately measured and recorded by the instrument—but to the question of whether the pattern of the readings can be directly and unequivocally related to truth or deception.

polymarker Generally, a kit for DNA TYPING that allows for the typing of more than one locus at a time, also known as multiplexing. The term is usually associated with the AmpliType PM+DQA1 kit, which allows typing of six loci. Although once commonly used, it is now being replaced with SHORT TANDEM REPEAT (STR) system typing.

polymerase An enzyme, such as that used in DNA TYPING, that catalyzes the polymerization of smaller units into larger ones. In the polymerase chain reaction (PCR) method of typing SHORT TANDEM REPEATS, DNA polymerase catalyzes the linkage of complementary bases to the half-strand of DNA to be copied.

polymerase chain reaction *See* PCR.

polymers Large molecules (macromolecules) that are created by linking together tens, hundreds, or thousands of subunits called monomers joined by strong chemical bonds. Polymers are common in nature and include PROTEINS (polymers of AMINO ACIDS) and carbohydrates (polymers of sugars). Cellulose is a glucose polymer that makes up about 90 percent of the weight of cotton, for example. Synthetic polymers, which were first produced in the mid-1800s, included cellulose nitrate and its derivative, celluloid, both of which are readily combustible and thus of limited commercial use. Bakelite, a rigid plastic polymer, was synthesized in 1907 by mixing together urea (the pungent compound that gives urine its characteristic odor) and formaldehyde. NYLON was introduced in 1939, to be followed by an explosion in the development and utilization of synthetic polymers. In forensic science, polymers are encountered as FIBER evidence (nylon, rayon, polyesters, and so on), in BUILDING MATERIALS (polyvinyl chloride [PVC] pipe), paints, plastics, ropes, and many other forms.

polymorphic/polymorphism Literally "many forms." In forensic serology and biology, this term is used to describe variants in blood that exist in different forms that are determined genetically. These variants include the ABO BLOOD GROUP SYSTEM, red cell ISOENZYME systems such as phosphoglucomutase (PGM), and the loci used in DNA TYPING.

pooled data Separate data sets or databases that are combined into one larger data set.

population genetics and databases The information and techniques that are used to estimate the frequency of types (typically blood or DNA types) in a population. For example, the B blood type of the ABO BLOOD GROUP SYSTEM occurs in approximately 3 percent of the population. Obviously not every person in the world has been typed, so this value is an estimate based on a database of people already typed and on statistical and genetic calculations.

poroscopy In FINGERPRINTS, the study and evaluation of the positions of pores within the print.

porous surface A surface that absorbs moisture or other liquids. Paper and wood are porous surfaces, whereas material such as tile and glass is nonporous. The nature of the surface is important when considering, for example, how to visualize FINGERPRINTS or how to interpret BLOOD-STAIN PATTERNS.

portrait parle See BERTILLONAGE.

postal inspection service See UNITED STATES POSTAL INSPECTORS SERVICE.

postcoital interval The amount of time that has passed since sexual intercourse or activity took place. Forensic estimates of this interval, also called time since intercourse (TSI), are usually based on the survival of sperm cells in the body cavity into which they were deposited and on the levels of acid phosphatase.

postmortem Occurring after the time of death, as in the postmortem interval or time since death.

postmortem examination See AUTOPSY.

postmortem interval (PMI) The time that has elapsed since death occurred. Contrary to popular belief, it is rarely possible to assign an exact time of death, and reliable estimates of the PMI require the use of multiple tools as opposed to one single method. Typically,

the MEDICAL EXAMINER is tasked with determining the PMI, but he or she is often aided by other forensic professionals, particularly in cases in which death occurred weeks or months before the body was discovered. For forensic purposes, the death event is divided into two processes, brain death and cell death (autolysis); the time of death generally refers to brain death. Techniques and processes used to estimate PMI include ALGOR MORTIS (rate of cooling), stiffening (RIGOR MORTIS), LIVOR MORTIS (pooling of blood at lowest points), DECOMPOSITION stage, and stomach contents.

postmortem submergence interval (PMSI) When a body is found in water, the amount of time that the body has been in the water. The PMSI is not necessarily the same as the POSTMORTEM INTERVAL (PMI), since a person may have died and later the body placed in water.

postural asphyxia Death by asphyxia (lack of oxygen) caused by the position of the body or compression of the chest and the resulting inability to expand it to fill the lungs with air. A person trapped in a crushed car may die of this.

potassium chlorate (KClO$_4$) An oxidant used in some types of incendiary devices or explosives.

powder See GUNPOWDER.

power of discrimination (P$_d$) See DISCRIMINATION INDEX.

precipitate A solid that forms in a solution as a result of a chemical or immunological reaction.

precipitin tests Immunological tests in which a positive result is evidenced by the formation of a precipitate or solid (also called an immunoprecipitate) that can be easily seen. In forensic applications the solid typically is the result of an immunological reaction between ANTIGENS found in a sample and ANTIBODIES contained in purified antisera applied in the test. Diffusion tests for species are an example of precipitin tests.

precision Degree of reproducibility. This term is commonly misused to mean ACCURACY. Accurate results are close to the true value; precise results are those that can be reproduced when the test or analysis is repeated.

precursors In the forensic context, the ingredients necessary to synthesize illegal or illicit drugs. For example, PHENYL-2-PROPANONE was once a common ingredient in the clandestine manufacture of METHAMPHETAMINE. However, once access to this substance was limited, this synthesis was replaced with others using different and more readily obtainable precursors. Similarly, MORPHINE is a precursor to HEROIN.

predator drugs *See* DATE RAPE DRUGS.

preparative chromatography A form of HIGH-PRESSURE LIQUID CHROMATOGRAPHY in which the goal is to clean or purify a sample for further analysis.

preponderance of evidence In civil trials, the criteria used to settle the case in favor of the defendant(s) or the plaintiff(s). It means that the majority of the evidence (however slim) supports the decision reached, but it is a more lenient standard than the "beyond reasonable doubt" standard used in criminal cases.

prescription drugs A common type of evidence encountered by forensic CHEMISTS, prescription drugs can be obtained fraudulently or otherwise diverted for illegal purposes. For the analysis of prescription medication, often the first task is a visual examination of the medication using the *PHYSICIAN'S DESK REFERENCE*. This is followed by a confirmatory test such as GAS CHROMATOGRAPHY/MASS SPECTROMETRY or INFRARED SPECTROSCOPY.

preservation of evidence A process that starts with collection of PHYSICAL EVIDENCE that encompasses proper packaging, storage, and documentation. If evidence is collected at a scene, the first steps in preservation are photography and documentation. Once this stage is complete, evidence is collected by using the method suitable to the specific type and a CHAIN OF CUSTODY document is started. During an analysis, all efforts should be made to preserve at least half of the sample for additional analysis by another forensic scientist.

pressure distortions In FINGERPRINTS, alterations of the pattern due to pressure applied to the finger or to curved or otherwise distorted surfaces on which the print is deposited.

presumptive tests Preliminary chemical tests widely used in the analysis of blood, drugs, GUNSHOT RESIDUE (GSR), and EXPLOSIVES. A presumptive test does not provide definitive identification; rather, it yields information useful for directing further analysis. Since most of these tests involve adding of a chemical reagent and looking for a resulting color change, these tests are sometimes referred to as color tests. Other terms used are *screening tests* and *spot tests*. Presumptive tests are subject to both FALSE POSITIVE and FALSE NEGATIVE results, which tend to be limited and recognized.

Tests for blood are based on the ability of the heme group in hemoglobin to act as a catalyst in chemical reactions that involve a color change. This ability is referred to as peroxidaselike activity in reference to biological enzymes that react with peroxide groups such as found in hydrogen peroxide (H_2O_2). The KASTEL MEYER test is widely used in this role. The LUMINOL test is unique in that the reaction does not cause a color change but rather results in the emission of light. Although sensitivities of these tests vary, usually it is possible to detect blood that has been diluted hundreds or thousands of times.

Perhaps in no other forensic area are preliminary color tests used more than in drug analysis. These tests are designed to narrow the possibilities when an unknown substance is delivered to the laboratory for identification. The other large group of screening tests targets the components of GSR and EXPLOSIVES, which share many common elements. The GRIESS TEST and DIPHENYLAMINE TEST react with nitrate and nitrite ions (NO_3^- and NO_2^-), found in both.

primary crime scene The location of a crime, such as a room where a murder occurred. This is in contrast to a secondary crime scene, which would be created if the body were dumped in a different location from the one where the killing occurred.

primary dentition Also called baby teeth, the 20 teeth that make up the dentition of a child or young adult and are later replaced by PERMANENT DENTITION. (See figure on page 204.)

primary explosive *See* HIGH EXPLOSIVE.

primary structure In proteins, the sequence of amino acids and the first of four levels of protein structure. The sequence of amino acids dictates the secondary structure, the three-dimensional twisting and orientations that occur as a result of the interactions among amino acids that are in close proximity. Tertiary structure is the shape that results from interactions of more distant amino acids, and quaternary structure refers to the arrangement of the subunits of a protein in space.

primary transfer A direct transfer or the first transfer that occurs. For example, if a person owns a dog, the dog sheds hairs onto furniture, and this is a primary transfer. If people sit on the furniture and as a result, dog hairs are transferred to their clothing, this is an example of a secondary transfer.

primer *See* PCR PRIMER.

primers (paint) Material that is applied as paint to surfaces such as bare metal in a car body or new plasterboard (Sheetrock)

in a wall before application of paint or other coatings. The primer can be used to seal a surface or otherwise treat it to ensure proper adhesion of subsequent layers.

primer/primer residue (firearms) A primary EXPLOSIVE that is used to ignite a PROPELLANT, which burns and creates gases that force a bullet forward. Primers consist of a shock-sensitive explosive (typically lead styphnate, also called lead trinitroresorcinate), an oxidizer (barium nitrate, $BaNO_3$), and a fuel such as antimony sulfide (Sb_2S_3). This is the same combination of ingredients used in COMBUSTION, and the ignition of a primer produces an intense flame that is directed through vents into the chamber of the CARTRIDGE CASE, where the propellant is stored. This flash ignites the propellant and causes the acceleration of the bullet down the barrel of the weapon. In addition to the initiator, fuel, and oxidizer, primers can contain many other ingredients, including sensitizers (that make the initiator more sensitive to shock), binders, and traces of other explosive materials. The compounds used in primers are important components of GUNSHOT RESIDUE. The two types of primers in use are CENTERFIRE and RIMFIRE (See figure on page 205).

principal focus With a LENS, such as used with a microscope, a term used to describe its behavior. If an object is placed an infinite distance away from a lens, the image appears at the principal focus point. Conversely, if an object is placed at the principal focus point, its image forms an infinite distance away from the lens. The symbol *f* is used to indicate the location of each principal focus, one on each side of the lens.

probability The likelihood of the occurrence of an event. In forensic science, the term is most often used in association with DNA types and their probability of occurrence in the population.

probability of discrimination (P_d) *See* DISCRIMINATION INDEX.

procaine (Novocain) A local anesthetic. Procaine hydrochloride (also known

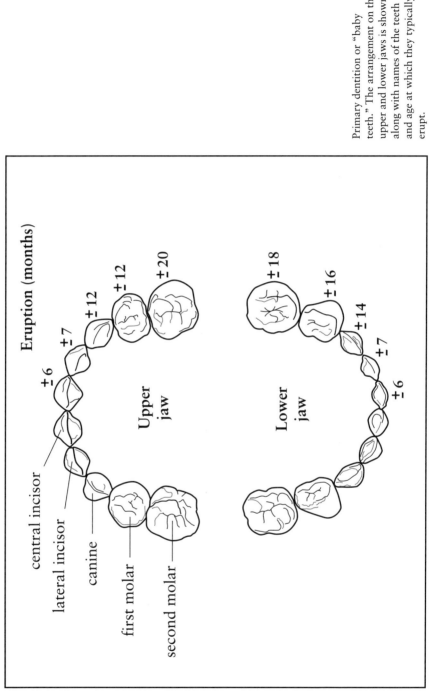

Eruption (months)

Upper jaw

Lower jaw

central incisor ±6
lateral incisor ±7
canine ±12
first molar ±12
second molar ±20

±18
±16
±14
±7
±6

Primary dentition or "baby teeth." The arrangement on the upper and lower jaws is shown, along with names of the teeth and age at which they typically erupt.

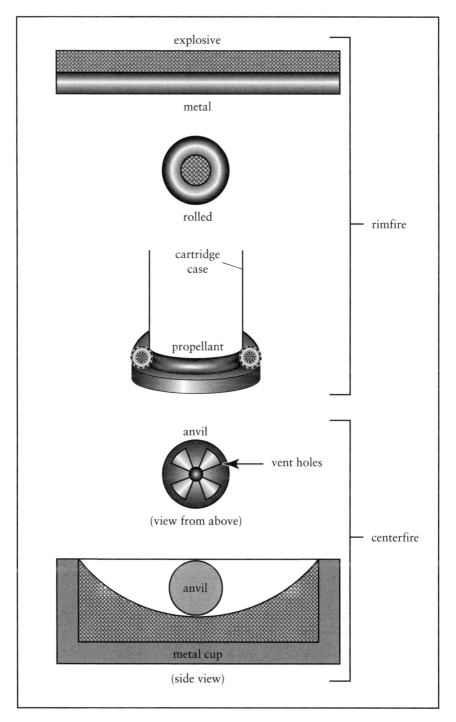

explosive

metal

rolled

rimfire

cartridge
case

propellant

anvil

vent holes

(view from above)

centerfire

anvil

metal cup

(side view)

The two different types of primers used in firearms, both shown in top and side views. The rimfire cartridge is used in small-caliber rounds such as .22 caliber; the centerfire is used in larger calibers.

as Novocain or allocaine) is a white powder that is sometimes used as a cutting agent (diluent) for cocaine and produces localized numbing.

product liability The legal procedures involved when a product causes, or is thought to have caused, injury. The injury may be physical (personal injury) or financial and can involve an individual or an organization. Product liability cases may involve a crime or criminal conduct, but most are civil cases, and many evolve into class-action lawsuits in which many complaints are combined into one. Perhaps the best known product liability cases of the last few years have involved the tobacco companies and payments for deaths and injuries caused by their products.

product tampering Adulteration or alteration of a product such as medicine or food. Product tampering attracted national attention in 1982, when seven people in Chicago died when cyanide powder was placed into Tylenol capsules. As a result of this and other incidents at the time, product packaging was changed dramatically to include safety seals and sealed packages. After September 11, 2001, the U.S. government became increasingly concerned about product tampering as a form of terrorist act, particularly in the area of adulteration of the food supply. The federal FOOD AND DRUG ADMINISTRATION is the principal responsible government agency and maintains a forensic facility in Cincinnati, Ohio.

professional associations See APPENDIX I.

proficiency testing The testing of laboratory analysts as part of obtaining or maintaining certification of a professional association. For example, to obtain certification from the AMERICAN BOARD OF CRIMINALISTICS, a person must complete written tests as well as laboratory proficiency testing in the area of specialization. Proficiency testing is part of QUALITY ASSURANCE/QUALITY CONTROL.

profiling Behavioral evidence (broadly defined) applied to the study of criminals at crimes and crime scenes. Many people are familiar through fiction with the FBI's Behavioral Science Unit, which is part of the National Center for the Analysis of Violent Crime (NCAVC) at the FBI Academy in Quantico, Virginia, formed in the early 1970s. Although there are no standards for profiling, the process normally includes examination and review of physical evidence and crime scenes as well as psychological evaluation of victims (VICTIMOLOGY). As in other forensic disciplines, a large part of profiling is seeking of patterns such as consistent elements of crimes, the MODUS OPERANDI, and any SIGNATURE BEHAVIOR shown during the commission of a crime.

projected blood pattern A type of bloodstain pattern created when blood is ejected with force, such as when a blow is struck.

propellant A low-EXPLOSIVE material used in AMMUNITION. Known informally as GUNPOWDER, propellant is placed into a cartridge case packed between the PRIMER and the projectile. BLACK POWDER (a mixture of charcoal, potassium nitrate, and sulfur) was used from ancient times until about the mid-1800s, when it was replaced by SMOKELESS POWDER. Modern smokeless powder propellants are either *single base,* consisting of NITROCELLULOSE or *double base,* consisting of cellulose nitrate and NITROGLYCERIN. As shown in the illustration, the impact of the firing pin on the primer causes a flash, which ignites the propellant. The rapid COMBUSTION that follows creates large volumes of hot expanding gas, resulting in high pressures that are confined within the barrel, forcing the bullet forward at high speeds.

propoxyphene See DARVON.

proteinase An enzyme that catalyzes the degradation of proteins by breaking the polypeptide bonds that exist between amino acids.

proteins Large complex molecules (MACROMOLECULES) formed by linking tens to thousands of AMINO ACID subunits. Proteins are biological POLYMERS and are also referred to as polypeptides. Because of the myriad attractions and repulsions of the subunits, proteins twist and fold into complex three-dimensional structures that are critical in determining their functions. The primary structure of any protein is the sequence of amino acids that constitute it; the secondary structure is defined by repeating subunits or sequences of amino acids along the chain. These structures can in turn interact with each other, forming clumps or globular shapes that define the tertiary structure. When several of these combine to form a functional protein, the quaternary structure results.

prosthesis A manufactured device used to replace something in the body. An artificial knee or limb is prosthesis, as are many dental implants such as bridges and crowns. Such implements can be useful in identification of a body, particularly in cases of mass disaster or other situations in which a body has been badly damaged or decomposed.

protofibril A microscopic substructure of HAIR composed of protein bundles called fibrils. Specifically, a protofibril consists of three coiled strands of a protein, and a group of protofibrils makes up a microfibril.

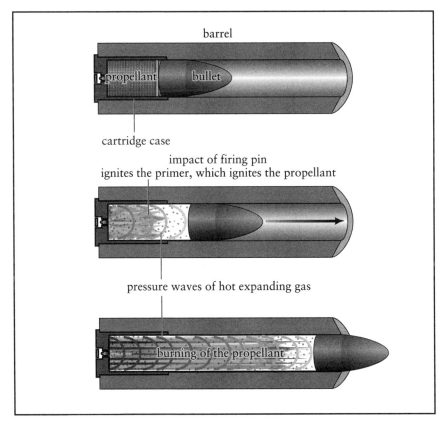

Exploitation of the burning of propellant in firearms to accelerate the projectile.

protonation The association or addition of a hydrogen cation (H^+) with another chemical species. The term is often used in association with acid–base interactions. For example, H_3O^+ is the protonated form of water.

proximal/proximate A term that refers to a point or structure nearest to the center of an attachment or to something nearest the center of a larger body or structure.

proximate cause The event or action nearest to the event in question. This phrase is most often used in relation to a death investigation, in which the proximate cause is the physiological event that precipitated the death. For example, if a stabbing victim arrives at the hospital and dies of shock, the shock is the proximate cause and the stabbing is the legal cause.

pseudogene A region of DNA that does not code for a protein (noncoding DNA) and is an inactive copy of an active gene.

pseudoscience Theories, ideas, or explanations that are represented as scientific but that are not derived from science or the scientific method. Pseudoscience often springs from claims or folk wisdom or selective reading without independent data collection or validation. Scientific statements are specific and well defined, whereas pseudoscience is vague and variable. One of the key differences between science and pseudoscience is that a scientific statement or theory is stated in such a way as to be FALSIFIABLE. In contrast, pseudoscientific statements are usually not falsifiable by means of objective experimental or observational evidence. Pseudoscience provides no room for challenge and tends to dismiss contradictory evidence or decide selectively which evidence to accept. Thus, pseudoscience is usually nothing more than a claim, belief, or opinion that is falsely presented as a valid scientific theory or fact.

pseudostationary phase *See* MICELLAR CAPILLARY ELECTROPHORESIS.

psilocyn and psilocybin Hallucinogens in mushrooms found principally in Mexico. In this respect, these compounds are similar to MESCALINE, which is obtained from peyote cactus. Psilocybin can also be synthesized from psilocyn, but most drug seizures are of the naturally occurring materials. In mushrooms, there is more psilocybin than psilocyn, but the psilocyn is about twice as potent. The drugs are taken by chewing the dried mushrooms, and the effects start within an hour and can last for four hours or more. These compounds are listed on Schedule II of the CONTROLLED SUBSTANCES ACT.

psychiatry, forensic, and psychology, forensic Disciplines involved with what is broadly categorized as behavioral evidence. Unlike physical evidence, behavioral evidence is intangible and subject to different perspectives and interpretations. Examples of the use of behavioral evidence include determining criminal responsibility and the validity of an insanity plea, assessing a person's danger to society and likelihood to reoffend, treating both victims and offenders, studying violent behavior, and evaluating eyewitness testimony. Behavioral evidence is used in civil as well as criminal cases in areas such as child custody and conflict resolution. Psychiatrists are physicians with an M.D. and additional training in behavior, neurology, and pharmacology and can prescribe medications. Psychologists have a college degree (usually an M.S. or Ph.D.) and do not have medical training. However, both are active in the justice system and perform many of the same functions. Related to these are neuropsychiatry, which blends neurology with psychiatry, and neuropsychology, which focuses on behavior as an indicator of brain and nervous system function.

psychological autopsy An informal term that usually refers to an evaluation of a person's psychological state and behavior as related to criminal activity. The subject of such an "autopsy" need not be dead; typically the usage refers to an analysis of past activities regardless of the state of the subject.

psychology, forensic *See* PSYCHIATRY, FORENSIC, AND PSYCHOLOGY, FORENSIC; PROFILING.

pubic hair combing A procedure used to collect evidence after a sexual assault. The goal is to collect hairs and fibers from the assailant.

pulp In a tooth, the central section, containing nerves, connective tissue, and blood vessels.

pupa/pupea An intermediate stage in an insect's metamorphosis important in forensic ENTOMOLOGY. In the blowfly, the pupa stage follows the egg and larval stages (informally called maggots) and is characterized by the formation of a hard exoskeleton around the larva. The adult fly eventually emerges from the pupa. The time required for this development is one tool used to estimate the POSTMORTEM INTERVAL.

purge and trap An analytical technique used in arson analysis and occasionally in toxicology. Generally, an inert gas such as helium is pumped through a sealed container in which the sample has been placed. The container may be at ambient temperature or heated. After passing through the sample, the gas stream, now containing volatile materials from the sample, is directed over trapping material such as solids containing charcoal. This trapping stage concentrates volatiles from the sample onto the trap material. At the end of the gas purging, the trap is flash heated to release the volatiles, which are then introduced into a gas chromatograph for analysis.

purine bases In DNA, the bases ADENINE (A) and GUANINE (G) are classified as purine bases. The classification is based on the skeleton of the molecules, in this case a double-ring structure. The other two bases in DNA (THYMINE ([T]) and CYTOSINE ([C])) are built in a single ring and are classified as PYRIMIDINE BASES.

putative suspect The person thought or alleged to have committed a crime.

putrefaction A stage in the DECOMPOSITION process. Putrefaction begins with a greening of the skin along with a surge in MICROBIAL DEGRADATION leading to bloating and purging of gases and fluids from the body. Putrefaction is marked by the characteristic foul odors of decomposition. The stage ends when all soft tissue has disappeared.

Pyrex A type of glass used extensively in laboratories because of its capacity to be rapidly heated and cooled without breaking.

pyridine An odoriferous solvent used for the analysis of inks, among other applications.

pyrimidine bases In DNA, the bases THYMINE (T) and CYTOSINE (C) are classified as pyrimidine bases. The classification is based on the skeleton of the molecules, in this case a single-ring structure containing five carbons and one nitrogen. The other two bases in DNA (ADENINE ([A]) and GUANINE ([G])) have two rings and are classified as PURINE BASES.

pyrocotton A propellant powder (single base) used in ammunition that contains 12.6 percent nitrogen.

pyrogram *See* PYROLYSIS/PYROLYSIS GAS CHROMATOGRAPHY.

pyrolysis/pyrolysis gas chromatography The word *pyrolysis,* which means "fire cutting," refers to the method in which a solid sample such as a fiber can be introduced into an instrument, usually a gas chromatograph. Normally, samples are dissolved in a solvent and injected into the instrument. In pyrolysis, the sample is rapidly heated, producing decomposition and the release of gaseous materials that can be directly introduced into the carrier gas stream of the gas chromatograph. The resulting output is sometimes called a pyrogram. Pyrolysis is not as reproducible as other methods of GAS CHROMATOGRAPHY and thus is not used often to obtain quantitative data on amounts and concentrations of individual components. However, the technique can be valuable in obtaining a qualitative "fingerprint" of a sample that can be used to compare separate fiber or paint samples to determine whether they are consistent or different.

pyruvate An organic molecule important in biochemical processes that can be found in latent fingerprint residues.

Quaalude *See* METHAQUALONE.

quadrupole mass spectrometer The most common type of MASS SPECTROMETER used as a detector for GAS CHROMATOGRAPHY. After exiting the chromatographic column, the sample molecule is introduced into the mass spectrometer, where it passes by a plate that has a positive charge. Ions (M) and fragments (F) are created by collision with a stream of electrons created by a tiny filament. The electrons are drawn toward the target that is positively charged because unlike charges attract. Conversely, since like charges repel, the positive ions are driven into a stack of focusing lenses by the positive charge on the repeller. The ions are focused into a tight beam that enters the quadrupole area, which consists of four metallic rods. Manipulation of the electrical fields allows the ions to be filters

so that only one mass is detected at any given time. The emerging ions are directed into an electron multiplier that amplifies the signal sent to the detector. *See also* MASS SPECTROMETRY.

qualifications A combination of education, training, and experience that allows a forensic scientist to testify as an expert witness in an area of expertise. The experts qualifications are introduced and reviewed during the process of VOIR DIRE, during which both sides have the opportunity to ask the expert questions and to judge whether he or she has qualifications that allow him or her to be accepted by the court as a reliable expert in some aspect of the case at hand. Qualifications are roughly divided into the categories of education (academic credentials), experience, on-the-job and continuing education,

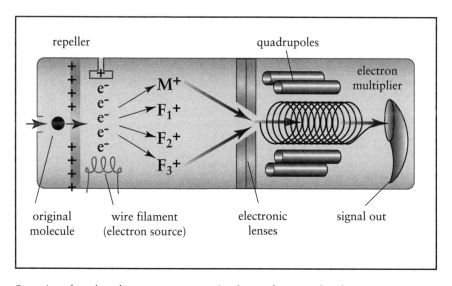

Operation of quadrupole mass spectrometer. See the text for a complete description.

membership in professional associations, publication record, and applicable certification and licenses. Although both the prosecution and the defense can ask questions during qualifications procedures, it is the judge who makes the final decision as to whether a person qualifies as an expert.

qualitative analysis and qualitative evidence Qualitative analysis involves observations and testing that identify the components or constituents of a sample. However, qualitative analysis does not determine the quantity of each component that is present. Similarly, qualitative evidence is evidence that is analyzed without producing quantitative data. Examples of qualitative evidence are those results obtained from the analysis of hairs, fibers, bullets, and cartridge casings, all of which are analyzed by microscopy. In many forensic applications, qualitative analysis is sufficient or serves as a starting point for further comparison and attempts at INDIVIDUALIZATION and linkage to a COMMON SOURCE. In other cases, qualitative analysis is the required starting point for a QUANTITATIVE ANALYSIS, since it is necessary to know what is present to determine how much of it is present.

quality assurance/quality control (QA/QC) An inclusive term for procedures and protocols used in forensic analysis to ensure that any data or results produced are accurate, reliable, and trustworthy. Often, the term *quality assurance* is used to describe the foundation established to ensure acceptable laboratory performance. In this usage, *QA* refers, for example, to training, documentation, laboratory policies and procedures, and METHOD VALIDATION. Quality control reflects all of the actual practices used to ensure the trustworthiness of data such as CONTROL SAMPLES, instrument logs, use of good laboratory practices (GLP), and proper documentation. Guidelines for QA/QC protocols are published by entities such as the National Institute of Standards and Technology (NIST) and the AMERICAN SOCIETY OF CRIME LABORATORY DIRECTORS (ASCLD). Many professional soci-

eties also publish or recommend procedures and guidelines that could be considered part of QA/QC.

quantitative analysis Testing that leads to determination of the amount of a given component or components in a sample. This is in contrast to QUALITATIVE ANALYSIS, in which only the identity of components is necessary, not amounts or concentrations. Quantitative analysis is most often an issue in TOXICOLOGY and DRUG ANALYSIS, in which the amount of an illegal substance present is critically important. For example, in the analysis of blood alcohol concentration the first stage is the definitive identification of ethanol (qualitative analysis), which is followed by an accurate determination of the quantity of alcohol present. A blood alcohol level below 0.08 percent in most states is not considered illegal whereas concentrations above 0.08 percent can result in a charge of driving under the influence of alcohol.

quantitative transfer A transfer of a sample from one container or vessel to another such that none is lost; a complete transfer.

quantum dot Also referred to as photoluminescent nanoparticles; used to visualize latent FINGERPRINTS. The types of quantum dots used for this purpose include CADMIUM SELENIDE and CADMIUM SULFIDE. A quantum dot is composed of a tiny (nano-) particle of a semiconductor material such as CdSe, CdS, and InAs that is luminescent when found in such small crystals. The nanoparticle is then coated or encapsulated with various materials that impart capabilities such as the ability to bind to certain surfaces or materials. Changing the size of the quantum dot changes its absorbance (excitation) and emission characteristics. This alteration increases their usefulness in cases when TIME-RESOLVED imaging is necessary or desirable for visualization of a fingerprint.

quartz wedge In microscopy, a tool that can be used to determine the BIREFRINGENCE of a fiber, in conjunction with a polarizing microscope and a MICHEL-LEVY CHART.

quasar light source   Another name for an ALTERNATE LIGHT SOURCE used in the visualization of latent FINGERPRINTS.

quaternary structure   The three-dimensional structure of a large protein molecule such as HEMOGLOBIN. It indicates the arrangement of the subunits of the protein in space and their contact with each other.

questioned documents (QD)   The specialty in forensic science that deals with suspicious, questioned, or damaged documents. Handwriting analysis is considered part of questioned document examination as well and should not be confused with GRAPHOLOGY. Analysis of questioned documents began to emerge as a forensic discipline around 1870, and an early practitioner was ALPHONSE BERTILLON. ALBERT S. OSBORN, who is considered to be a pioneer of document examination in the United States, wrote *Questioned Documents* in 1910, along with a later revision. These books are still cited in the field, and Osborn's sons continue to be active examiners. The principal professional associations for document examiners are the American Society of Forensic Document Examiners (ASQDE) and the Questioned Document section of the AMERICAN ACADEMY OF FORENSIC SCIENCES (AAFS). Together, the ASQDE, the AAFS, and the Canadian Society of Forensic Science formed the American Board of Forensic Document Examiners (ABFDE) in 1977 to provide professional certification of document examiners. In addition to handwriting analysis, QD includes any type of printed document such as copies, typed documents, faxes, and computer printouts, as well as the examination of inks.

questioned sample   The sample in question, the one that has been collected as evidence. For example, a questioned fingerprint would be one collected at a crime scene; it would be compared to standards or exemplars collected from suspects or other persons who may have deposited it.

Quincy   *Quincy M.E.* was a television show that ran from 1976 to 1983. Jack Klugman played the lead role, as a MEDICAL EXAMINER in the Los Angeles County Coroner's office.

quinine   An alkaloid plant extract that is occasionally encountered as a CUTTING AGENT in drugs such as HEROIN. It has a bitter taste and was used to treat malaria.

R

racemic mixture A sample of a single compound that contains a mix of STEREOISOMERS. A mixture of *d*-cocaine and *l*-cocaine is a racemic mixture.

racial determination The determination of racial origin of evidence such as HAIR or bones based on MORPHOLOGY, measurements, or other characteristics.

radial diffusion When a liquid sample is placed into a circular well that has been punched in a porous medium such as gel, this liquid diffuses outward into the gel in a roughly circular pattern called radial diffusion. Radial diffusion is exploited in some IMMUNOLOGICAL TESTS for species, which are also described as radial immunodiffusion. An example is the OUCHTERLONY TEST.

radial fractures The roughly linear fractures that form at an impact site in GLASS and move outward. These are accompanied by CONCENTRIC FRACTURES, which form circular shapes around the impact.

radial loop A loop pattern in FINGERPRINTS that opens toward the thumb.

radioimmunoassay (RIA) A type of IMMUNOASSAY in which the label attached to the drug is radioactive and is detected by the emission of radiation.

radiology, forensic (radiography, forensic) The application of medical and dental radiology (X-ray techniques) to forensic work. Forensic radiology is considered a branch of forensic medicine that is related to but separate from forensic PATHOLOGY. Although the most frequently used tool in

the field is the familiar X-ray film, all radiological techniques are used, including ultrasound, magnetic resonance imaging (MRI, based on NUCLEAR MAGNETIC RESONANCE [NMR]); fluorescent imaging, computed tomography (CT) scans, and nuclear imaging techniques in which radioactive materials are used.

RAM The abbreviation for the dye combination Rhodamine 6G, Ardrox, and MDB. RAM is used to visualize latent fingerprints.

Raman spectroscopy A technique related to INFRARED SPECTROMETRY (IR) that is used in forensic science, principally with MICROSCOPY, for DRUG ANALYSIS. Unlike traditional infrared spectroscopy, which relies on measuring the amount of infrared radiation that is absorbed, Raman spectroscopy is based on the quantity of light that is scattered. Since this scattering behavior is specific to a given molecule, Raman spectroscopy (as does IR spectroscopy) produces a unique spectrum for each compound. Thus, Raman spectroscopy can be used to identify molecules or to perform qualitative analysis of complex mixtures.

To be infrared-active, a compound must have a dipole moment, which in simple terms is an imbalance or unevenness in the electron cloud that surrounds the atoms of a molecule. However, for Raman activity, these bonds must be polarizable by incoming electromagnetic energy. Furthermore, the degree of polarizability must change as the bonds stretch and contract. Interestingly, this leads to a "mutual exclusion" property—a bond that is IR-active is Raman-inactive, and vice versa. Raman spectroscopy has been used for explosives and drugs as well as for mixtures.

random match (random man not excluded, RMNE) A measurement most often used in DNA TYPING (but applicable to other types of evidence) that indicates how likely a given combination of frequencies is to occur in a population. For a bloodstain typed for several different DNA loci, knowing how common that particular combination of types is important. In other words, if a "random man" were selected from the same population, how likely would it be that the man would have the same combination of types and thus not be excluded as a possible source? The value of this probability of a random match is determined by using POPULATION GENETICS and POPULATION FREQUENCIES. Assume that a stain is typed for the first four DNA loci of the CODIS system and that the combination of frequencies (in the U.S. Caucasian population) is 0.080, 0.068, 0.041, and 0.080. The likelihood of finding this combination in a random selection from that same population is expressed as the product of these four values, or 1.78×10^{-5}. This means that this combination of types would be found in about 18 people in 1 million. Another way to phrase it would be to say that in the same population, there would be 18 chances in 1 million that any one man (or person) selected at random would have exactly the same combination of types.

rape kit (sexual assault kit) One of several commercially available kits used to collect evidence in sexual assault cases. In rapes, physical evidence is crucial since there are rarely witnesses and trials often boil down to one person's testimony against another's. Thus, the evidence collected from the victim by using the sexual assault kit is crucial in determining the truth. Generally, sexual assault kits consist of a whole blood sample, swabs of any dried secretions, and swabs of the vagina, genital area, thighs, anus, and mouth. In addition, smears on slides are made from these swabs, and all must be air-dried. A vaginal rinse may also be collected. Any visible HAIRS and FIBERS are collected and the victim may be asked to undress over a large sheet of paper so that any other

TRACE EVIDENCE that falls off in the process can be collected and preserved. All clothing worn also becomes part of the physical evidence. *See also* PERK KIT.

Rayleigh scattering A form of scattering of light that occurs without a change in wavelength. If light of a wavelength of 483 nm is scattered by interaction with a sample, the Rayleigh scattering is also 483 nm. This phenomenon is exploited in RAMAN SPECTROSCOPY.

rayon One of the first synthetic fibers manufactured. Rayon is not based on synthetic polymers; rather, it is classified as a REGENERATED FIBER because it is made from cellulose, which is derived from plant material. The cellulose is obtained from a material such as cotton and then further processed to make the new fiber material.

RDX A high EXPLOSIVE that is a component of C4. The chemical name of the compound is cyclotrimethylenetrinitramine, and it is also known as cyclonite and hexogen. The origin of the abbreviation is unclear, variously reported as *"Royal Demolition Explosive"* or *"Research Department Explosive."* RDX is a secondary high explosive, meaning that it is shock-insensitive and must be detonated by a primary high explosive. It is the principal component (~90 percent) of C4, a military explosive that is moldable and is sometimes referred to generically as PLASTIC EXPLOSIVE or PLASTIQUE. (See figure on page 216.)

reaction order In a process or chemical reaction, the order relates to the kinetics or rate of a reaction. The rate of a first-order reaction depends solely on the concentration of one component. Radioactive decay and decomposition reactions are examples of first-order reactions. Similarly, the rate of a second-order reaction depends on the concentration of two reactants.

real image In MICROSCOPY, an image that forms on a screen or other surface as the result of the convergence of light rays. Lenses and lens combinations can be used

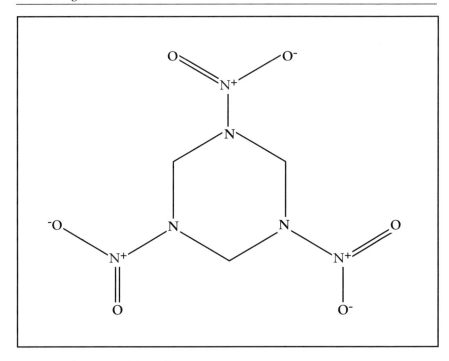

The chemical structure of the explosive RDX.

to form and manipulate a real image. *See also* VIRTUAL IMAGE.

reannealing   *See* HYBRIDIZATION.

receiver operating characteristics (ROC)   A mathematical relationship applied in forensic science, particularly in cases of computer-aided comparisons and identifications. A typical ROC application would be to plot the rate of FALSE POSITIVE findings obtained by a given technique on the *x*-axis and the rate of FALSE NEGATIVE findings for the technique on the *y*-axis.

recessive   A genetic allele or trait that is not expressed unless it is HOMOZYGOUS. For example, if there are two alleles of a gene and X is dominant and x is recessive, the x variant is expressed only when the person's genotype is xx. If it is Xx XX, the dominant form X is expressed. *See also* CODOMINANT.

recoil operation   *See* BLOWBACK.

recombinant DNA   A segment of DNA that is removed from one organism and inserted into the DNA of another. Recombinant DNA technology is widely used in biotechnology and is sometimes referred to as genetic engineering.

reconstruction   The use of physical evidence and, on occasion, other information to recreate the event sequence that produced it. It is analogous to an archaeologist's use of artifacts recovered from an excavation to study the lives and cultures of the people who used them. The difference is the time frame: forensic scientists recreate recent events.

reconstruction evidence   Evidence and findings that are gleaned from a crime scene RECONSTRUCTION.

recross   *See* REDIRECT.

recurving ridges Fingerprint ridge pattern in which a curve that approximates a circle is observed. An ARCH pattern is not recurving; a WHORL is.

redact A process used to prepare images for publication or viewing. This term is often used in forensic science also to refer to the process of digitally modifying crime scene photos such that they will be less inflammatory when presented to the jury but still accurately convey the scene.

redirect In courtroom testimony, a witness for the prosecution, for example, is first questioned by the prosecutor during the direct examination. The defense attorney then cross-examines the witness. This testimony can be followed by additional questions by the prosecutor in what is called the redirect. The defense can then have a recross, and so on.

Red lake C A fluorescent powder that has been used to visualize latent FINGERPRINTS.

redox An abbreviation for OXIDATION/REDUCTION, a type of chemical reaction in which electrons are transferred from the oxidized species to the reduced species.

reducing agent A chemical compound that promotes the reduction of another compound. In an OXIDATION/REDUCTION reaction, one compound is oxidized and one is reduced. The compound that is oxidized is the reducing agent.

reduction A chemical process that is the companion of oxidation. Such paired reactions are referred to as OXIDATION/REDUCTION or REDOX, and one cannot occur without the other, just as acid–base reactions must occur together. Reduction can be defined, depending on the situation, gain of electrons, the gain of hydrogen, or the loss of oxygen. If CO_2 is converted to CH_4, the carbon dioxide is reduced. *See also* REINSH TEST.

reference collections Large groups of specimens that are used for examination and identification of evidence. Examples of reference collections would be SOILS, PAINTS, HAIRS, FIBERS, GLASS, and POLLENS.

reference points In the examination and comparison of latent prints, a point or points selected on one print that serve as a reference point. The locations of other features are characterized in relation to the reference point.

reflected light/reflected illumination Lighting that is reflected off the surface of a specimen or subject. In microscopy, a stereomicroscope uses reflected illumination in which light from a lamp above the specimen reflects off it and is collected by the lens. This contrasts with TRANSMITTED LIGHT, which is used in a biological microscope.

reflection grating *See* DIFFRACTION GRATING.

refractive index (RI) A quantity (PHYSICAL PROPERTY) that measures the bending of light as it travels from one medium into another. Mathematically, the refractive index is defined by the following formula:

$$RI = \frac{velocity\ of\ light\ in\ a\ vacuum}{velocity\ of\ light\ in\ a\ medium} = \eta$$

The symbol η also stands for the refractive index. The speed of light in a vacuum is the maximum achievable, so the RI is a quantity that is greater than 1.00, and it represents the degree to which light bends when passing from one media to another. (See figure on page 218.)

regenerated fibers Fibers such as rayon and triacetate that are made by processing of cellulose derived from plants. Thus, although technically such fibers are synthetic, they are not manufactured from synthetic polymers such as polyesters and nylon.

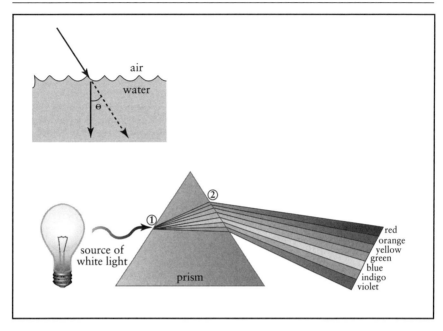

Refractive index. Note the bending of the light as it passes from air to water.

Reinsh test A presumptive test used to identify the presence of heavy metals such as arsenic in a sample of a body fluid. To perform the test, a strip or coil of copper is boiled in acid to clean the surface and then placed into the sample. Metals (arsenic, mercury, bismuth, and antimony) if present in high concentrations form a dark gray coating on the copper. This is an example of an OXIDATION/REDUCTION reaction.

relative error Generally, the error in an analysis relative to the mean. For example, if a cocaine sample has a true concentration of 100.0 units and an analysis yields a value of 90.0 units, the relative error is 10/100 or 0.1. This quantity is most often reported as a percentage, which would be 10 percent in this example.

relative retention time In a chromatographic analysis such as GAS CHROMATOGRAPHY or HIGH-PRESSURE LIQUID CHROMATOGRAPHY, the RETENTION TIME of the compound of interest calculated relative to the retention time of a standard compound.

repetitive DNA Repeated sequences of very similar or identical BASE PAIR (BP) SEQUENCES. An estimated 30 percent or more of human DNA consists of these repeats. These repetitive sequences are sometimes differentiated by length; SINES are short interspersed repeats 100–500 BP in length; LINES are long interspersed repeats that can be several thousand base pairs long. *Alu* repeats are abundant SINES. Repetitive DNA sequences are targeted in the SHORT TANDEM REPEAT (STR) DNA typing techniques.

representative sampling The process of obtaining samples for testing that accurately reflect the composition of the bulk material in question. This is an issue anytime evidence is collected and subject to forensic analysis. For example, an analyst who needs to compare a cotton fiber found at a crime scene with jeans recovered from a suspect must obtain a representative sample of fibers from the jeans since fibers vary in appearance, depending on where on the jeans they are collected.

request standards Samples of handwriting that are requested of a person for the purpose of QUESTIONED DOCUMENT examination. These samples are collected while the writer is being observed so there is no question as to their origin.

resins Most often, a term that refers to absorbent materials used in SOLID PHASE MICROEXTRACTION, a sampling protocol exploited in toxicology and arson investigation. Resins are selected on the basis of the compounds that are of interest.

resolution A term that has two uses in forensic science. First, it can refer to the resolution of a digital image, photo, copy, or printout in dots per inch (dpi). The greater the dpi number, the sharper the image. Second, resolution is the degree of separation that is achieved in a chromatographic separation as in GAS CHROMATOGRAPHY. The greater the separation between components achieved by the chromatographic column, the higher the resolution. For any two adjacent peaks in a CHROMATOGRAM, the resolution can be calculated by the following formula:

$$R = \frac{t_{r_2} - t_{r_1}}{\frac{1}{2}\left(W_2 - W_1\right)}$$

where tr_1 and tr_2 are the RETENTION TIMES of the two peaks and the W values are the widths. If $R = 1.5$, there is very little overlap (~0.1 percent or less) and the peaks are said to have baseline resolution.

restriction enzyme An ENZYME that is used to locate a specific segment of DNA and to break the DNA chain at that point. These enzymes are often informally referred to as molecular scissors and are used in restriction fragment length polymorphism (RFLP) DNA TYPING techniques.

restriction fragment length polymorphism *See* RFLP.

retention index A calculated quantity that is used to standardize RETENTION TIMES in relation to a common reference point. To calculate a retention index, the retention time of the compound in question is divided by that of the reference compound (often a straight chain hydrocarbon). The resulting retention index can be evaluated against tables of standard retention indices.

retention time (t_r) In a chromatographic analysis such as GAS CHROMATOGRAPHY or HIGH-PRESSURE LIQUID CHROMATOGRAPHY, the time at which a component of a mixture exits from the chromatographic column. A compound with a short retention time emerges from the column more quickly than a compound with a longer retention time.

reticular dermis One of two layers into which the structure of human skin can be divided. It is the lower layer and contains connective tissues.

reversed phase In HIGH-PRESSURE LIQUID CHROMATOGRAPHY, analytical conditions in which the stationary phase (the chromatographic column) is composed of nonpolar materials and the mobile phase (the solvent) is composed of a polar solvent or solvents.

revolver A type of handgun or pistol in which the cartridges are housed in a rotating cylinder. Revolvers can hold five to seven cartridges and include calibers from .22 to .45. The forensic examination of revolvers follows the same general protocols as those for any rifled weapon. *See also* AMMUNITION; BULLETS; CALIBER; CARTRIDGES; FIREARMS.

RFLP (restricted fragment length polymorphism) An approach to DNA TYPING largely but not completely displaced by the SHORT TANDEM REPEAT (STR) typing procedures that are included in the CODIS system. As in STR typing, RFLP typing targets TANDEM REPEATS in the DNA; however, they are longer than those used in STR.

Rh factors (Rh blood group) A blood group system that, like the ABO BLOOD GROUP SYSTEM, is based on antigens located on the surface of the red blood cell. Rh type is best known for potential effects on babies. To simplify, if a mother who is Rh-positive (Rh⁺) gives birth to a child who is Rh-negative (Rh⁻), her body produces antibodies to the Rh⁻ factors. If she conceives another child who is also Rh⁻, severe complications are possible. However, Rh was rarely used in forensic applications, primarily because of difficulties in typing it in stains and because of the advent of DNA TYPING.

Rhodamine A group of dyes that are used to enhance the visualization and development of LATENT FINGERPRINTS. Rhodamine dyes are often coupled with CYANOACRYLATE fuming (Superglue) techniques. Rhodamine dyes include Rhodamine 6G (also called R6G) and Rhodamine B.

rhodizonate test A PRESUMPTIVE TEST for GUNSHOT RESIDUE that targets lead that originates not in the bullet, but in the PRIMER. Often the test is performed by hot pressing a garment such that any residue is transferred to photographic paper placed beneath it. After the reagent, made by using sodium rhodizonate and a tartrate buffer, is applied, lead residues turn reddish pink.

ricin A highly toxic protein that can be isolated from the common castor bean (*Ricinus communis*) plant. After ingestion, the victim usually experiences nausea, vomiting, and tightness in the chest and many die within 36 hours of respiratory failure, depending on the dosage and means of delivery.

ricochet A bullet fired from a gun that contacts a surface, causing the flight path to be altered.

ridge characteristics (ridge detail) Detailed features of the ridges found on the hands and feet, most particularly for forensic purposes, on the fingers. Collec-

tively, these details are referred to as MINUTIAE, and they are the key to fingerprint analysis. Once the fingerprint analyst has been provided with prints to compare, he or she selects a set of ridge characteristics to examine on both prints. The characteristics, their number, direction, location, and location relative to each other, are all part of this comparison. This process is referred to as a POINT COMPARISON, after which the analyst determines that the compared prints match or do not. Currently, there is no universal standard for the minimal number of comparison points required in order for a match to be pronounced.

ridge count The number of ridge features found in a designated area of a LATENT FINGERPRINT. This term is most often used in conjunction with automated fingerprint identification systems (AFIS).

ridge endings The ending or termination points of a ridge feature in a LATENT FINGERPRINT.

rifle A high-powered FIREARM designed to be fired from the shoulder with two hands. Rifles have long barrels and impart high muzzle velocity to the bullets fired. Types of rifles include single-shot, lever-action (seen in western movies), bolt action, semiautomatic, and automatic. Representative calibers of rifles include 30–06 and .223. Assault rifles are automatic rifles that include the famous M-16, AK-47, and Uzi weapons developed by the U.S., Soviet, and Israeli militaries, respectively. These rifles can be fired in fully automatic or semiautomatic mode.

rifling *See* LANDS AND GROOVES.

rigor mortis The stiffening of a body that occurs shortly after death and that can be used to estimate the POSTMORTEM INTERVAL. Rigor begins to set in two to six hours after death, starting in the small muscles of the jaw and progressing through the trunk and out to the arms and legs. The stiffness remains for two to three days and

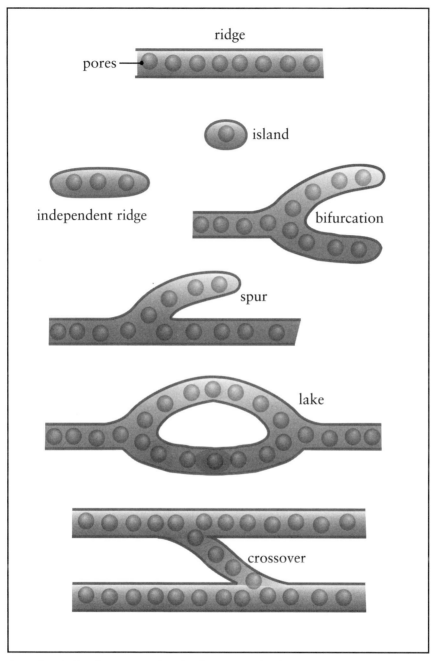

Examples of ridge characteristics found in fingerprints. The dots represent pores in the skin.

then releases in the reverse order. The rate of stiffening is temperature-dependent: colder temperatures accelerate it and warm temperatures slow it. The rate can also be influenced by physical activity before death. Running or strenuous activity immediately before death increases the rate of stiffening. Rigor mortis is not the same as cadaveric spasm, which can lock a joint in the position it was in at the moment of death.

rim fire primer (rim fire primer) A type of PRIMER used in smaller-caliber AMMUNITION such as .22. In a rimfire design, the initiator is contained inside a roll of metal that wraps around the base of the cartridge.

Ritalin (methylphenidate) *See* STIMULANTS.

RMNE *See* RANDOM MATCH.

ROC *See* RECEIVER OPERATING CHARACTERISTICS.

rohypnol (flunitrazepam) A drug that is legal in Europe but not in the United States. It produces a sedative effect and can lead to short-term amnesia. As a DATE RAPE DRUG, it is often slipped into an alcoholic beverage.

rolled ink fingerprints Fingerprints that are collected by inking fingers and rolling them on a paper fingerprint card. Although the technique is still used, increasingly fingerprints are obtained digitally.

rope and cordage An inclusive term referring to ropes or other materials that may be used in the commission of a crime. Ropes can be used as restraints for victims or as LIGATURES. Ropes may leave impressions in skin, and fibers from rope may also be collected as physical evidence.

Roxburgh model A model for describing the variability of fingerprints that was proposed in 1933. The model is unique in that it used polar coordinates rather than simple rectangular coordinates to map features on the print.

RSD (%RSD) Percentage relative standard deviation, also known as the coefficient of variation (CV). The %RSD is calculated by dividing the standard deviation of a set of data by the mean and multiplying it by 100. The quantity is used to gauge precision or reproducibility of replicate measurements.

RTX An abbreviation for ruthenium tetroxide, a toxic compound that has been researched as a means of visualizing LATENT FINGERPRINTS.

Ruhemann purple (RP) The characteristic purple produced by the contact of NINHYDRIN with latent prints. It is produced by the reaction of the ninhydrin with amino acids.

Rule 702 *See* FEDERAL RULES OF EVIDENCE.

rules of discovery *See* DISCOVERY.

Ruybal test *See* COBALT THIOCYANATE.

Sacco-Vanzetti case A landmark case in the history of early firearms examination. On April 20, 1920, a robbery took place in South Braintree, Massachusetts, in which five men robbed a paymaster, killing him and the guard. A month later, two men, Sacco and Vanzetti, were charged with the crime. A .38 pistol was recovered from Vanzetti, but it could not be conclusively tied to any of the evidence recovered. From the bodies, .32 ACP bullets were recovered, and Sacco had a .32 pistol in his possession when arrested. Numerous experts, including CALVIN GODDARD, testified, and the trial was surrounded by controversy. The men were convicted and executed; later review of the forensic work upheld the conclusions.

sacrum *See* APPENDIX V.

safety fuse A cord fuse used to set off high explosives without using electricity. Safety fuse is composed of a core of BLACK POWDER wrapped with protective layers, including waterproofing.

safety glass Special types of glass engineered to be stronger than normal window glass and/or fabricated to shatter into small pieces with a minimal number of sharp edges. Tempered glass is a type of safety glass that is strengthened by special heating processes; glass in windshields is laminated and strengthened with plastic.

saline/isotonic saline A solution of sodium chloride (NaCl) in distilled water that has the same concentration of this salt as in blood, 0.9 percent (weight/volume [w/v]).

saliva A common form of BODY FLUID evidence that has become more useful as DNA TYPING techniques have improved. Saliva can also be used in forensic TOXICOLOGY as it is considered to be a filtrate of the blood, similar to URINE. Three pairs of salivary glands produce saliva. Unlike stains of BLOOD and SEMINAL FLUID, saliva stains are usually colorless and hard to detect. The most common PRESUMPTIVE TEST for saliva involves the detection of the enzyme amylase, which catalyzes the breakdown (specifically the hydrolysis) of STARCH in food to glucose. Amylase is found in other body fluids, but the concentration in saliva is generally the highest; however, this means that the intensity of the response must be considered. *See also* PHADEBAS REAGENT AND TESTS.

saltpeter An old name for the salt potassium nitrate, KNO_3. It was an ingredient in early forms of GUNPOWDER.

salts Ionic compounds such as NaCl (table salt) that are made up of metals electrostatically bonded to nonmetals. Potassium nitrate, KNO_3, is also a salt; the K^+ is ionically bonded to the nitrate group, which carries a negative charge.

sampling *See* REPRESENTATIVE SAMPLING.

SANE Abbreviation for SEXUAL ASSAULT NURSE EXAMINER, a certification obtained through the International Association of Forensic Nurses (IAFN).

saponification Informally, the type of reaction that forms a soap; specifically it is the hydrolysis of an ester by an aqueous base. The substance ADIPOCERE is created by a saponification reaction of human fat.

satellite DNA Sections of repetitive DNA segments common in human DNA. Many of these sites are characterized by TANDEM REPEATS of the BASE PAIR sequence, meaning that sequences of bases are replicated. For example, the base sequence TTGA in tandem repeat would be TTGATTGATTGA, . . ., and so on. Current forensic DNA TYPING targets satellite DNA and tandem repeats.

satellite spatter Smaller blood droplets that are ejected from a larger-volume drop when that drop hits a surface. The thin taillike portion of the satellite points backward toward the parent drop.

Savannah River Technology Center Located in Savannah River, Georgia, a Department of Energy (DOE) facility that supplies technical support and assistance to the FBI in areas related to nuclear materials.

scallop pattern A type of edge pattern seen in blood drops and bloodstain patterns. The greater the height from which a droplet falls, the more exaggerated and elongated the scalloping is.

scanning electron microscope (SEM, SEM-EDX, SEM-EDS, SEM-XRD) Apparatus employed in a form of MICROSCOPY that uses beams of electrons instead of beams of light and is able to achieve very high magnification, up to 1,000,000X (magnification by a factor of 1 million). SEM also has high depth of field, meaning that a large portion or depth of the image remains in focus no matter how high the magnification. This contrasts with traditional light microscopy, in which the higher the magnification, the shallower the depth of focus. SEM and the associated X-RAY TECHNIQUES (electron dispersive spectroscopy [also called energy dispersive] and X-ray diffraction [XRD]) are used for ELEMENTAL ANALYSIS.

A simple schematic of an SEM system is shown in the figure. Inside a vacuum chamber, an electron gun supplies a tightly focused beam of incident electrons that interact with the sample. Although many types of interactions result, it is the emission of backscattered and secondary electrons that is used to create an image. On the sample's surface, elements with higher atomic numbers (*see* APPENDIX III) scatter more incident electrons and appear brighter than elements that have lower atomic numbers. This difference is discernible in the final image. Secondary electrons, which are actually emitted from the sample rather than scattered, are used to obtain information about surface features (topography). To generate an image, the electron beam is moved over the surface, scanning it as the backscattered and secondary electrons are collected. The image is a display that shows the relative intensity of the electrons collected at a given location. Older systems used cathode ray tubes (CRTs, similar to older televisions and to CRT computer monitors) to display the image, whereas newer systems typically incorporate digital imaging. Since the signal is related only to electron detection and not to detection of light as in traditional microscopy, the image is not colored. However, color (called false coloring) can be added to improve the visualization.

Schiff test *See* PERIODIC ACID–SCHIFF TEST.

science Broadly speaking, science is a method of study of the natural world and the universe using observation, experimentation, and experience. Science attacks problems and seeks understanding by using experimentation and observation based on measurable criteria. Temperature, pressure, and volume are measurable quantities, whereas criteria such as "worthiness" are subjective and not measurable. Other qualities that distinguish science from other areas as well as from PSEUDOSCIENCE include the use of HYPOTHESIS AND THE SCIENTIFIC METHOD and the principle of FALSIFIABILITY, which requires that any scientific finding or theory be stated specifically enough that it can be tested, replicated, and proved false or true.

scientific method *See* HYPOTHESIS AND THE SCIENTIFIC METHOD.

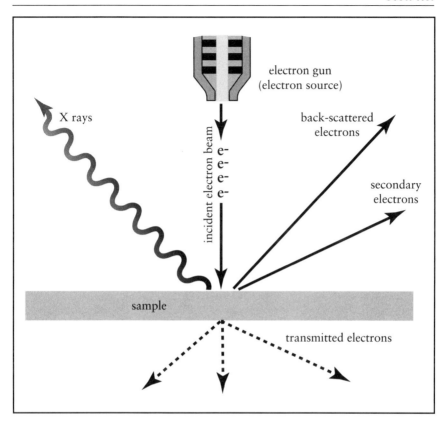

Schematic of a scanning electron microscope (SEM).

Scientific Working Group (SWG)   In forensic science, a group of forensic scientists in a particular specialty who work together to study methods, practices, and other aspects of that specialty. SWGs are sponsored by the Federal Bureau of Investigation (FBI); examples include SWG-DAM [DNA] analysis) and SWGFAST (latent fingerprints analysis).

scopolamine   A BARBITURATE that can be used as a sedative or as a poison. Scopolamine was originally obtained from extraction from plants of the Solanaceae family, with a chemical synthesis described in 1956. The synthetic version is sometimes referred to as Atroscine. It has the same molecular formula ($C_{17}H_{21}NO_4$) as COCAINE but shares none of its physiological properties. Scopolamine is listed on Schedule II of the CONTROLLED SUBSTANCES ACT. It has also been used during the course of narcoanalysis, known informally as the use of truth serums, an aspect of forensic DECEPTION ANALYSIS. Courts do not generally accept testimony that arises from the use of truth serums, including scopolamine.

Scotland Yard   A term that is used to describe the Metropolitan Police, the city police department of London. The main office is located in a complex called New Scotland Yard.

Scott test   *See* COBALT THIOCYANATE.

scrape cutter A type of machining tool that is used to carve out the RIFLING pattern in the barrel of a firearm.

scrapings Samples taken from underneath the fingernails of victims of murders, assaults, and sexual assaults, among other crimes. Fingernail scrapings are included in many RAPE KITS and are routinely collected at AUTOPSY when foul play may have occurred. Scrapings can be collected from suspects as well.

screening tests *See* PRESUMPTIVE TESTS.

scuff marks Marks on materials such as shoe soles that result from contact and rubbing with another surface or material. SKID MARKS are a type of scuff mark.

search patterns The walking patterns used to find, document, and collect evidence at a crime scene. The pattern selected depends on the unique characteristics of the scene. For a single person working outdoors, the outward spiral, either clockwise or counterclockwise, is a good choice, whereas a single person working indoors may elect to use a point-to-point method, in which the searcher moves between key locations within a room. Large outdoor scenes are best searched by several people using a strip or modified strip pattern. Searchers stand at arm's length from each other and walk in tandem, thoroughly and efficiently covering a large area.

sebaceous glands One of the three types of glands in SKIN that contribute to the composition of sweat and in turn to the composition of LATENT FINGERPRINTS. These glands, which are found in association with hair follicles as well as on the face, secrete oil. The material secreted from the sebaceous glands is called sebum.

sebum *See* SEBACEOUS GLANDS.

secondary crime scene A scene related to a crime, but not the location where the crime itself occurred. If a person is killed in one location and dumped in another, the dump site is a secondary scene.

secondary explosive *See* EXPLOSIVES; HIGH EXPLOSIVE.

secondary structure In proteins, which are large macromolecules, the secondary structure is the folding and twisting of the molecule in space that results from HYDROGEN BOND interactions. In DNA, the twisting of the individual strands is an example of secondary structure.

secondary transfer A transfer of physical evidence once removed from the original transfer of interest. For example, a burglar who breaks into a home with a cat may get cat fur on his or her clothing, a primary transfer. A secondary transfer would occur if the cat fur were then transferred to the car seat on which the burglar later sat.

second-level detail In a fingerprint, features such as MINUTIAE, scars, friction ridge paths, and creases (such as the "life line" on the palm) are classified as second-level details.

second-order reaction A chemical reaction in which the rate (speed) depends on the concentration of two reagents.

secretors The roughly 80 percent of the population who secrete antigens from the ABO BLOOD GROUP SYSTEM and other GENETIC MARKERS into their BODY FLUIDS. Before the use of DNA TYPING, secretor status was an important consideration in forensic SEROLOGY, particularly in sexual assault cases. As was discovered in the 1930s, a person's secretor status is under genetic control and is an inherited characteristic. Depending on the population studied, approximately 75–85 percent are secretors. In a secretor, body fluids that contain the ABO substances are SEMEN, SALIVA, vaginal fluid, and, to a lesser extent, URINE, SWEAT, and gastric juices.

secretory glands Generally, the three glands that contribute substances to sweat and thus to LATENT FINGERPRINTS: the SEBACEOUS GLANDS, ECCRINE SWEAT GLANDS, and APOCRINE GLANDS. *See also* SKIN.

Secret Service (United States Secret Service, USSS) A branch of the U.S. Treasury Department that is responsible for the protection of federal officials and for the investigation of counterfeiting crimes. The Secret Service was created in 1865 to battle counterfeiting after the Civil War. After the assassination of President William McKinley in 1901, their duties expanded to include fraud and presidential protection. The Secret Service has a significant forensic capability, emphasizing, but not limited to, support of counterfeiting investigations.

seismology, forensic Seismic monitoring and analysis of seismic signals to detect unnatural events such as underground nuclear tests and accidents such as the explosion aboard the Russian submarine *Kursk* on August 12, 2000.

selected ion monitoring (SIM) A technique used in MASS SPECTROMETRY to increase the sensitivity of the analysis of a selected compound or small set of compounds. In the most common form of mass spectrometer operation, the mass detector scans through a large range of atomic masses, collecting fragment ions at a given mass for only a fraction of a second before moving to the next mass. The wider the mass range scanned, the less time spent collecting ions of each individual mass. When SIM is employed, the detector collects data at only a few masses or a single mass. As a result, the detector collects ion fragments at the targeted masses for longer times, increasing the sensitivity and allowing for detection of smaller quantities. However, since only a few fragments are collected, identification of unknown compounds, which requires a large range of mass fragments, is more difficult.

selective partitioning The separation of mixtures of compounds. For example, if oil and water are placed into a jar and shaken, they separate, forming two phases. If sugar is added and the mixture shaken again, the sugar enters the water layer since it dissolves in water but not in oil. This is an example of selective partitioning based on solubility.

selectivity A term that refers to what certain procedures or instrumentation detect. A highly selective procedure, such as an immunoassay, responds to only one compound or a small set of compounds. A less selective procedure may react with or detect many compounds. Selectivity is not the same as SENSITIVITY, which is a measure of the smallest concentration of a compound or species of interest that can be detected.

selenium (Se) A chemical element that is used, among other applications, to coat drums in copy machines.

semen The fluid ejaculated by a man during sexual intercourse. Older PRESUMPTIVE TESTS for seminal fluid targeted such compounds. A single ejaculation typically consists of a few milliliters (two to six milliliters) of fluid that, as does BLOOD, has a serum component (seminal plasma) and a cellular component (SPERM cells or spermatozoa). The number of cells found in a milliliter of the ejaculate is variable but is usually in the range of 100,000,000. The cellular component constitutes about 10 percent of the volume of the ejaculate. Semen is a thick milky liquid that dries as a crusty, somewhat shiny material that acquires a slight yellowish tinge as it ages.

semiautomatic weapon A firearm that automatically ejects a spent cartridge and inserts a fresh one into the firing chamber. This is accomplished by exploiting the hot gases created when the cartridge is fired. *See also* BLOWBACK.

seminal acid phosphatase (SAP) An enzyme found in seminal fluid. Other secretions such as vaginal fluid contain acid phosphatase, although vaginal acid phosphatase (VAP) can be distinguished from SAP under certain analytical conditions. Regardless, the SAP PRESUMPTIVE TEST is useful for screening large items and for detecting semen stains in areas where they are difficult to see. *See also* P30.

SEMTEX A high EXPLOSIVE that is a combination of PETN and RDX.

sensitivity *See* SELECTIVITY.

sensitizer A compound or compounds added to the PRIMERS used in ammunition to make them more shock-sensitive, thus ensuring detonation and firing of the cartridge when the trigger is pulled and the hammer strikes the primer.

sequence of strokes In handwriting, the sequence in which pen or pencil strokes are used to create a letter. For example, the printed letter *A* can be made by first creating a shape Λ followed by a line across the center. Alternatively, one of the slanted lines can be made first, then the center line, and finally the last slanted line. This sequence can be useful in identification of forgeries since a forger may not use the same sequence as the person whose handwriting is being forged.

serial dilution The sequential dilution of a sample or other solution by powers of 10. For example, an undiluted ANTISERUM can be diluted by a factor of 10 (diluted 1/10); then the diluted sample itself is diluted again by 1/10. The final solution is 1/100th as concentrated as the original. Continued dilutions are serial dilutions.

serial number restoration The process of revealing serial numbers that have been filed, scraped, polished, or otherwise obliterated. Although restoration is most common in FIREARMS cases in which serial numbers on weapons have been removed, any metal object with a stamped serial number such as an engine block, tool, or equipment can be treated in a similar manner. When a serial number is stamped into a metal object, the force of compression is transferred to the metal below the indentation, causing imperceptible damage and strain. The depth of this strain can be several times the depth of the original impression. This damage to the metal makes it more susceptible to attack by oxidizing agents such as strong acids.

seriation Placement of something in a natural series. The technique of SERIAL DILUTION is an example of seriation, in which samples are diluted in a series, each one 1/10th as concentrated as the previous sample.

serology, forensic Serology is a subdivision of IMMUNOLOGY, the science that focuses on the reactions between antigens and antibodies such as are found in the ABO BLOOD GROUP SYSTEM. The word *serology* arises from the serum portion of the BLOOD, where the antibodies are found. Forensic serology generally refers to ABO blood group typing and isoenzyme typing. As DNA TYPING has all but replaced traditional forensic serology, the more inclusive term *forensic biology* is being used to describe characterization of blood and body fluids. In addition to DNA work, forensic biology includes disciplines such as forensic ENTOMOLOGY and ECOLOGY.

serum *See* BLOOD.

serum proteins POLYMORPHIC proteins that are found in the serum portion of BLOOD. Serum proteins are GENETIC MARKER SYSTEMS in the serum portion of blood that were occasionally used in the past to help individualize bloodstain and some types of body fluid evidence. HEMOGLOBIN is one example of a serum protein that has more than one form (POLYMORPHIC) and that can be typed by using simple gel ELECTROPHORESIS techniques. Others that have been used in forensic serology include haptoglobin (Hp), transferrin (Tf), and group-specific component (Gc). Since the introduction of DNA TYPING techniques, serum protein typing is no longer used.

sex determination (gender determination, sexing) The analysis of blood, bloodstain, body fluid, or skeletal remains to determine the sex of the person from whom the evidence originated. Although current DNA TYPING methods make this procedure relatively easy for blood and body fluids, this was not always the case.

The sexing of skeletal remains the responsibility of forensic ANTHROPOLOGISTS and relies on the MORPHOLOGY and size of bones. Of most use in determining sex are the pelvic bones and structures and the cranium (skull). Often, bones have been scattered or scavenged and all that remain are fragments and pieces. In such cases, analysis of the remaining portions can be performed and compared to large collections to provide the best possible sexing of the remains. The program FORDISC, created by the Department of Anthropology at the University of Tennessee, Knoxville, is widely used for this purpose.

sexology, forensic Generally considered to be a part of forensic psychology and psychiatry, a discipline the focuses on criminal or deviant sexual behavior.

sexual asphyxia (autoerotic death) An accidental death that is a form of STRANGULATION. Usually the victim is a male who uses a noose or other type of ligature to decrease blood flow to the brain to increase sexual pleasure.

sexual assault A violent attack on an individual that has a sexual motive or component. Sexual assault includes rape, both heterosexual and homosexual; child rape and abuse; and elder abuse. Sexual assault can also be committed during other crimes such as burglary and murder. *See also* PERK KIT; RAPE KIT.

sexual assault kit *See* PERK KIT; RAPE KIT.

sexual assault nurse examiner *See* SANE.

shading In handwriting, a difference in pressure applied to the writing instrument on an upstroke as compared to a downstroke. Differences in shading can be useful in detecting forgery, since a forger may not be able to reproduce the shading used by the writer.

shaving marks Marks that can be encountered on bullets fired from a REVOLVER. If the bullet scrapes against the flange at the base of the chamber as fired, the resulting marks are called shaving marks since the metal can be described as having been shaved off.

Sheppard, Sam A physician convicted of murdering his wife in 1953. The trial in 1954, as were the LINDBERGH KIDNAPPING case before and the O. J. SIMPSON case to follow, was called the "trial of the century." However, it was to become a series of three trials, one involving the United States Supreme Court. Although Sheppard was eventually released from prison, the continuing controversy was the inspiration for the 1960s television series *The Fugitive* and the 1993 movie of the same name.

shock wave A pressure wave (sensed as sound) produced by something moving through the air at a speed faster than the speed of sound. In forensic science, the term is generally used in association with FIREARMS. Many RIFLES and some smaller weapons fire BULLETS that travel so fast that they produce a shock wave or sonic "boom." Silencing such guns often involves decreasing the propellant so that the bullet exits at a slower speed, eliminating the characteristic cracking sound.

shoe prints (shoe impressions, footprints, footwear impressions) A type of impression evidence commonly encountered in indoor as well as outdoor CRIME SCENES. Footwear impressions can be two-dimensional (such as a shoe print seen on a dusty floor) or three-dimensional, such as a shoe print in mud or snow. The former are treated in much the same way as latent FINGERPRINTS; the latter are treated similarly to TIRE IMPRESSIONS.

shored exit wound An exit wound created by a bullet that has an appearance that is due to clothing or other support. Shored exit wounds generally appear smaller than might otherwise be expected. For example, if a person is shot though the back of the neck and the exit wound is through a tight necktie, the tie can support and cradle the skin, lessening the damage done as the bullet exits.

short tandem repeats (STRs) The current methods of DNA TYPING, which are used almost exclusively, revolve around short tandem repeats (STRs), which consist of only three to seven base pairs and total fragment lengths of about 400 base pairs. Since they are small, degradation is not as serious a problem as it is for longer fragments and polymerase chain reaction (PCR) can be used to amplify the sample. STRs are abundant in DNA, and although each locus has only a few variants, typing several loci at once dramatically increases the discriminating power. Thirteen loci have been selected as the standard in the United States; they include loci such as TH01, D7S1S820, and D3S1358. Typing three systems leads to discrimination power of about one in 5,000, and typing all 13 pushes this number into the trillions. An added advantage is that commercial STR kits provide a method to sex samples, a procedure that was cumbersome and difficult with techniques such as staining BARR BODIES. The amelogenin gene, which codes for dental pulp, is found on both the X and Y chromosomes but is six base pairs shorter on the X. Women, who are XX, show one type, since both variants are the same; men (XY) show two, one shorter than the other. An added advantage to sexing a stain or other evidence is that the results eliminate approximately half the population as a potential source with one test.

STRs have become a standard typing test in most forensic labs and the data collected are being integrated into a national database system coordinated by the FBI. The CODIS SYSTEM stores DNA typing data from convicted felons and sexual assault cases and has been useful in linking crimes and identifying assailants.

shotgun A type of firearm that fires collections of pellets rather than a single projectile. Shotguns do not have rifled barrels, and they fire shotgun shells, which are a distinctly different type of AMMUNITION from that used by RIFLES and HANDGUNS. Shotguns are identified by their GAUGE.

signal-to-noise ratio (SNR) *See* NOISE.

signature/signature analysis An individual's characteristic written identification, which is used on legal (and all manner of other) documents. Most signatures are made in cursive; some individuals print theirs. On the other extreme, many signatures are illegible but are still considered to be an individualized and authentic representation of one person. Other signatures acquire flourishes and stylistic traits that are not transferred to a person's normal writing. For these reasons, the signature is considered to be a specialized case of handwriting, and one of the most frequently contested elements of a questioned document.

signature behavior In criminal profiling, a signature consists of actions that are consistent across all crimes in a series and is an element that is unique to one perpetrator. This is in contrast to the modus operandi (MO), which may change or evolve. Identification of a signature can help investigators determine whether the same person likely committed several crimes or whether crimes are unrelated.

significant differences *See* EXPLAINABLE DIFFERENCES.

significant figures In reading data from an instrument, the number of digits that are certain plus one. For example, in reading a typical needle-style bathroom scale, the dial is marked with lines at each pound. If weight falls halfway between the lines for 130 pounds and 131 pounds, the weight is read as 130.5 pounds. The first three digits are certain since the reading is clearly greater than 130 but less than 131, but since the dial is not calibrated for anything smaller than one pound, the 0.5 pound is an estimate and thus an uncertain digit. The value 130.5 has four significant digits. Significant figure considerations are crucial when reporting data obtained from instrumentation, and rules for handling significant figures must be followed in any calculation based on or related to data obtained from an instrument.

sign of birefringence/sign of elongation A characteristic of materials such as FIBERS that have different optical properties along their center axis (parallel to it) as opposed to across it (perpendicular). In fiber analysis, a positive sign of elongation means that the fiber has a higher REFRACTIVE INDEX (RI) in the parallel direction and a negative value means that the RI is greater perpendicular to the long axis.

silencers FIREARMS accessories used to reduce the noise associated with firing a weapon. Silencers can be commercially produced or homemade and can be added to pistols and rifles. When ammunition is fired in a weapon, two processes can contribute to the sound produced. As PROPELLANT ignites, it produces a pressure wave of hot expanding gas that propels the bullet down the barrel. When this wave of high pressure reaches the atmosphere, a loud sound, called the report, is always heard. The sound is also referred to as the muzzle blast. The second sound is a sonic boom that results when the velocity of the projectile exceeds the speed of sound, 1,100 feet per second (roughly 770 miles per hour at room temperature). The small sonic boom resembles a sharp cracking sound. To alleviate the cracking sound, all that needs to be done is to reduce the muzzle velocity of the bullet to below 1,100 feet per second. To silence the muzzle blast, alterations to the barrel are required.

silver halides Halides are the elements that are in Group 8 of the Periodic Table of Elements (*see* APPENDIX III); silver halides are compounds such as silver chloride (AgCl) and silver bromide (AgBr). Silver chloride is a light-sensitive material that is exploited in some methods of LATENT FINGERPRINT visualization.

silver nitrate (AgNO₃) A salt that is used as a part of fingerprint visualization processes such as PHYSICAL DEVELOPER.

silver physical developer *See* PHYSICAL DEVELOPER.

Simpson, O. J. A former football star and celebrity acquitted of the brutal murder on the night of June 12, 1994, of his former wife, Nicole Brown Simpson, and Ronald Goldman. The case was noteworthy, even notorious, for many reasons, including the forensic aspects, which focused on CRIME SCENE documentation, CHAIN OF CUSTODY, and DNA TYPING results. Like the LINDBERGH KIDNAPPING trial in 1932 and the SAM SHEPPARD murder trial of 1954, the Simpson trial was hailed as the "trial of the century."

SINES *See* REPETITIVE DNA.

single action A type of firearm action in which hammer is manually cocked back before the trigger is pulled.

single base powder A type of SMOKELESS POWDER (PROPELLANT for AMMUNITION), which consists of nitrocellulose (made by treating wood shavings or cotton with nitric and sulfuric acids, HNO₃ and H₂SO₄), diphenylamine, and other additives. The amount of nitrogen in the nitrocellulose is reported by weight: "guncotton" is 13.3 percent nitrogen by weight. To prepare the single base powder, an emulsion is made with solvents, is dried, and is further treated before it is ground or otherwise powdered for use.

single diffusion IMMUNODIFFUSION technique in which only one component, either an antigen or an antibody, is mobile in the gel.

single nucleotide polymorphism (SNP) In DNA, a variation in a single NUCLEOTIDE at a given location that shows polymorphism across a population.

singleton DNA POLYMORPHISM observed in a single NUCLEOTIDE (SINGLE NUCLEOTIDE POLYMORPHISM) that is observed in less than 5 percent of the population.

sinsemilla A form of MARIJUANA that has a high concentration of the active ingredient TETRAHYDROCANNABINOL.

skeletal identification *See* AGE ESTIMATION; ANTHROPOLOGY, FORENSIC; STATURE ESTIMATION.

skeletal measurement   A process used by forensic anthropologists to estimate stature of a person, age, population, or sex. Procedures used for obtaining these measurements must be carefully followed to allow for use of databases and programs such as FORDISC. *See also* ALLOMETRY.

skeletal remains   A body that has been completely defleshed, so that only bones and teeth remain.

skeleton   *See* APPENDIX V.

skeletonized bloodstain   A bloodstain pattern that can occur if a blood drop is allowed to dry partially and is then wiped. Since the blood tends to be thinner at the edge of the drop, this portion dries first, leaving liquid in the middle. This liquid can be wiped away, leaving a hollowed area in the center.

skeleton wipe   A type of bloodstain pattern that can occur if a stain partially dries and is then wiped. Portions of the blood that are wet can be wiped away, whereas portions that have already dried remain, creating hollowed or skeleton areas.

sketches   Hand drawings; usually refers to crime scene documentation or drawings done by police or forensic artists.

skid   In forensic applications, a mark created by tires; it is also a type of mark that can be created on the surface of a bullet as it skids along a surface.

skin   The largest organ of the body, the structure of which is critical in the way fingerprints are formed. It consists of two layers, the DERMIS and the EPIDERMIS, which can further be divided into layers referred to as strata. Examples include the STRATUM CORNEUM and STRATUM GRANULOSUM. Structures such as pores, hair follicles, and secretory glands are found in the skin as well. The FINGERPRINT entry is accompanied by a figure showing the various layers and structures of skin.

skull   *See* APPENDIX VI.

slant   In handwriting analysis, the angle of the writing that is characteristic of a person's style and can be useful in detecting forgeries.

SLICE (spectral library for identification and classification engine)   A database system used to store data obtained by energy dispersive X-ray spectrometry (EDS). The program was developed for the FBI specifically for forensic applications.

slippage   A term that has two meanings in the forensic context. Skin slippage marks one of the phases of decomposition of a body. In this stage, the skin slips away from the underlying tissue. The other meaning is used in firearms analysis, in which slippage is a type of mark that can be imparted to a bullet.

slurry explosive   A type of explosive made by creating a slurry or gel in which the explosive compounds are suspended.

small particle reagent technique (SPR)   A group technique used to visualize latent FINGERPRINTS. In this approach, tiny particles with various properties are suspended in a solution that is applied to the print. The particles adhere to the oils in the print and help make it visible.

smear   A type of bloodstain pattern in which blood on a surface is smeared by moving contact with another surface or object.

smokeless powder   A low EXPLOSIVE that is used as the propellant in AMMUNITION and as a component of pipe bombs and other homemade explosive devices. Smokeless powder was a military innovation introduced in the late 1800s to replace BLACK POWDER, which created large amounts of smoke that obscured battlefields. As a propellant, black powder is now used principally in historical weapons owned by collectors and hobbyists. There are two kinds of smokeless powder, single base and double base. Single base powder consists of nitrocellulose (made by treating wood shavings or cotton with nitric and sulfuric acids, HNO_3 and H_2SO_4), diphenylamine, and other additives. The amount of nitrogen in the nitrocellulose is reported by weight: "guncotton" is 13.3 percent nitrogen by weight. To prepare the single base powder, an emulsion is made with solvents, is dried, and is further treated before it is ground or otherwise powdered for use. Double base powder consists of nitrocellulose and nitroglycerin along with various additives. In addition to use as propellants, smokeless powders are encountered in the forensic context as a component in an INCENDIARY DEVICE or a homemade bomb such as a pipe bomb.

smudge A type of bloodstain pattern in which a small volume of blood is smeared and displaced from a larger drop while it is wet.

smudge ring *See* BULLET WOUNDS.

Snow Print Wax A spray-on material that is used to cast impression evidence such as TIRE PRINTS or SHOE PRINTS made in snow.

SNP *See* SINGLE NUCLEOTIDE POLYMORPHISM.

SNR (signal-to-noise ratio) *See* NOISE.

social work, forensic Social work applied to areas of civil and criminal law. Examples include evaluation of competency, child custody, child or elder neglect, and abuse.

soda lime glass The most common type of glass, used in windows, bottles, and many other products. It contains silicates, aluminum, quartz, sodium, and magnesium.

sodium D line An intense yellow color of ELECTROMAGNETIC RADIATION emitted by sodium that is in the excited state. For example, when sodium is placed in a flame or a sodium lamp, the characteristic yellow color is emitted. The D line is at 589.3 nm.

sodium rhodizonate *See* RHODIZONATE.

soil A complex mixture of minerals, organic material, botanical and animal products and debris, and anthropogenic (human-made) materials that is naturally found on the Earth's surface. Although thousands of minerals have been identified, most soil samples contain only a few, typically 20 or fewer. Examples of common minerals are dolomite, $CaMg(CO_3)_2$, and talc $Mg_3(Si_4O_{10})(OH)_2$. The chemical and physical characteristics of soil vary with location, both along the surface (horizontally or laterally) and below it (vertically). Given this variability, soil samples taken six inches or a foot apart may show significant differences, as can two soils taken from the same location but different depths.

sole/outsole The portion of a shoe that has contact with the walking surface and the source of impressions that are made by this contact. The sole is the bottom of a shoe.

solid phase extraction and solid phase microextraction A group of related techniques used to extract compounds from water, other liquids, or gases. It has developed as an alternative to SOLVENT EXTRACTION for sample preparation. In forensic science, solid phase methods are used for sample preparation in ARSON cases, DRUG ANALYSIS, and TOXICOLOGY. A simple illustration of solid phase extraction is the use of charcoal filters to purify drinking water. Adsorption of compounds on a SORBENT is similarly exploited in solid phase extraction, when the sample, usually water, is passed over the sorbent to remove the compounds of interest. The sorbent is then rinsed to remove unwanted materials. The compounds of interest are then desorbed (eluted) off the sorbent by using heat or a solvent. This solvent extract is then concentrated down to a small volume that is ready for analysis, most often by using GAS CHROMATOGRAPHY or HIGH-PERFORMANCE LIQUID CHROMATOGRAPHY. A recent innovation is miniaturization by placing sorbents into capillary tubes or coating them onto thin plates. This is called solid phase microextraction (SPME), which is now commonly used for the analysis of arson evidence via a HEADSPACE technique.

solubility testing A technique used in the forensic analysis of PAINT, FIBERS, PLASTICS, EXPLOSIVES, and DRUGS. The solvents used vary with the material being tested. For drugs and explosives, testing water solubility can be useful. For paints, common solvents used include concentrated acids such as nitric and hydrochloric (HNO_3 and HCl), acetone, methyl ethyl ketone (MEK), and dimethylformamide (DMF). For fibers, solvents are generally organics such as DMF and nitromethane. Plastics, which are synthetic polymers with much in common with synthetic fibers, can also be tested with different solvents for classification and comparison purposes.

solvent extraction A method frequently used in forensic CHEMISTRY and TOXICOLOGY to isolate compounds of interest from a complex sample matrix. The principle of solvent extraction is based on SELECTIVE PARTITIONING between two different phases based on solubility. In solvent extraction, a sample such as urine can be extracted with an organic solvent such as chloroform or methylene chloride (dichloromethane) to isolate compounds in the organic layer selectively. Other solvents commonly used are hexane, acetone, alcohols, and ethers. Acid-base extractions are also extensively used to isolate drugs that have acidic and basic forms. In recent years, SOLID PHASE EXTRACTION techniques have replaced many traditional solvent extractions since they can often yield a "cleaner" extract with fewer extraneous materials. This characteristic is particularly valuable in toxicology, as the sample matrix itself, particularly blood, complicates solvent extraction.

somatic Referring to the body or a body structure. For example, the somatic origin of a hair refers to the part of the body where it originated, such as the scalp, beard, armpit, or leg.

SOP An abbreviation for *standard operating procedure*. An SOP can describe anything from the way evidence is logged in to the way a bloodstain is analyzed. Use of SOPs is an important part of QUALITY ASSURANCE/QUALITY CONTROL.

sorbent A material that is used in analytical techniques such as SOLID PHASE MICROEXTRACTION (SPME). The sorbent selectively absorbs vapors from a sample, such as arson debris, and as a result preconcentrates it. The sorbent is then chemically or thermally desorbed to release the absorbed materials into an instrument such as a GAS CHROMATOGRAPH.

sound spectrograph *See* VOICE PRINT.

Southern blotting *See* BLOTTING.

souvenir A term in profiling that refers to an object that is taken by a perpetrator from a crime scene or victim.

spalling A type of fire damage seen in concrete and similar surfaces. As an example, when concrete is exposed to extreme heat, water trapped inside the material can boil and force its way to the surface, resulting in physical damage such as cracking and crumbling.

spatter Usually, patterns of blood deposited on structures or objects during a violent event.

specialist *See* GENERALIST.

speciation The process of determining the species from which a body fluid stain originated. Blood, saliva, and semen are the body fluids most commonly tested for species. Before the widespread adoption of DNA TYPING using PCR amplification, immunological tests such as IMMUNODIFFUSION and crossed-over electrophoresis, an IMMUNOELECTROPHORESIS method, were employed for speciation. These tests involve the use of antisera created to react with blood or body fluids of other species such as human, cat, or dog. In DNA typing procedures using PCR, one step of the procedure requires determining how much human (higher primate) DNA is present in the recovered sample. In so doing, the test confirms the presence or absence of human DNA.

specific gravity A measure of the DENSITY of a liquid as compared to water. It is calculated by dividing the density of the substance by the density of water, which is often taken as 1.00 g/mL, although density does depend on temperature.

spectral transition The transition of an element or compound to an unstable excited state by the absorption of ELECTROMAGNETIC RADIATION. Broadly speaking, this is the basis of all SPECTROSCOPY; it is often represented by the generic notation M + photon => M^*, in which M represents the material of interest and M^* indicates that material in the excited state.

spectrofluorometer (spectrofluorometry) An instrumental technique that uses ELECTROMAGNETIC ENERGY (EMR or radiation) to promote an atom, compound, or molecule to an excited state and detects the radiation that is emitted (FLUORESCENCE) when the species returns to the original state (ground state). What constitutes an excited state varies with the type of spectrometry being used; however, when the target species takes the final step to relax to the ground state, electromagnetic energy is released. Since the energy gap is smaller than that initially jumped, the energy of the emitted radiation is lower, and thus a longer wavelength is observed. This process is fluorescence, which can be exploited to make very sensitive instrumental detectors. Fluorescent species, whether natural or derived, are referred to as fluorophores and fluorescent detection is used in automated DNA TYPING procedures. Another common forensic use of fluorescence is in the analysis of INORGANIC materials using X-ray fluorescence (XRF).

spectrophotometry (spectroscopy, spectrophotometer) Generically, the study of the interaction of ELECTROMAG-NETIC ENERGY with matter. The interaction can be measured and used to determine what is present (QUALITATIVE ANALYSIS) and in what quantity (QUANTITATIVE ANALYSIS). The simplest form of spectrophotometry and also the first developed is COLORIMETRY, in which the visible portion of electromagnetic energy is used. Spectrometers consist of an energy (light) source, a mechanism or device to filter the source energy and select the wavelength(s) of interest, a device or method to hold the sample, and a detector system, which converts electromagnetic energy (light) to a measurable electrical current. Other kinds of spectrometry widely used in forensic science include INFRARED SPECTROPHOTOMETRY, ATOMIC ABSORPTION, and ULTRAVIOLET/VISIBLE spectrometry.

spermine A compound found in high concentrations in seminal fluid and the basis of the presumptive BARBERIO TEST for seminal fluid.

sperm (sperm cell, spermatozoa) The cellular component of semen, the male ejaculate fluid. A typical ejaculation contains on the order of 10^8 sperm cells per milliliter, with three to four milliliters ejaculated. Sperm cells are the male sex cells and carry 23 chromosomes that pair with those from the female egg when fertilized. Microscopic identification of intact sperm cells is a conclusive test result for presence of semen; however, in stains and swabs, finding intact sperm is not always easy.

splash A generic type of BLOODSTAIN PATTERN created in the same way water splashes.

spoliation The destruction of evidence, whether by incorrect preservation or deliberate action. The term is often used to describe decomposition or degradation of biological evidence such as BLOOD and body fluid stains.

spot tests *See* PRESUMPTIVE TESTS.

spurious minutiae The creation or visualization of false features such as pores or ridge characteristics in a finger-print. Incorrect collection techniques and contaminants such as dust, for example, may create such features.

squalene A large hydrocarbon molecule that is a precursor of cholesterol and also a component of the residue deposited in a latent fingerprint.

stabilizer Chemical compound added to PROPELLANTS such as SMOKELESS POW-DER to prevent easy or unintentional igni-tion.

stab wound A puncture wound that is characterized by being deeper than it is long. This contrasts with a cut or slice, which is generally shallow and long.

staged crime scene A crime scene that is altered to conceal evidence or obscure what happened. For example, a murder scene can be staged to appear as if the killing was the result of a burglary when in fact the murder was the purpose of the intrusion.

stage micrometer A tool mounted on the stage of a microscope that provides a scale and is used to measure microscopic features.

staining A family of techniques used in microscopy to aid in the visualization of otherwise difficult to see features. For example, sperm cells are often stained with a pair of reagents collectively called Christmas tree stain to make the cells eas-ier to see. Some of these stains have been adapted to other uses such as in the visual-ization of latent FINGERPRINTS.

standard addition A method used to determine the quantity of a given sub-stance present in a sample. In a standard addition quantitative analysis, a known amount of the target substance is added to the sample in increasingly larger portions. From a standard addition curve, a plot of amount of substance added versus signal, the original concentration in the sample, can be determined.

standard deviation A measure of the spread or reproducibility of replicate tests. The standard deviation is the average deviation of all samples from the mean value and is symbolized by s for small sample sets and the letter sigma (σ) when larger data sets are used. If the same sam-ple is analyzed 10 different times, for example, the standard deviation provides a measure of how closely the results of the 10 samples agree. The formula for stan-dard deviation, assuming a small data set, is given by

$$s = \sqrt{\frac{\sum_i \left(x_i - \bar{x}_i\right)}{n-1}}$$

where n is the total number of repeated measurements, x_i is an individual mea-surement, and $\bar{x}_i$ is the average value of all the measurements.

standardization The process of assur-ing that a method or instrument will pro-duce accurate and reproducible results under specified standard operating condi-tions. Standardization is a vital part of QUALITY ASSURANCE/QUALITY CONTROL and extends from very simple procedures such as assuring that balances obtain accurate weights to performing the most complex DNA analysis. For example, when a police officer stops someone he or she suspects of driving under the influ-ence of alcohol, a BREATH ALCOHOL test is usually administered by using an instrument designed specifically for that purpose. It is crucial that the instrument be properly calibrated and tested by using known standards of alcohol to ensure that the results the officer obtains are trustworthy.

standard methods Procedures for analysis of evidence that have been tested, validated, approved, and used routinely in the laboratory. The use of standard meth-ods is critical to overall QUALITY ASSUR-ANCE/QUALITY CONTROL to ensure that results obtained are as accurate and reli-able as possible. The use of standard methods also facilitates comparison of results between labs.

standard of care   The normal or standard procedures, requirements, and so on, that are applied to a given profession or practice.

standards   Reagents or materials of known composition that are used in an analysis. These are used both as positive controls and as references for QUANTITATIVE ANALYSIS. A special group of standards are those prepared by the National Institute of Standards and Technology (NIST), which are characterized by multiple methods and are generally considered to be the best and most reliable standards available. *See also* SAMPLES (CONTROLS).

Starburst dendrimer (PAMAM)   A particular type of CDS QUANTUM DOT (nanoparticle) used to visualize FINGERPRINTS. The dot is bound chemically to the DENDRIMER (branched chemical structure), which in turn attaches to residues in the fingerprint.

starch   A POLYMER of glucose, a simple sugar, that has three principal roles in the forensic science context. First, starches, particularly cornstarch, which is inexpensive and widely available, are used as CUTTING AGENTS for drugs that are white powders such as COCAINE. Starch can also be a component of dust and thus a form of TRACE EVIDENCE. In both cases, the identification of starch is made easier by its distinctive appearance under a POLARIZED LIGHT MICROSCOPE (PLM). Finally, starch–iodine is a common staining agent and has been used as part of the amylase test for the presence of SALIVA.

starch-gel electrophoresis   A form of electrophoresis in which the separation of the components of interest is accomplished in a gel derived from a STARCH material such as AGAROSE. The gel is poured in a thin slab into which the sample is inserted. Separation is accomplished by applying an electrical field across the gel, causing in the movement of charged species.

static load   A term in forensic engineering that refers to the load placed on a structure such as a pillar or support beam that results from the weight of materials atop it. The term *static* refers to a load that does not result from movement, as compared to a DYNAMIC LOAD created by wind.

stationary phase   In CHROMATOGRAPHY, the phase that does not move. In GAS CHROMATOGRAPHY (GC), for example, the stationary phase is coated on the inside of the column.

statistics and probability   Although often lumped together, terms that are not identical. Both are considered to be branches of mathematics; however, *statistics* generally refers to the analysis of collections of data, whereas PROBABILITY is the study of the likelihood of certain events or occurrences.

stature estimations   Considered primarily a task of forensic ANTHROPOLOGY, the determination of the likely height of an individual on the basis of skeletal remains. This is accomplished by measurements of the long bones such as those found in the arm and the leg (the femur, for example). On the basis of these measurements and databases of information, the anthropologist can estimate height. Stature estimates are difficult if not impossible if only partial skeletal remains are recovered.

stellate pattern   A term that has two common meanings in forensic PATHOLOGY. When severe burns are seen, stellate can refer to a pattern of sharp, starlike projections emanating from a split in the skin. More commonly, it refers to features seen in contact gunshot wounds to the head. When the barrel of a gun is placed in contact with the skin over the skull, firing the weapon results in ejection of gases into the skin and an outward rupture that can create a roughly star-shaped injury in the skin. *See also* BULLET WOUNDS.

stereobinocular microscope   A type of microscope that uses two eyepieces, relatively low magnification, and reflected light. This term is generally taken to be synonymous with *stereomicroscope* or *stereoscopic microscope*. Stereomicroscopes are used for tasks such as identifying and sorting TRACE EVIDENCE such as HAIRS, FIBERS, and SOIL.

stereoisomers Compounds that have identical molecular formulas and orders of connection of atoms but differ in the three-dimensional arrangement of these atoms. If two stereoisomers are mirror images of each other, they are called ENANTIOMERS; otherwise they are referred to as DIASTEREOISOMERS.

stereomicroscope *See* STEREOBINOCULAR MICROSCOPE.

stereoscopic microscope *See* STEREOBINOCULAR MICROSCOPE.

sternum *See* APPENDIX V.

sticking marks Markings that are sometimes found on medicine tablets that are the result of material's sticking to the punch used to create the tablet.

Stielow case A key case in the early history of firearms examination. In 1917, Charles Stielow had been accused and was later convicted of shooting his landlord and his housekeeper. An expert at his trial stated that Stielow's pistol had fired the fatal shots, and Stielow was sentenced to die. Shortly thereafter, another man confessed to the crime, prompting a reexamination. No evidence of a conclusive match of the bullet to Stielow's pistol was found. Critical in the reversal was the work of Charles Waite, who went on to collect information about the rifling characteristics of firearms in the United States and in Europe. In 1924, he started a private laboratory called the Bureau of Forensic Ballistics staffed by pioneers such as CALVIN GODDARD, John Fisher, and Philip Gravelle. Goddard went on to greater fame in cases such as the Saint Valentine's Day Massacre.

stimulants Substances that stimulate the central nervous system (CNS), producing sensations of wakefulness, decreased fatigue, decreased appetite, and general well-being. COCAINE, amphetamine, methamphetamine, Ritalin (methylphenidate), caffeine, and nicotine fall into this category. In higher doses, many stimulants can also act as hallucinogens.

stippling *See* BULLET WOUNDS; STELLATE PATTERN.

stipulation Agreement reached voluntarily by the prosecution and the defense relating to some point in a case. When a forensic scientist completes an analysis and writes a report, often the defense stipulates to the report as fact rather than calling the expert to testify.

stoichiometry In chemistry, the tools used to perform quantitative calculations by using chemical reactions and formulas.

Stoney and Thorton model A detailed model that attempted to address the individuality of FINGERPRINTS and issues such as variability in repeated printings of the same finger statistically.

STP Shorthand slang for the drug methyldimethoxymethyl-phenethylamine, a HALLUCINOGEN.

STR *See* SHORT TANDEM REPEAT (STR).

strangulation Death by asphyxia that is brought about by compression of the neck. The cause of death is lack of blood flow to the brain. Strangulation can be performed manually with the hands (throttling) or by use of a LIGATURE. HANGING is a form of ligature strangulation in which body weight is used to generate the compression.

stratum corneum One of several layers in the SKIN. The stratum corneum is the outermost layer of the EPIDERMIS and the one exposed to the elements. The name refers to the cornified or horny layer. Epithelial cells are converted to keratin and shed from this layer.

stratum germinativum One of several layers in the SKIN. This layer is more commonly known as the basal cell layer and is the innermost layer of the EPIDERMIS. When considered together with the STRATUM SPINOSUM, the combination is referred to as the Malpighian layer or the STATRUM MALPIGHII.

stratum granulosum One of several layers in the SKIN; also called the granular layer. It is found in the EPIDERMIS between the outermost horny layer (STRATUM CORNEUM) and the STRATUM SPINOSUM.

stratum lucidum One of several layers in the SKIN and a transitional layer between the granular layer of the epidermis (STRATUM GRANULOSUM) and the outermost horny layer (STRATUM CORNEUM). The name refers to the translucent appearance of the layer.

stratum malpighii One of several layers in the SKIN and the name for the combination of the STRATUM SPINOSUM and STRATUM GERMINATIVUM (basal cell layer.) Also called the Malpighian layer, it is part of the EPIDERMIS.

stratum spinosum One of several layers in the SKIN. This layer is found in the EPIDERMIS atop the basal cell layer (STRATUM GERMINATIVUM) and the granular layer, the STRATUM GRANULOSUM.

stray light In SPECTROPHOTOMETRY and related techniques, excessive or unwanted light that adversely alters the light of interest in the analysis.

stress marks *See* CONCHOIDAL LINES.

striae grooves A term that refers to grooves in a mark created by an inked writing device such as a ball point or felt tip pen. These are grooves in the pen stroke that do not contain any ink.

striations Marks that are made in a surface as the result of motion of one surface across another. Scratching, sliding, and scraping motions all can produce striations on one or both surfaces. In the case of metal-to-metal contact, the softer metal yields to the harder metal and shows striations.

striker (striker pin, striking pin) In a gun, the striker is the small metal point that slams into the PRIMER when the trigger is fired. It can leave a distinctive mark on the primer. This term is considered to be an older term for FIRING PIN.

stringency conditions In DNA TYPING using PCR techniques, the chemical composition and conditions necessary to ensure that DNA molecules HYBRIDIZE.

string matcher *See* ELASTIC MATCHING.

strychnine An extremely poisonous substance that can be extracted from the seeds of the plant *Strychnos nux vomica* or synthesized. It is a large molecule ($C_{21}H_{22}N_2O_2$) that can be found in the form of salts, many of which are bitter tasting. It is an ingredient in rodent poisons.

student *t* value A statistical value used in tests that compare means, among other uses. The *t* value for a given test is selected from a table and depends, for example, on the size of the data set and the desired level of certainty that can be associated with the test result.

Study in Scarlet, A The first story written by Sir Arthur Conan Doyle in the Sherlock Holmes series. It was published in 1887.

stylistics, forensic A discipline closely related to forensic LINGUISTICS, which concentrates on the style of speech (oral or written) characteristic of a group or individual. For example, the speaking style of a person from New England can be quite distinct from that of a person from Texas though both are English speakers.

sublimation A change in physical state that passes directly from solid to vapor without passing through a liquid phase. Dry ice (frozen carbon dioxide) sublimes, as does solid iodine. Iodine fuming was once widely used (and still is used) to visualize LATENT FINGERPRINTS.

subpoena A document issued by a court that requires someone to appear on a specified date and time to provide testimony in the matter at hand. The word means literally "under [*sub*] penalty [*poena*]," referring to penalties that will be applied if the person fails to appear. Forensic scientists routinely receive subpoenas to testify to their findings on evidence they have analyzed.

substrate Generally, the material on which something else is layered or deposited; a lower or supporting layer. For example, if a bloodstain exists on jeans, the substrate is the denim.

succession and succession patterns A natural process that involves advancement along defined pathways. Ecological succession, a natural process that is considered in locating CLANDESTINE GRAVES, occurs when a patch of ground is disturbed to dig a hole. When the hole is filled in and the dirt replaced, all existing plants have been destroyed, leaving essentially bare earth. Ecological succession occurs as grass or moss colonizes the bare soil, to be followed later by larger plants. Similarly, forensic entomology takes advantage of the succession of insect species that can colonize a body after death.

sucrose Table sugar, a disaccharide (two-sugar) molecule that consists of glucose and fructose.

sugars (saccharides) Carbohydrates that have the general formula $C(H_2O)_n$— hence the term *carbo-* (from carbon) *hydrate* (water). Glucose (blood sugar), for example, has the molecular formula $C(H_2O)_6$. In forensic science, the principal importance of sugars is in the area of DRUG ANALYSIS, because sugar is frequently used as a CUTTING AGENT. Sugars that are detected include table sugar (sucrose), fructose, xylose, glucose, galactose, lactose, and maltose.

suicide Death caused by purposeful actions or omissions by the victim. Suicide is considered to be a MANNER OF DEATH, along with accidental, homicidal, and indeterminate.

supercritical fluids Fluids under very high pressures that are occasionally used as part of EXTRACTION processes. When they are used this way, the process is called supercritical fluid extraction; carbon dioxide is a commonly used compound for such extractions.

Super Glue A cyanoacrylate glue that is used to develop and preserve latent FINGERPRINTS. It is usually applied by fuming after the addition of a basic compound such as sodium hydroxide. *See also* CA FUMING.

superimposition The process of laying one image or photo over another. In forensic science, this most commonly refers to a technique used in forensic anthropology and forensic art in which the image of a skull is superimposed over a picture of a person's face to determine how the features align.

surface analysis Chemical, instrumental, or microscopic analysis of a surface. This is opposed to what is usually referred to as bulk analysis, in which all of a material is tested. SCANNING ELECTRON MICROSCOPY is an example of a surface analysis technique.

surfactant A compound such as soap that breaks up the surface tension of water by disrupting the hydrogen bonds that form between water molecules.

surveying, forensic Measuring and mapping of features (natural and human-made) that may be involved in legal matters. The emphasis in such procedures is on showing relative positions and locations.

swabbing A process used to collect many types of forensic samples such as suspected BLOOD and GUNSHOT RESIDUE. To collect the sample, a cotton swab is wetted with distilled water or saline solution and gently rubbed across the area containing the sample.

swaging   A machining method used to create LANDS AND GROOVES (RIFLING) in the barrels of firearms.

sweat   *See* ECCRINE SWEAT.

sweat pores   Small openings in the SKIN through which sweat is released. The pattern of pores can be used as part of the analysis of LATENT FINGERPRINTS.

swipe/swipe pattern   A type of blood spatter (bloodstain) pattern created by a swiping motion through an existing wet stain. Often swipe patterns are created by hands or hair.

synergistic effect   A combination of two or more effects in which the result is greater than what would be expected by summing the individual effects. Drugs that are mixed can have a synergistic effect; for example, alcohol increases the drowsiness produced by medications such as antidepressants.

synthetic   A material that is human-made rather than derived from existing mineral, plant, or animal sources. Cotton is a natural fiber, which is derived directly from the cotton plant; polyester is a synthetic fiber that is made from material synthesized from petroleum. The conversion of cotton plant to fiber requires no chemical conversions.

T

$t_{1/2}$ The symbol for HALF-LIFE.

taggants Materials or compounds that are placed in a product that can be chemically identified and linked back to a source, time of manufacture, and other characteristics. As an example, taggants were once placed in many inks and when detected could be used to identify the ink by the company that made it and the period it was made.

Takayama test *See* HEMOCHROMOGEN TEST.

tandem repeat In DNA, a location on the strand that has sections containing repetitions of the same BASE SEQUENCE. For example, the sequence GTCTAGTC-TAGTCTAGTCTA is a tandem repeat of the sequence GTCTA.

tape lifts A technique of evidence collection used to recover LATENT FINGER-PRINTS, FIBERS, HAIRS, and other types of transfer and TRACE EVIDENCE. To remove fibers from a surface such as clothing, clear tape can be rolled into an inside-out (sticky side up) loop. This loop is then pressed against the surface, to pick up fibers and other clinging matter.

taphonomy The systematic study of "death assemblages" and DECOMPOSITION applied to forensic science. Taphonomy is a subdiscipline of anthropology and archaeology that has found increasing use in forensic cases over the past 10–15 years. Strictly defined, taphonomy focuses on the process of death and the aftermath that ultimately led to the fossilization of a remains, but in the forensic context, the time frame of interest is much shorter.

Taq polymerase An enzyme that catalyzes the addition of nucleotides to DNA and a key component in PCR amplification and DNA TYPING.

Tardieu's spots *See* PETECHIAL HEMOR-RHAGE/SPOTS/TARDIEU'S SPOTS.

target analyte The compound or element that is the subject of a chemical analysis. In an analysis for GUNSHOT RESIDUE, the target analytes are the elements barium, antimony, and lead.

TATP (triacetone triperoxide) An EXPLOSIVE synthesized from acetone and hydrogen peroxide.

TCD (thermal conductivity detector) A detector used in some types of GAS CHROMATOGRAPHY. It is based on the thermal conductivity of a carrier gas such as helium, which does not change until another substance is mixed in with the helium. Such a change indicates that a compound has been detected.

Technical Working Group (TWG) A predecessor of the SCIENTIFIC WORKING GROUP.

teeth and tooth marks *See* BITE MARKS; TOOTH.

Teichman test *See* HEMATIN TEST.

telogen phase The final phase of the growth cycle of HAIR. It is during this phase that hair sheds naturally. Most hairs found as evidence, unless forcibly removed, are in the telogen phase.

TEM *See* TRANSMISSION ELECTRON MICROSCOPY.

tempered glass   *See* SAFETY GLASS.

templates   *See* HYBRIDIZATION.

tent pattern/tented arch   One of the main FINGERPRINT patterns, in which the ridge pattern enters from the side and travels in an arching pattern but lacks a central core or delta pattern. The tented arch has a distinctive peak, as seen in a tent. *See also* ARCH; PLAIN ARCH; TENTED ARCH.

terminal velocity   In forensic science, usually the maximal speed of blood traveling through air, which is about 25 feet per second.

tertiary structure   One of four levels of structure in large protein molecules. The term refers to the way the molecules fold and twist in space on the basis of interactions between amino acids in the chain.

test impression   An impression of an imprint or mark made by an object such as a tire or shoe. Test impressions are compared to questioned impressions, such as might be found at a crime scene.

tetrahydrocannabinol (THC)   A member of a class of organic compounds called cannabinoids and the active ingredient in MARIJUANA and derivatives such as HASHISH. In marijuana, there are more than 50 cannabinoids, of which THC (a HALLUCINOGEN) is considered to be the most important active one. It is found in the oily resin, flowering tops, and leaves of the plant, with the highest concentrations found in the oil. Concentrations of THC in marijuana have been steadily increasing as growers improve the quality of their crop by using selective breeding. THC is a thick, oily liquid that is actually the result of degradation, and so its concentration increases as the plant or extract ages. It was first isolated from marijuana in 1964.

tetramer   A tetranucleotide or a segment of DNA that has four base pairs that repeat; a four-nucleotide TANDEM REPEAT.

tetramethylbenzidine (TMB)   A PRESUMPTIVE TEST for BLOOD made by dissolving TMB in glacial (concentrated) acetic acid. As in other similar tests such as the BENZIDINE and KASTEL MEYER tests, TMB changes color in a reaction that is catalyzed by HEMOGLOBIN. The TMB test is an alternative to the older benzidine test, which has been abandoned because of the carcinogenic properties of that compound.

tetranucleotide repeat   A section of DNA composed of a sequence of four NUCLEOTIDES that are repeated. For example, a DNA section with the BASE SEQUENCE AATGAATGAATG contains three repetitions of a tetranucleotide.

tetryl   Shorthand name for 2,4,6-trinitrophenylmethylnitramine, a high EXPLOSIVE.

TH01   A gene locus classified as a SHORT TANDEM REPEAT (STR) that is typed in current DNA TYPING procedures. It is one of the 13 loci included in the CODIS system.

THC   *See* TETRAHYDROCANNABINOL.

thermal cycler   A device used in PCR amplification of DNA. A thermal cycler carefully controls temperature through successive heating and cooling, which is essential to copying DNA sections.

thermal luminescence   The emission of ELECTROMAGNETIC RADIATION such as visible or infrared (IR) light that is caused by excitation with thermal energy. Metal heated "red hot" is an example of thermal luminescence.

thermionic detector   *See* NITROGEN PHOSPHORUS DETECTOR.

thermograph/thermographics   An imaging technique in which heat (infrared radiation) is detected and transformed into a visible signal. Simple night vision goggles that detect body heat exploit this technique.

thermoplastic Material that becomes fused to a surface such as paper when heated. The TONERS used in copiers and laser printers are thermoplastics.

thin layer chromatography (TLC) A simple form of CHROMATOGRAPHY that separates mixtures by exploiting solvents traveling over a solid support phase and the resulting chemical interactions between the solvent, solid support, and molecules of interest. As shown in the figure, tiny spots of dissolved sample are placed in a line across the bottom of a plate or paper. The plate is coated with a thin layer of a silica or related powdery material. The line is called the origin. The plate is then placed into a shallow solvent bath so the level of the solvent is below that of the origin. The solvent can be as simple as water (used for ink analysis) or may consist of two or more organic solvents such as benzene, ethanol, or acetonitrile.

Capillary action draws solvent up the plate in the same way that water is drawn up into a paper towel. As the solvent encounters the sample, some or all of it dissolves and begins moving along with the solvent "front" as it creeps up the plate. Some components in the mixture interact with the silica material, causing it to fall behind components that interact less. Eventually, all separable components are spread out in spots across the plate. Any components that are not soluble in the solvent remain at the origin. In the case of ink analysis, spots are clearly visible; however, in many cases a developer must be sprayed or otherwise applied to visualize the spots.

three-dimensional impression An impression that has depth as well as pattern. For example, a shoe on a dusty floor creates a two-dimensional (flat) pattern in the dust, but the same impression made in mud has depth and is thus three-dimensional.

throttling STRANGULATION that is accomplished by the hands and not by use of a LIGATURE such as a rope or cord.

thymine (T) One of four NUCLEOTIDE bases that compose DNA and RNA (ribonucleic acid). Because of its molecular structure, thymine associates with ADENINE (A), and the two are referred to as complements of each other.

time of death *See* POSTMORTEM INTERVAL.

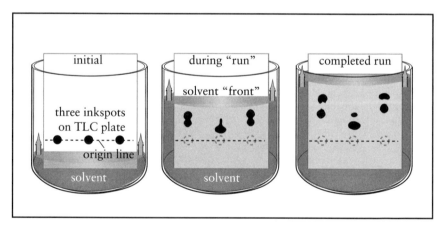

The process of thin layer chromatography. Capillary action draws the solvent upward, allowing it to interact with components in the sample. Components separate on the basis of the extent of their interaction with the powder relative to the solvent.

Thymine (T)

The structure of thymine (T), one of the four bases found in DNA and the complement of the base adenine (A).

time of flight (TOF) A type of MASS SPECTROMETER (MS) in which ions are separated by the time they take to cover a large distance. Smaller ions travel faster and reach the detector first; the time required to cover that distance can be correlated to the mass of the ion.

time resolved imaging/time resolved spectroscopy A technique of collecting images over a period to collect different information. The technique is exploited in detection of FINGERPRINTS by using special treatments such as QUANTUM DOTS. For example, a fingerprint may be on a surface that fluoresces under ultraviolet (UV) light so brightly that the fingerprint image is washed out. By using treatments that phosphoresce (a process similar to fluorescence but longer-lasting), this can be prevented by delaying collection of phosphorescent data until after the fluorescence has faded.

time since intercourse (TSI) *See* POST-COITAL INTERVAL.

tin (Sn) A heavy metal element most commonly encountered in GUNSHOT RESIDUE ANALYSIS.

tire impression A form of IMPRESSION EVIDENCE. A tire print in soil, mud, or snow is created by compression and is sometimes referred to as compression evidence or as an indentation. Tire prints can also be created without compression, as when black tire marks are left on the concrete floor of a garage. When indented tire prints are discovered, they must be protected and preserved until documentation and CASTING can be performed. Further analysis is conducted on the cast.

tire track The path made by a tire.

titer The relative strength of a solution, most often of an ANTISERUM. It is determined by successive dilution and testing against the target ANTIGEN.

TNT (2,4,6-trinitrotoluene) A secondary high EXPLOSIVE used extensively in

World War II as a military explosive. It is prepared by treating toluene (a common solvent used as a paint thinner) with nitric and sulfuric acids (HNO_3 and H_2SO_4). A related compound, 1,3,5-trinitrotoluene, can also be prepared; it is often used as a standard in forensic analyses. A secondary high explosive, TNT must be detonated by a primary high explosive and thus is relatively shock-insensitive.

TOF *See* TIME OF FLIGHT.

toners Materials that are used in laser printers and copiers to produce the image or document. Toners are powdery materials that adhere to static charges on the paper that are then fixed by heat and pressure to the paper.

toolmarks A form of IMPRESSION EVIDENCE created when a metal tool has contact with softer metal or materials such as wood or paint. The most common source of toolmarks evidence are burglary cases. Tools can create indentation or compression imprints, scraping or STRIATION marks, or a combination of the two. Through two factors—grinding during manufacture and WEAR PATTERNS—toolmark impressions can often be linked to a specific tool if the quality of the impression is good enough and the suspect tool is recovered.

tooth and teeth Forensically, among the most important components of the body because of durability. As shown in the figure (opposite page), the tooth nestles in the gum (gingiva) and is composed of a central pulpy cavity, the DENTIN, and the CEMENTUM. The exposed upper surfaces are coated with enamel, and it is this ceramic material that gives teeth their durability. The figure on page 248 illustrates how the teeth are organized in the mouth, a pattern that is important in BITEMARK evidence. Lingual surfaces are closest to the tongue; the buccal surface is beside the cheek.

total internal reflectance *See* INTERNAL REFLECTANCE.

toxicity The degree of harm that a substance can do to an organism. Toxicity is a function of the substance, the amount present, and the span of time over which it is ingested. The adage "The dose makes the poison" applies to toxicity; even common substances such as aspirin and table salt can be toxic if a large enough amount is ingested over a short period. The size of a person also plays a role. Taking three aspirin tablets may give an adult a stomachache but kill an infant.

toxicology (toxicology, forensic) The analysis of drugs and POISONS in BLOOD and body fluids. The forensic toxicologist works with biological samples and must consider absorption of the material, distribution in body tissues, metabolic conversions (METABOLITES), movement through the system (PHARMACODYNAMICS and PHARMACOKINETICS), persistence of the drug, and modes of excretion. One of the largest sources of samples for forensic toxicologists is collected for BLOOD ALCOHOL.

The history of toxicology traces back to the late 1700s and early 1800s, when incidents of ARSENIC poisoning were rampant. James MARSH, an English chemist, developed a reliable test for arsenic in body tissues that was first used in a legal setting in 1840. In that case, M. B. ORFILA, an Italian credited with being the father of forensic toxicology, performed the analysis. Currently, forensic toxicology can be divided into three areas: postmortem toxicology, drug testing (such as urine screening for employment or participation in athletics), and human performance toxicology (including blood alcohol analysis). A forensic toxicologist deals with tissue and organ samples, BLOOD, URINE, bile, VITREOUS FLUID, and, if gaseous poisons or drugs are suspected, lung tissue.

TPOX A gene locus classified as a SHORT TANDEM REPEAT (STR) that is typed in current DNA TYPING procedures. It is one of the 13 loci included in the CODIS system.

trace analysis An informal term usually applied to either TRACE EVIDENCE ANALYSIS (typically microscopic examina-

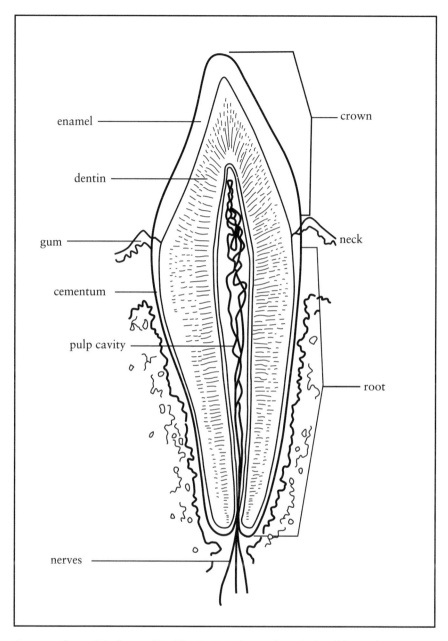

enamel

dentin

gum

cementum

pulp cavity

nerves

crown

neck

root

Anatomy of a tooth in the gum line. The dentin makes up the majority of the tooth's structure.

tion) or to analysis of a sample for very low concentrations of substances. Typically, substances with parts-per-million (ppm, milligrams per liter of water) concentrations are considered to be found in trace quantities.

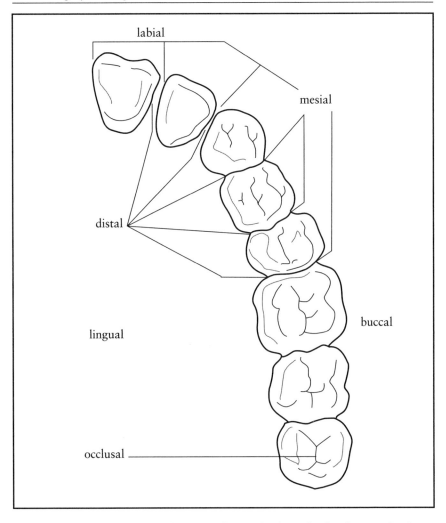

Terms associated with location of the teeth in the mouth. The *occlusal* surfaces are the chewing surfaces; *buccal* refers to the cheek and *lingual* to the tongue.

traced forgery/tracings In QUES-TIONED DOCUMENTS cases, a forgery of writing such as a signature that is accomplished by tracing the original writing.

trace element analysis A process similar to TRACE ANALYSIS except that the sub-stances of interest are metals such as lead (Pb) or arsenic (As).

trace evidence Physical evidence that is found in very small or trace quantities. Common forms of trace evidence include HAIR, FIBER, SOIL, DUST, and GLASS. The primary tool for the analysis of trace evidence is the MICROSCOPE. The term is often used synonymously with TRANSFER EVIDENCE.

track width In a vehicle such as a car or truck, the distance from one center wheel to the other center wheel mounted on the same axle.

traffic accident reconstruction A type of forensic ENGINEERING involving the study of automobile accidents and related accidents involving motorcycles, trucks, pedestrians, bicycles, and other vehicles. Reconstructions can be used in civil or criminal cases and can become crucial when there are no witnesses. Tools for reconstruction include measurements, damage assessment, skid marks, and calculations.

trailer In an arson fire, a trail of debris or other material that leads to the POINT OF ORIGIN of the fire.

traits Expressed characteristics that have variants controlled by the laws of heredity and genetics. Eye color is a trait, as is ABO BLOOD GROUP type.

trajectory Path of travel or flight, often of a blood spatter or a bullet.

tranquilizers A class of drugs used to reduce anxiety and to assist in sleeping. Examples of tranquilizers include BENZO-DIAZEPINES such as Valium (diazepam) and BARBITURATES.

transferability A term sometimes used to describe the ease of transfer of material. For example, shed head HAIRS and DUST have high transferability relative to that of carpet FIBERS still woven into a carpet.

transfer evidence (trace evidence) Materials that are transferred from one person or place to another person or place. The evaluation of transfer evidence is the heart of traditional forensic science and often is conducted by using MICROSCOPY. Edmund LOCARD (1877–1966) has been credited with developing what has become

known as LOCARD'S EXCHANGE PRINCIPLE, paraphrased as "Every contact leaves a trace." Locard was particularly interested in everyday materials such as dust, which are among the most common forms of transfer evidence. Other ubiquitous types of materials found as transfer evidence are HAIRS, FIBERS, GLASS, SOIL, and paint chips.

transfer pattern A type of BLOOD SPATTER (bloodstain) pattern. If a person who has bloody hands places a palm on a wall, the palm print is considered a transfer pattern.

transient evidence Evidence that is easily lost or destroyed or is liable to decomposition.

transition A term usually applied to DNA. A transition is replacement in the DNA sequence of one base for another in the form of guanine (G) to adenine (A), A to G, cytosine (C) to thymine (T), or T to C.

transmission electron microscopy (TEM) A microscopic technique that uses electrons instead of light to create images. TEM uses equipment similar to that used for SCANNING ELECTRON MICROSCOPY. As in standard light MICRO-SCOPY, the incident radiation (in this case, an electron beam) can interact with the sample by reflection or transmission. In light microscopy, this is referred to as transmitted light or reflected light. Similarly, in electron microscopy, reflected electrons (backscattered electrons, as well as others) are used to create an image in SEM, whereas transmitted electrons are used in TEM. Thus, for TEM studies, samples must be thin enough to allow for electron transmission.

transmitted light In microscopy, light that originates below and passes through a sample. Accordingly, such samples must be thin enough to transmit light. *See also* ILLUMINATION.

transportation disasters Large-scale disasters such as airline crashes or railroad mishaps, which are investigated by the National Transportation Safety Board (NTSB). If criminal or terrorist actions are suspected, forensic scientists work closely with forensic engineers and other specialists in the investigations.

transposon In DNA, a short sequence that can jump and insert itself into new positions in the molecule.

trash mark In QUESTIONED DOCUMENTS, imperfections, scratches, dirt, or other foreign matter on the glass surface of a photocopier. These marks can be imaged and placed on the copies and can be used to link a copy to another copy or to a specific copier.

tread In shoes and tires, the portion that has contact with a surface. Tread patterns are CLASS CHARACTERISTICS; WEAR PATTERNS or accidental characteristics can be acquired to assist in individualization.

tread wear indicator A small rubber bar embedded in a tire to allow measurement of wear in the tread. The indicator is 1/16 inch high and is placed in the grooves of the tread; when the tread wears down to this level, it indicates that the tire should be replaced.

tremor In handwritten documents, a shaking pattern seen in the lines that form the letters. This can be the result of conditions such as intoxication, illness, or age.

triangulation A measuring technique used at crime scenes to determine distances and locations. In triangulation, the position of an object is determined relative to two fixed points, creating a triangle.

tricyclic antidepressants *See* BENZODIAZEPINES.

trier of fact The party responsible for making a decision or finding on a matter before a court. The trier of fact may be the jury and the judge, or just a judge if there is no jury present. However, for a question related to a point of law such as admissibility of evidence or testimony, the trier of fact is the judge.

trigger pull The amount of force needed to pull the trigger of a firearm, usually measured in pounds.

trinucleotide repeat A section of DNA composed of a sequence of three NUCLEOTIDES that are repeated. For example, a DNA section with the BASE SEQUENCE AATAAT AAT contains three repetitions of a trinucleotide.

trophy An object that a perpetrator takes from a victim or crime scene.

TSI Time since intercourse.

TTI An abbreviation for *transmitting terminal identifier* that is printed on the top line of a FAX document sent from one machine (the transmitting FAX) to the receiving one.

twelve-point rule In fingerprint comparisons, debate continues as to how many individual points (features or MINUTIAE) on one print must match before a match is declared. The twelve-point rule is a historical one that is widely accepted but is not coded into law in the United States. However, other countries do specify the number of points that must be compared and matched before a print can be positively linked to another or to a specific person.

two-dimensional impression An impression such as a fingerprint on glass that lacks depth. This is in contrast to a three dimensional impression, such as a shoe print in mud.

Tylenol tampering In Chicago in 1982, powdered CYANIDE was added to several bottles of Tylenol, resulting in the death of seven people. The perpetrator was never caught. As a result, product packaging was redesigned to include seals, and PRODUCT TAMPERING was classified as a federal offense.

typefaces A collection of printed letters with the same pattern; similar to font styles used in computer word processors. Typewriters typically have the Pica or Elite typeface.

Type I error A classification of error used in statistical nomenclature. A Type I error is an incorrect exclusion or a FALSE NEGATIVE. It is the incorrect rejection of the NULL HYPOTHESIS.

Type II error A classification of error used in statistical nomenclature. A Type II error is an incorrect inclusion or a FALSE POSITIVE. It is the incorrect acceptance of the NULL HYPOTHESIS.

U

ulnar loop A fingerprint pattern named for the direction toward which it opens. The ulna is the lower arm bone that is aligned roughly with the smallest finger. Accordingly, an ulnar loop pattern flows out toward the ulna.

ultraviolet light and ultraviolet spectroscopy (UV/VIS) In the electromagnetic spectrum (ELECTROMAGNETIC RADIATION AND THE ELECTROMAGNETIC SPECTRUM), ultraviolet (UV) energy lies on the high-energy side of visible (VIS) light. Since many of the analytical techniques that utilize the UV range also work in the visible range, the acronym *UV/VIS* is often used in their description. The UV range encompasses the wavelength range of 200 to 400 nm, at which point the energy can be detected visually as a violet light, hence the name *ultraviolet*. The range is further subdivided into UVA (320–400 nm), UVB (280–320 nm), and UVC, the highest-energy UV light, 200–280 nm. These designations are found on sunscreen products that are designed to prevent UV damage to skin. Many organic compounds can absorb UV/VIS light, and this behavior is the basis of UV/VIS spectrophotometry. In addition, UV illumination, known informally as black light, has many forensic applications. In QUESTIONED DOCUMENT analysis, UV light can be used to stimulate FLUORESCENCE and to reveal different inks or erasures and OBLITERATIONS. Similarly, drugs such as LSD fluoresce under UV light. Finally, microscopes can be designed to work with UV light, although images must be converted to the visible range to be recorded.

uncertainty A range, value, or other quantity that expresses the range of a value. For example, in a drug analysis the forensic chemist may perform a quantitative analysis for COCAINE in a white powder and return a result such as 25.0 percent cocaine, ±1.0 percent. The one percent value is the uncertainty. Specific procedures are used to determine the uncertainty value.

United States Postal Inspection Service The investigative branch of the Post Office that deals with mail fraud, letter and package bombs, mailing of controlled substances, trafficking in child pornography by using the mail, counterfeit stamps, and related crimes. The roots of the postal inspection service can be traced back to the first postmaster, Benjamin Franklin, and the system now includes five labs around the United States. Not surprisingly, a large contingent of their work is in the area of QUESTIONED DOCUMENTS and FINGERPRINT analysis. Other units within the labs include physical evidence, chemistry, photography, digital evidence (forensic COMPUTING), and POLYGRAPH evidence, postal inspectors were deeply involved in the ANTHRAX mailing attacks in late 2001.

United States Secret Service *See* SECRET SERVICE.

upper explosive limit (UEL) The highest concentration of a flammable or explosive mixture in air that will explode. Above this concentration, the mixture is too "rich;" meaning that there is too much fuel for the oxygen available. *See also* LOWER EXPLOSIVE LIMIT.

urea A compound found in URINE and known for its distinctive odor. One presumptive test for urine involves heating a stain to determine whether the odor of urea is detected.

urea nitrate An explosive material that can be manufactured from fertilizer. It is classified as an improvised explosive since it can be made from common ingredients.

urine/urine testing/urinalysis A body fluid frequently used in forensic TOXICOLOGY, occasionally as physical evidence, and a source of cellular material for DNA TYPING. When a stain that may be urine is encountered, the simplest PRESUMPTIVE TEST is based on smell. On heating, urine gives off a distinctive characteristic smell. There are also gel-based diffusion techniques that can be used. With the advent of DNA typing, it is sometimes possible to type the cellular material excreted in urine; however, the sample must be fairly concentrated for this to be an option. Urine contains urea, creatinine, uric acid, and a number of ionic materials such as chloride (Cl^-) and phosphate (PO_4^{3-}). By far the most important role of urine in forensic science is in the area of TOXICOLOGY, in which it is a preferred medium for detection of many drugs and poisons. Traces of drugs and metabolites remain in the urine much longer than in BLOOD, and it is generally easier to screen a urine sample than a blood sample quickly.

urobilinogen A compound found in fecal material that can be detected in PRESUMPTIVE TESTS for feces.

vaculoes *See* BUBBLE RING; CORTICAL FUSI.

vacuum sweeping A method of collecting trace evidence from large areas such as rooms and cars. Sweepings have been employed since the early part of the last 20th century, using specialized attachments and systems designed to trap materials such as DUST, HAIRS, and FIBERS. Although thorough, sweepings are indiscriminant and collect a large amount of evidence, most of which is usually irrelevant.

vaginal acid phosphatase (VAP) A type of phosphatase enzyme found in vaginal secretions that originates in the cells that line the vagina.

vaginal swabs Swabs taken as part of a RAPE KIT in sexual assault cases. Portions of the swabs are smeared on microscope slides and allowed to dry, and both swabs and slides are submitted to a laboratory for analysis. On the slides, the analyst looks for sperm cells characteristic of SEMEN; however, a lack of visible sperm cells does not indicate that intercourse has not occurred. Further tests are performed on the swabs to detect the presence of SEMEN by using the prostate-specific antigen (PSA) (P30) test. The swabs can also be used as a source of DNA for DNA TYPING.

validation The process of determining the accuracy, precision, and limitations of a forensic analysis or procedure. Validation is an integral part of QUALITY ASSURANCE/QUALITY CONTROL and takes place at many levels. Methods such as a DNA TYPING procedure or the analysis of a drug are validated by repeat analysis of different types of samples (not casework) under controlled conditions. The method can be validated internally (within the lab) or among many different laboratories. Similarly, validation can be applied to a single analyst, who must show that he or she is capable of performing the test correctly and obtaining acceptable results. Even procedures such as documentation can be validated. The goal of validation is to ensure that reliable results are obtained when proper procedures are followed.

Van Deemter equation An equation used to describe the separation process in CHROMATOGRAPHY, expressed as

$$H \approx A + \frac{B}{\mu_x} + C\mu_x$$

where H is a measure of efficiency of separation called a height equivalent of a theoretical plate. This term originated in distillation, in which a greater number of condensation plates corresponds to better separation of components found in the mixture being distilled. In the Van Deemter equation, A is the term that describes the paths materials that are flowing in a chromatographic column take, B refers to diffusion processes, and C refers to the time materials introduced into the column require to equilibrate.

Van Urk's test *See* ERLICH'S TEST.

variable number of tandem repeats (VNTRs) In DNA, a locus that has a base pair sequence that repeats a variable number of times. Thus, variation within a population does not lie in the bases found, but in the frequency with which a given sequence is repeated. Both current STR and older RFLP techniques target VNTR sites.

vehicle identification number (VIN) A unique identifying number assigned to every car and truck made. It is stamped into metal on several locations in the vehicle and can be used to identify where and when the vehicle was produced.

velocity Speed; how quickly a certain distance is covered by a moving object. Velocity is important in analysis of BLOODSTAIN PATTERNS and FIREARMS.

ViCAP (Violent Criminal Apprehension Program) A database of information maintained by the FBI that contains information on homicides and other violent crimes. The program is useful for linking seemingly unrelated crimes through similarities in crime scene patterns, behavior, and physical evidence. Participation by police agencies is voluntary, and the FBI provides the software needed.

victimology In forensic behavioral sciences and profiling, a detailed study of the victims of crimes undertaken to reveal information about the perpetrator.

videography, forensic Techniques of video recording used at CRIME SCENES to supplement (but not replace) photography and other traditional forms of documentation. Videotapes and video digital recordings offer the added advantage of sound, so the person recording can also provide a narrative that will later be transcribed. The video can also record evidence in place and in context as well as record the process of photographing and collecting evidence.

video spectral comparison A technique used in the analysis of inks. It operates in the infrared (IR) region.

virtual autopsy A relatively new technique in which data obtained from a traditional dissection autopsy are obtained noninvasively by using medical imaging techniques such as X rays, computed tomography (CT) scanning, and positron emission tomography (PET) scanning.

virtual image In microscopy, an image that cannot be seen unless the viewer is looking through some type of lens. It is not formed by converging light traces. *See also* LENSES; REAL IMAGE.

Vitali's test An older PRESUMPTIVE TEST for drugs that involves adding a drop or two of fuming nitric acid (HNO_3) to a small amount of the material in question. Any color change is noted, and then the sample is evaporated to dryness, when the sample is dry, a second color change may be noted. Finally, alcoholic potassium hydroxide (KOH) is added, and any additional color changes are recorded. Vitali's test is used for MORPHINE and HEROIN (yellows), LSD (brownish purple), and MESCALINE (reddish browns).

vitreous humor The fluid inside the eyeball that is frequently used for postmortem toxicological analyses. Since this fluid is contained and fairly isolated, it degrades more slowly than other fluids such as BLOOD and URINE and thus is usually preferred in cases in which any significant decomposition has begun.

VNTR *See* VARIABLE NUMBER OF TANDEM REPEATS.

voiceprint A graph of the frequency structure of a voice produced by a sound spectrograph is called a spectrogram, sound spectrogram, or voiceprint. The voiceprint was developed in 1941 at Bell Labs by Lawrence Kersta. Since all people have different vocal structures, it was thought that each voice pattern is unique. However, voiceprints have never had the acceptance that FINGERPRINTS have, and voiceprint analysis remains controversial.

voice stress analysis Use of a sound spectrograph (VOICEPRINT) to detect signs of stress in the speaker. As are voiceprints in general, it is a controversial practice not universally accepted as valid or reliable.

void In bloodstain patterns, a void is an area that is somehow shielded and thus does not show staining that would otherwise be expected. Also called a shadow pattern, a void pattern can be seen when a person's body, shoe, hand, and so on, intercept BLOOD in flight, preventing it from reaching a surface such as a wall or floor.

void volume In CHROMATOGRAPHY, the volume of liquid or gas that a column can contain.

voir dire "To Speak the truth," a term sometimes incorrectly spelled *voie dire.* Prospective jurors as well as expert witnesses are subject to this process. The *voir dire* is a preliminary investigation by the court as to qualifications, suitability, and, in the case of forensic scientists, expertise and competency in the matter at issue.

Vollmer, August (1876–1955) Vollmer was an American police officer instrumental in linking university scientists to the analysis of evidence and in developing forensic science laboratory systems in the United States. In 1907, he recorded the first use of scientific analysis of evidence in a case in the United States, involving the analysis of BLOOD, SOIL, and FIBERS. He is best known for establishing the first forensic science laboratory in the United States while he served as chief of police in Los Angeles in 1923–24. Back in Berkeley, California, in 1930, he established the first CRIMINOLOGY and CRIMINALISTICS program (police science) in the United States at the University of California, Berkeley.

vomitus (vomit) Disgorged stomach contents and evidence that are commonly found at death scenes. Vomit can be important for toxicological analysis and detection of drugs and poisons and occasionally as TRANSFER EVIDENCE.

V-shaped pattern A burn pattern often seen in fires, where the V points upward from the point of ignition. This occurs as hot gases and smoke rise and spread.

Vucetich, Juan (Ivan) (1858–1925) Vucetich, an Argentinean, was a pioneer in the use of FINGERPRINTS in criminal investigation and is credited with the first case solved by using them. In 1892, a woman murdered her two sons and, in an attempt to deflect suspicion, injured herself. However, Vucetich was able to locate her bloody fingerprint on a doorjamb, leading to a confession. Vucetich had read Sir Francis GALTON's book *Finger Prints* and had created a large collection of prints by 1891. This early collection still included BERTILLONAGE measurements, although fingerprints would soon supplant them. By 1896, he had developed and instituted a fingerprint classification system that is still extensively used in Latin America.

vWa A gene locus classified as a SHORT TANDEM REPEAT (STR) that is typed in current DNA TYPING procedures. It is one of the 13 loci included in the CODIS system.

W

wadding A plug made of plastic, cardboard, paper, or felt that is used to separate the propellant from the shot in SHOTGUN AMMUNITION. An older term for wadding is *overpowder wadding*. In newer ammunition, a one-piece plastic cup-and-wad combination is often used. The wadding material is blown out of the barrel when the ammunition is fired, and the position of the wadding at a shooting scene can sometimes be helpful in estimating distances and locations of the shooter and victim.

Wagenaar test *See* ACETONE-CHLOR-HEMIN TEST.

Walker test A presumptive test for GUNSHOT RESIDUE useful for clothing. The Walker test detects the nitrite ion (NO_2^-) found in PROPELLANTS in AMMUNITION. The test works by transferring the residues to inactivated photographic paper that has been pretreated with sulfanilic acid and 2-naphthylamine (or similar reagents). Development of an orange-red color is indicative of gunshot residues.

walk-through At a crime scene, an initial procedure undertaken to determine the scope of a scene and to gain an overview of the areas to be searched in detail.

Wallner lines *See* CONCHOIDAL LINES.

Warren Commission A commission appointed after the assassination of President John F. Kennedy in 1963. It was headed by Chief Justice Earl Warren of the United States Supreme Court. The findings of a lone assassin were controversial but were supported by the evidence and forensic testing done at the time.

watermark A physical mark made in some types of paper that is created by thinning of the material in a set pattern. Watermarks can be seen when the paper is held up to light, since the thinner areas allow more light to pass than thicker ones do. Watermarks can be used to identify and in some cases date paper. Some word processors also can create digital watermarks in printed documents.

wave A type of cast-off bloodstain pattern that is created when blood is still being pumped. For example, if an artery is severed in the neck, blood is ejected under alternating spurts of high pressure corresponding to the heartbeat. This process can create a distinctive wavelike pattern on a wall.

wavelength *See* ELECTROMAGNETIC ENERGY and the ELECTROMAGNETIC SPECTRUM.

wavenumber An alternative method of expressing the wavelength of ELECTROMAGNETIC RADIATION. It is used in INFRARED SPECTROMETRY; measurements are expressed in inverse centimeters (cm^{-1}).

wear patterns Patterns acquired in or on an object as a result of normal usage. Wear patterns are found in objects ranging from clothing to tools and can be distinctive enough to allow INDIVIDUALIZATION.

weasel words Derisive slang used for statements that are not specific and leave wide room for interpretation.

weathering Degradation and other changes brought about by exposure to the environment. In ARSON, WEATHERING refers to the loss (through evaporation) of lighter hydrocarbons found in accelerants such as gasoline.

wet chemistry A term used to describe older chemical analyses that do not rely on instrumentation. Examples of wet chemical analyses include gravimetric analysis and many types of volumetric titrations. In a gravimetric analysis, for example, the quantity of a given ion such as silver (Ag^+) in a solution is determined by combining it with a solution that results in the formation of an insoluble material that precipitates out of solution. The solid is then dried and weighed, and the concentration of silver in the original solution is determined by calculation. No instrumentation other than a balance is required. Most PRESUMPTIVE TESTS used in forensic science would be considered wet chemistry.

wet origin impression An impression, usually from footwear, created when a person steps into something wet such as a puddle of water or mud and then steps on another surface, leaving a shoe print made by deposition of the water or mud.

wheelbase In a car or other vehicle, the distance between the front and rear axles. This measurement is used in traffic ACCIDENT RECONSTRUCTION.

wheel gun Slang for a revolver-type handgun or a reference to an older type of RIFLE that used a wheel as part of the mechanism for lighting the POWDER.

whorl/whorl pattern One of the main FINGERPRINT patterns, in which the ridge pattern is roughly circular and has a central core.

wildlife forensics *See* FISH AND WILDLIFE SERVICE.

wild type In DNA, the nucleotide or more generally the variant that occurs with the greatest frequency in the population.

William of Orange (1650–1702) An English nobleman and king who is purported to be one of the first government officials in the Western world to accept FINGERPRINTS and palm prints as identification.

Williams, Wayne A prime suspect in the Atlanta (Georgia) child murders of the early 1980s. The case was notable for its dependence on FIBERS and HAIRS as the principal forms of physical evidence.

wipe pattern A type of bloodstain pattern produced when a bloody object such as a hand, clothing, or hair is wiped across a surface such as a wall.

Wood's lamp A type of ultraviolet lamp used primarily to look for SEMEN stains.

working range/linear range Analytical instruments respond to compounds according to concentration: the higher the concentration, the higher the response. The linear range of an instrument is the range in which a plot of concentration versus response is linear and thus predictable.

wound ballistics The study of wound patterns and forms, principally by forensic PATHOLOGISTS, to determine such elements as what type of weapon was used, what movement took place during the struggle, when the wounds were inflicted, and whether wounds might have been defensive or self-inflicted.

X chromosome A sex chromosome and location of the AMELOGENIN GENE locus. A female has the XX GENOTYPE and a male the XY genotype.

xenobiotic (xenobiotic substance) A substance introduced into the body such as a drug or poison that is not normally present; a foreign substance. This term is used in forensic TOXICOLOGY and PATHOLOGY.

xenon arc lamp A light source that produces radiation in the range of 250 nm (ultraviolet [UV]) to about 600 nm. The visible range is ~400–700 nm. It can be used to induce fluorescence.

X-ray crystallography *See* X-RAY DIFFRACTION.

X-ray diffraction (XRD) A technique used to identify solid crystalline compounds. Crystals are characterized by their having a regular and orderly structure with distances separating the atoms on the order of the size of the wavelengths of X-ray radiation. As a result, X rays can interact with the atoms in the crystals and cause the bending (diffraction) of the radiation in ways that are characteristic of the crystal. The principle of diffraction is based on constructive and destructive interference patterns created by reflection and scattering of the radiation within the crystal. XRD is particularly valuable for the analysis of GUNSHOT RESIDUE (GSR) and explosives, which often contain compounds such as potassium nitrate (KNO_3) and potassium chlorate (K_2ClO_4).

X-ray emission An emission from an atom that can occur when an electron is lost. (See figure on page 260.) *See also* X-RAY FLUORESCENCE.

X-ray fluorescence (XRF) When x RAYS interact with the electrons surrounding an atomic nucleus, they can cause inner shell electrons to be ejected completely from the atom. This process creates instability that is remedied when an electron from an outer shell moves in to fill the vacancy. When an electron "falls" into an empty orbital, it must release energy and does so in the form of an X ray. The emitted X ray is of a longer wavelength (lower energy) than the original incident X ray. If a filling electron originates one shell away, the process is referred to as alpha (α) transition; from two shells away, as beta (β) transition; and from three shells, as gamma (γ) transition. Because more than one electron can be ejected and more than one electron can change shells during the refilling process, each atom can emit more than one wavelength of radiation. The pattern of emitted X rays is characteristic of a given element.

There are two principal methods for implementation of XRF, energy-dispersive and wavelength-dispersive. These techniques are sometimes referred to as energy-dispersive spectroscopy (EDS) and wavelength-dispersive spectroscopy (WDS). In EDS, the detection system differentiates X rays emitted from elements on the basis of the different energies produced. In WDS, this differentiation is achieved by separating and detecting the various wavelengths emitted from the sample. Another variant of XRF is the electron microprobe. The fluorescence is produced not by X rays but by bombardment of the sample surface with a beam of focused electrons, similar to what occurs in SCANNING ELECTRON MICROSCOPY.

X-ray photoelectron spectroscopy (XPS) A form of SPECTROSCOPY in

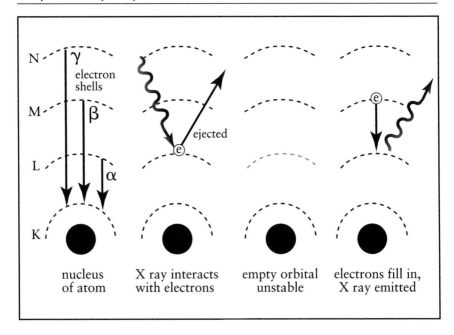

N
γ
electron
shells

M
β

L

α

K

ejected

e

empty orbital
unstable

electrons fill in,
X ray emitted

nucleus
of atom

X ray interacts
with electrons

The process of X-ray emission. When an X ray interacts with an electron in one of the shells, it can be ejected, creating an unstable atom. Electrons from above empty space fall in to replace it, causing the emission of X-ray radiation.

which atoms that are bombarded with energy (in this case, X rays) eject electrons. Such electrons are called photoelectrons, since they are ejected as a result of the absorption of photons. XPS can be used to characterize the composition of surfaces.

X rays and X-ray analysis A family of instrumental techniques that exploit the interaction of X rays with matter. Broadly defined, these techniques include forensic RADIOLOGY, fluoroscopy, and analytical techniques such as X-ray fluorescence (XRF) and X-ray diffraction (XRD). Many related techniques exist; XRF and XRD are the most widely used in forensic applications. X rays are a highly energetic form of ELECTROMAGNETIC ENERGY characterized by high frequencies and short wavelengths.

X rays, as can any form of electromagnetic radiation, can be partially absorbed by the sample. This is the basis of diagnostic medical and dental X rays that are produced by exposing the body to a short burst of energy. In general, the heavier the element (the larger its atomic number), the more X-ray radiation it absorbs. Thus bones, which contain large amounts of elements such as calcium (atomic number 20), absorb more radiation than soft tissues containing large amounts of lighter elements such as hydrogen (atomic number 1) and oxygen (16). Fluoroscopy, such as that used in airport luggage screeners, also depends on X-ray absorption. In this technique, emerging X rays are directed to a screen that fluoresces, giving off visible light that can be focused and viewed on a screen. In addition to absorption, X rays can cause elements to emit radiation of longer wavelengths (lower energy); this effect is exploited in XRF. The X rays can also be scattered and bent (diffracted). X-ray diffraction (XRD) is used to study the crystal structure of materials. Diffraction techniques were used to help decipher the double-helix structure of DNA.

Xylocaine *See* LIDOCIANE.

Y–Z

yaw Along with pitch and roll, a characteristic angle of something in flight such as a bullet. The yaw angle of a bullet in flight is the angle between its center axis and the direction of flight. If it is moving forward in level flight but the nose is pointed slightly off-center, that angle between the line of flight and the tip of the nose is the yaw angle.

Y chromosome A sex chromosome and location of the AMELOGENIN GENE locus. A female has the XX GENOTYPE and a male the XY genotype.

Zwikker test *See* DILLIE-KOPPANYI TEST.

APPENDIXES

Appendix I: Bibliographies and Web Resources

Print
Organized by forensic discipline

Accounting
Bologna, G. J., and R. J. Lindquist. *Fraud Accounting and Forensic Accounting New Tools and Techniques.* New York: John Wiley and Sons, 1995.

Analytical Chemistry
Buffington, R., and M. K. Wilson. *Detectors for Gas Chromatography: A Practical Primer.* Avondale, Pa.: Hewlett-Packard Inc, 1987.

Dean, J. A. *Analytical Chemistry Handbook.* New York: McGraw Hill, 1995.

Eiceman, G. A., and Z. Karpus. *Ion Mobility Spectrometry.* Boca Raton, Fla.: CRC Press, 1995.

Fritz, J. S. *Analytical Solid-Phase Extraction.* New York: Wiley-VCH, 1999.

Harris, D. C. *Quantitative Chemical Analysis.* 4th ed. New York: W. H. Freeman and Company, 1995.

Jenkins, R. "X-ray Fluorescence Spectrometry." Vol. 99. In *Chemical Analysis: A Series of Monographs on Analytical Chemistry and Its Applications.* New York: John Wiley and Sons, 1988.

Skoog, D. A., D. M. West, F. Holler, F. James, and S. R. Crouch. *Analytical Chemistry, An Introduction.* 7th ed. Orlando, Fla.: Harcourt College Publishers, 2000.

Yinon, J., ed. *Forensic Applications of Mass Spectrometry.* Boca Raton, Fla.: CRC Press, 1995.

Anthropology and Archaeology
Fairgrieve, S. I., ed. *Forensic Osteological Analysis: A Book of Case Studies.* Springfield, Ill.: Charles C. Thomas, 1999.

Maples, W. R., and M. Browning. *Dead Men Do Tell Tales, The Strange and Fascinating Cases of a Forensic Anthropologist.* New York: Doubleday, 1994.

Morse, D., J. Duncan, and J. Stoutamire, eds. *Handbook of Forensic Archaeology and Anthropology.* Tallahassee, Fla.: Rose Printing Co., 1983.

Nafte, M. *Flesh and Bone: An Introduction to Forensic Anthropology.* Durham, N.C.: Carolina Academic Press, 2000.

Pickering, R. B., and D. C. Bachman. *The Use of Forensic Anthropology.* Boca Raton, Fla.: CRC Press, 1997.

Reichs, K. J., ed. *Forensic Osteology.* 2d ed. Springfield, Ill.: Charles C. Thomas, 1998.

Ubelaker, D., and H. Scammell. *Bones: A Forensic Detective's Casebook.* New York: HarperCollins, 1992.

Arson and Related Topics
Redsicker, D. R., and J. J O'Connor. *Practical Fire and Arson Investigation.* 2d ed. Boca Raton, Fla.: CRC Press, 1997.

Art
Taylor, K. T. *Forensic Art and Illustration.* Boca Raton, Fla.: CRC Press, 2001.

Blood Evidence and Blood Spatter Patterns
James, S. H., and W. G. Eckert, eds. *Interpretation of Bloodstain Evidence at Crime Scenes.* 2d ed. Boca Raton, Fla.: CRC Press, 1999.

Blood-Spatter Pattern Analysis

Bevel, T., and R. M. Gardner. *Bloodstain Pattern Analysis with an Introduction to Crime Scene Reconstruction.* Boca Raton, Fla.: CRC Press, 1997.
James, Stuart H., ed. *Scientific and Legal Applications of Bloodstain Pattern Interpretation.* Boca Raton, Fla.: CRC Press, 1999.
Wonder, A. Y. *Blood Dynamics.* San Diego, Calif.: Academic Press, 2001.

Computer Forensics

Casey, E. *Digital Evidence and Computer Crime: Forensic Science, Computers, and the Internet.* London: Cambridge University Press, 2000.

Crime Scenes

Fisher, B. A. J. *Techniques of Crime Scene Investigation.* 6th ed. Boca Raton, Fla.: CRC Press, 2000.
Lee, H. C., T. Palmbach, and M. T. Miller. *Henry Lee's Crime Scene Handbook.* San Diego, Calif.: Academic Press, 2001.

Deception Analysis

Hall, H. V., and D. A. Pritchard. *Detecting Malingering and Deception: Forensic Deception Analysis.* Delray Beach, Fla.: St. Lucie Press, 1996
Turvey, B. E. *Criminal Profiling: An Introduction to Behavioral Evidence Analysis.* San Diego, Calif.: Academic Press, 1999.

DNA Typing

Inman, K., and N. Rudin. *An Introduction to Forensic DNA Typing.* 2d ed. Boca Raton, Fla.: CRC Press, 2002.

Engineering

Brown, J. F., and K. S. Osborn. *Forensic Engineering: Reconstruction of Accidents.* Springfield, Ill.: Charles C. Thomas, 1990.
Carper, K. L., ed. *Forensic Engineering.* 2d ed. Boca Raton, Fla.: CRC Press, 2001.
Milton, P. *In the Blink of an Eye, The FBI Investigation of TWA Flight 800.* New York: Random House, 1999.
Noon, R. K. *Engineering Analysis of Vehicular Accidents.* Boca Raton, Fla.: CRC Press, 1994.
Shepherd, R., and J. D. Frost, eds. *Failures in Civil Engineering: Structural, Foundation, and Geoenvironmental Case Studies.* Reston, Va.: American Society of Civil Engineers, 1995.

Entomology (Insects)

Byrd, J. H., and J. L Caster, eds. *Forensic Entomology: The Utility of Arthropods in Legal Investigations.* Boca Raton, Fla.: CRC Press, 2001.

Fingerprints

Lee, H. C., and R. E. Gaensslen. *Advances in Fingerprint Technology.* Boca Raton, Fla.: CRC Press, 1994.
Lee, H. C., and R. E. Gaensslen. *Advances in Fingerprint Technology.* 2d ed. Boca Raton, Fla.: CRC Press, 2001.

Firearms and Gunshot Residue

Schwoeble, A. J., and D. L. Exline. *Current Methods in Forensic Gunshot Residue Analysis.* Boca Raton, Fla.: CRC Press, 2000.

Footwear, Shoeprints, and Tire Impression Evidence

Bodziak, W. J. *Footwear Impression Evidence*. New York: Elsevier, 1990.
McDonald, P. *Tire Imprint Evidence*. Boca Raton, Fla.: CRC Press, 1993.

General Textbooks and References

De Forest, P. R., R. E. Gaensslen, and H. C. Lee. *Forensic Science: An Introduction to Criminalistics*. New York: McGraw Hill, 1983.
Eckert, W. E., ed. *Introduction to Forensic Sciences*. 2d ed. Boca Raton, Fla.: CRC Press, 1992.
James, S. H., and J. J. Nordby, eds. *Forensic Science: An Introduction to Scientific and Investigative Techniques*. Boca Raton, Fla.: CRC Press, 2003.
Lane, B. *The Encyclopedia of Forensic Science*. London: Headline Book Publishing, 1992.
Nickell, J. N., and J. F. Fisher. *Crime Science: Methods of Forensic Detection*. Lexington: University Press of Kentucky, 1998.
Nordby, J. J. *Dead Reckoning—The Art of Forensic Detection*. Boca Raton, Fla.: CRC Press, 2000.

Geology

Murray, R. C., and J. C. F. Tedrow. *Forensic Geology*. Englewood Cliffs, N.J.: Prentice Hall, 1992.

Law

Becker, R. F. *Scientific Evidence and Expert Testimony Handbook*. Springfield, Ill.: Charles C. Thomas, 1997.
Black's Law Dictionary. 6th ed. St. Paul, Minn.: West Publishing Company, 1990.
Moenssens, A. A., J. E. Starrs, C. E. Henderson, and F. E. Inbau. *Scientific Evidence in Civil and Criminal Cases*. 4th ed. Westbury N.Y.: The Foundation Press, Inc., 1995.

Miscellaneous

Gerber, S. M., ed. *Chemistry and Crime: From Sherlock Holmes to Today's Courtroom*. Washington, D.C.: American Chemical Society, 1983.
Kaye, B. H. *Science and the Detective, Selected Reading in Forensic Science*. Weinheim, Germany: VCH Press, 1995.
Kurland, M. *A Gallery of Rogues, Portraits in True Crime*. New York: Prentice Hall General Reference, 1994.
Lee, J. A. *The Scientific Endeavor: A Primer on Scientific Principles and Practice*. San Francisco: Addison Wesley Longman, 2000.
Saferstein, R., and S. M. Gerber, eds. *More Chemistry and Crime: From Marsh Arsenic Test to DNA Profile*. Washington, D.C.: American Chemical Society, 1997.

Pathology and Taphonomy

Brogdon, B. D. *Forensic Radiology*. Boca Raton, Fla.: CRC Press, 1998.
DiMaio, D. J., and V. J. M. DiMaio. *Forensic Pathology*. Boca Raton, Fla.: CRC Press, 1993.
Dix, J., and R. Calaluce. *Guide to Forensic Pathology*. Boca Raton, Fla.: CRC Press, 1999.
Dix, J., and M. Graham. *Time of Death, Decomposition, and Identification, An Atlas*. Boca Raton, Fla.: CRC Press, 2000.
Haglund, W. D., and M. H. Sorg, eds. *Forensic Taphonomy. The Postmortem Fate of Human Remains*. Boca Raton, Fla.: CRC Press, 1997.
Knight, B. *Simpson's Forensic Medicine*. 10th ed. London: Edward Arnold, 1991.
Spitz, W. U., ed. *Spitz and Fisher's Medicolegal Investigation of Death*. 3d ed. Springfield, Ill.: Charles C. Thomas, 1993.

Psychology, Psychiatry, Profiling, Etc.; Behavioral Evidence

Sweet, J. J., ed. *Forensic Neuropsychology Fundamentals and Practice.* Lisse, Netherlands: Swets and Zeitlinger Publishers, 1999.

Turvey, B. E. *Criminal Profiling: An Introduction to Behavioral Evidence Analysis.* San Diego, Calif.: Academic Press, 1999.

Questioned Documents

Ellen, D. *The Scientific Examination of Documents, Methods and Techniques.* 2d ed. London: Taylor and Francis, 1997.

Hilton, O. *Scientific Examination of Questioned Documents.* Rev. ed. Boca Raton, Fla.: CRC Press, 1993.

Koppenhaver, K. M. *Attorney's Guide to Document Examination.* Westport, Conn.: Quorum Books, 2002.

Toxicology

Levine, B. *Principles of Forensic Toxicology.* Washington, D.C.: American Association of Clinical Chemistry, 1999.

Trace Evidence/Hairs/Fibers

Houck, M. *Mute Witness: Trace Evidence Analysis.* San Diego, Calif.: Academic Press, 2001.

Robertson, J., ed. *Forensic Examination of Fibers.* London: Ellis Horwood, 1992.

Web Resources

General Information

Armed Forces Institute of Pathology. www.afip.org
Zeno's Forensic Page. www.forensic.to/forensic.html
Reddy's Forensic Page. www.forensicpage/com
Nikon's Microscopy University. www.microscopyu.com

Professional Organizations and Societies

American Academy of Forensic Psychology (AAFP, www.abfp.com)
American Academy of Forensic Sciences (AAFS, www.aafs.org)
American Academy of Psychiatry and the Law (AAPL, www.emory.edu/AAPL)
American Association of Physical Anthropologists (AAPA, www.physanth.org)
American Board of Criminalistics (ABC, www.criminalistics.com/ABC/A.php)
American Board of Forensic Anthropology (ABFA, www.csuchico.edu/anth/ABFA/)
American Board of Forensic Document Examiners (ABFDE, www.abfde.org/)
American Board of Forensic Entomologists (ABFE, www.missouri.edu/~agwww/entomology/)
American Board of Forensic Odontologists (ABFO, www.abfo.org)
American Board of Forensic Toxicology (ABFT, www.abft.org)
American Board of Medicolegal Death Investigators (ABMDI, www.slu.edu/organizations/abmdi)
American Board of Questioned Document Examiners (ABQDE, www.asqde.org)
American Chemical Society (ACS, www.chemistry.org)
American Society for Testing Materials (ASTM, www.astm.org)
American Society of Crime Laboratory Directors (ASCLD, www.ascld.org)
Association for Crime Scene Reconstruction (ACSR, www.acsr.com)
Association for Environmental Health and Science (AEHS, www.aehs.com)
Association of Certified Fraud Examiners (ACFE, www.cfenet.com/home.asp)

Association of Firearm and Toolmark Examiners (AFTE, www.afte.org)
California Association of Criminalists (CAC, www.cacnews.org)
Canadian Association of Forensic Sciences (CAFS, www.cafs.ca)
College of American Pathologists (CAP, www.cap.org)
Forensic Sciences Society (FSS, www.forensic-science-society.org.uk)
International Association of Bloodstain Pattern Analysis (IABPA, www.iabpa.org)
International Association of Forensic Nurses (IAFN, www.iafn.org)
International Association of Forensic Toxicologists (IAFT, www.tiaft.org)
International Association of Identification (IAI, www.theiai.org)
National Association of Forensic Engineers (NAFE, www.nafe.org)
National Association of Medical Examiners (NAME, www.thename.org)
National Association of Traffic Accident Reconstructionists and Engineers (NATARI, www.natari.org)
Society for Forensic Toxicologists (SOFT, www.soft-tox.org)

Appendix II: Common Abbreviations and Acronyms

The following table contains abbreviations commonly used in forensic science. Except for a few instances, chemical names are not included.

Abbreviation	Meaning
AA or AAS	Atomic absorption spectroscopy
AAFS	American Academy of Forensic Sciences
AAR	Amino acid racemization
Ab	Antibody
ABC	American Board of Criminalistics
ABFA	American Board of Forensic Anthropology
ABFE	American Board of Forensic Entomologists
ABFO	American Board of Forensic Odontologists
ABO	ABO blood group system
ACFE	Association of Certified Fraud Examiners
ACP	Acid phosphatase
ADA	Adenosine deaminase
AEHS	Association for Environmental Health and Science
AFIS	Automated fingerprint identification system
AFLP	Armed Forces Laboratory of Pathology
AFLP	Amplified fragment length polymorphism
AFTE	Association of Firearm and Toolmark Examiners
Ag	Antigen
AK	Adenylate kinase
ALS	Alternate light source
AP	Acid phosphatase
ARF	Anthropological Research Facility ("Body Farm")
As	Arsenic
ASCLD	American Society of Crime Laboratory Directors
ASQDE	American Society of Questioned Document Examiners
ASTM	American Society for Testing Materials
ATF	Bureau of Alcohol, Tobacco, and Firearms (federal)
ATR	Attenuated total reflectance
BAC	Blood alcohol concentration
BSA	Bovine serum albumin
BZ	Benzoylecgonine
CAB	Civil Aeronautics Board (federal)
CAC	California Association of Criminalists
CA II	Carbonic anhydrase
CAT	Computerized axial tomography
CCSC	Criminalistics certification study committee
CDC	Centers for Disease Control (federal)
CE	Capillary electrophoresis
CFC	Chlorofluorocarbon
CFN	Clinical forensic nursing
CGE	Capillary gel electrophoresis
CI	Chemical ionization
CIEF	Capillary isoelectric focusing
CIL	Central Identification Laboratory (US Army)
CNS	Central nervous system

CO	Carbon monoxide
CODIS	Combined DNA Indexing System
CPA	Certified public accountant
CSA	Controlled Substances Act
CSI	Crime scene investigation
CZE	Capillary zone electrophoresis
DA	District attorney
DE	Diatomaceous earth
DEA	Drug Enforcement Agency (federal)
DNA	Deoxyribonucleic acid
DOJ	Department of Justice (federal)
DOT	Department of Transportation (federal)
DRE	Drug recognition expert
DRIFTS	Diffuse reflectance infrared Fourier transformation spectroscopy
EAP	Enzyme acid phosphatase
EAP	Erythrocyte acid phosphatase
EDS	Energy dispersive spectroscopy
EDTA	Ethylenediamine tetraacetic acid
EDXRF	Electron diffraction X-ray fluorescence spectroscopy
EI	Electron impact ionization
ELISA	Enzyme linked immunoassay
EMIT	Enzyme multiplied immunoassay
EMP	Electron microprobe
EMR	Electromagnetic radiation
EP	Electrophoresis
EPA	Environmental Protection Agency (federal)
ESD	Esterase D
ESDA	Electrostatic detection apparatus
FAA or FAS	Flame atomic absorption spectroscopy
FBI	Federal Bureau of Investigation
FDA	Food and Drug Administration (federal)
FID	Flame ionization detector for gas chromatography
FISH	Forensic information system for handwriting
FISH	Fluorescent in-situ hybridization sex determination
FPIA	Fluorescent polarization immunoassay
FSS	Forensic science service (UK)
FTC	Federal Trade Commission
FTIR	Fourier transform infrared spectroscopy
GAO	General Accounting Office (federal)
GC	Gas chromatography
Gc	Group specific component
GC/MS or GC-MS	Gas chromatography-mass spectrometry
GKE	General Knowledge Examination
GLO I	Glyoxalase I
GLP	Good laboratory practice
GSR	Gunshot residue
Hb	Hemoglobin
HLA	Human leukocyte antigen
Hp	Haptoglobin

HPD	Heavy petroleum distillates
HPLC	High performance liquid chromatography

IABPA	International Association of Bloodstain Pattern Analysis
IAFIS	Integrated automatic fingerprint identification system
IAFN	International Association of Forensic Nurses
IAI	International Association of Identification
IBIS	Integrated ballistics identification system
IC	Ion chromatography
ICP	Inductively coupled plasma
ICP-AES	Inductively coupled plasma-atomic emission spectroscopy
ICP-MS	Inductively coupled plasma-mass spectroscopy
IEF	Isoelectric focusing
IMS	Ion mobility spectrometry
IOC	International Olympic Committee
IR	Infrared region of the electromagnetic spectrum
IRS	Internal Revenue Service (federal)

LD_{50}	Lethal dose 50
LPD	Light petroleum distillates
LSD	Lysergic acid diethylamide

MALDI	Matrix assisted laser desorption/ionization
ME	Medical examiner
MECC or MEKC	Micellular electrokinetic capillary chromatography
MO	Modus operandi
MPD	Medium petroleum distillates
MRI	Magnetic resonance imaging
MS	Mass spectrometry
MSP	Microspectrophotometry
MtDNA or mDNA	Mitochondrial DNA

NAA	Neutron activation analysis
NAFE	National Association of Forensic Engineers
NASA	National Aeronautics and Space Administration
NASH	Natural, accidental, suicidal, homicidal (cause of death)
NC	Nitrocellulose
NCAVC	National Center for the Analysis of Violent Crime
NCFS	National Center for Forensic Sciences
NCIC	National Crime Information Center
NFPA	National Fire Protection Association
NG	Nitroglycerin
NHTSA	National Highway Transportation Safety Administration
NIJ	National Institute of Justice (federal)
NMR	Nuclear Magnetic Resonance
NPD	Nitrogen phosphorus detector for gas chromatography
NTSB	National Transportation Safety Board (federal)

OTC	Over the counter; drugs and medicines available without a prescription

P/M	Parent to metabolite ratio
P2P	Phenyl-2-propanone
PCP	Phencyclidine
PCR	Polymerase chain reaction

Pd .	Power of discrimination
PD	Physical developer
PDA	Personal digital assistant
PDQ	Paint data query
PDR	Physician's Desk Reference
PGM	Phosphoglucomutase
PLM	Polarizing light microscopy
PMI	Post mortem interval
PSA	Prostate specific antigen
QA/QC	Quality assurance/quality control
RBC	Red blood cell
RCMP	Royal Canadian Mounted Police
RFLP	Restricted fragment length polymorphisms
RI	Refractive index
RIA	Radioimmunoassay
RMNE	Random man not excluded
SANE	Sexual assault nurse examiner
SAP	Seminal acid phosphatase
SEC	Securities and Exchange Commission (federal)
SEC	Size exclusion chromatography
SEM	Scanning electron microscopy
SIM	Selected ion monitoring
SPF	Spectrofluorometry
SPME	Solid phase microextraction
STR	Short tandem repeats
TEM	Transmission electron microscopy
Tf	Transferrin
THC	Tetrahydrocannabinol
TLC	Thin layer chromatography
TNT	2,4,6-trinitrotoluene
TOF	Time of flight
USPS	United States Postal Service
USSS	United States Secret Service
UV	Ultraviolet region of the electromagnetic spectrum
V/V	Volume to volume ratio
VAP	Vaginal acid phosphatase
ViCAP	Violent criminal apprehension program
VIS	Visible region of the electromagnetic spectrum
VNTR	Variable number of tandem repeats
W/V	Weight to volume ratio
WBC	White blood cell
WDS	Wavelength dispersive spectroscopy
WTC	World Trade Center
XRD	X-ray diffraction spectroscopy
XRF	X-ray fluorescence spectroscopy

Appendix III

Periodic Table of the Elements

Legend:
- 1 — atomic number
- H — symbol
- 1.008 — atomic weight

Numbers in parentheses are atomic mass numbers of radioactive isotopes.

1	2	3	4	5	6	7	8	9	10	11	12	13	14	15	16	17	18
1 H 1.008																	2 He 4.003
3 Li 6.941	4 Be 9.012											5 B 10.81	6 C 12.01	7 N 14.01	8 O 16.00	9 F 19.00	10 Ne 20.18
11 Na 22.99	12 Mg 24.31											13 Al 26.98	14 Si 28.09	15 P 30.97	16 S 32.07	17 Cl 35.45	18 Ar 39.95
19 K 39.10	20 Ca 40.08	21 Sc 44.96	22 Ti 47.88	23 V 50.94	24 Cr 52.00	25 Mn 54.94	26 Fe 55.85	27 Co 58.93	28 Ni 58.69	29 Cu 63.55	30 Zn 65.39	31 Ga 69.72	32 Ge 72.59	33 As 74.92	34 Se 78.96	35 Br 79.90	36 Kr 83.80
37 Rb 85.47	38 Sr 87.62	39 Y 88.91	40 Zr 91.22	41 Nb 92.91	42 Mo 95.94	43 Tc (98)	44 Ru 101.1	45 Rh 102.9	46 Pd 106.4	47 Ag 107.9	48 Cd 112.4	49 In 114.8	50 Sn 118.7	51 Sb 121.8	52 Te 127.6	53 I 126.9	54 Xe 131.3
55 Cs 132.9	56 Ba 137.3	57–71*	72 Hf 178.5	73 Ta 180.9	74 W 183.9	75 Re 186.2	76 Os 190.2	77 Ir 192.2	78 Pt 195.1	79 Au 197.0	80 Hg 200.6	81 Tl 204.4	82 Pb 207.2	83 Bi 209.0	84 Po (210)	85 At (210)	86 Rn (222)
87 Fr (223)	88 Ra (226)	89–103‡	104 Rf (261)	105 Db (262)	106 Sg (263)	107 Bh (262)	108 Hs (265)	109 Mt (266)	110 Ds (271)								

*lanthanide series

57 La 138.9	58 Ce 140.1	59 Pr 140.9	60 Nd 144.2	61 Pm (145)	62 Sm 150.4	63 Eu 152.0	64 Gd 157.3	65 Tb 158.9	66 Dy 162.5	67 Ho 164.9	68 Er 167.3	69 Tm 168.9	70 Yb 173.0	71 Lu 175.0

‡actinide series

89 Ac (227)	90 Th 232.0	91 Pa 231.0	92 U 238.0	93 Np (237)	94 Pu (244)	95 Am (243)	96 Cm (247)	97 Bk (247)	98 Cf (251)	99 Es (252)	100 Fm (257)	101 Md (258)	102 No (259)	103 Lr (260)

The Chemical Elements

element	symbol	a.n.	element	symbol	a.n.	element	symbol	a.n.	element	symbol	a.n.
actinium	Ac	89	einsteinium	Es	99	mendelevium	Md	101	samarium	Sm	62
aluminum	Al	13	erbium	Er	68	mercury	Hg	80	scandium	Sc	21
americium	Am	95	europium	Eu	63	molybdenum	Mo	42	seaborgium	Sg	106
antimony	Sb	51	fermium	Fm	100	neodymium	Nd	60	selenium	Se	34
argon	Ar	18	fluorine	F	9	neon	Ne	10	silicon	Si	14
arsenic	As	33	francium	Fr	87	neptunium	Np	93	silver	Ag	47
astatine	At	85	gadolinium	Gd	64	nickel	Ni	28	sodium	Na	11
barium	Ba	56	gallium	Ga	31	niobium	Nb	41	strontium	Sr	38
berkelium	Bk	97	germanium	Ge	32	nitrogen	N	7	sulfur	S	16
beryllium	Be	4	gold	Au	79	nobelium	No	102	tantalum	Ta	73
bismuth	Bi	83	hafnium	Hf	72	osmium	Os	76	technetium	Tc	43
bohrium	Bh	107	hassium	Hs	108	oxygen	O	8	tellurium	Te	52
boron	B	5	helium	He	2	palladium	Pd	46	terbium	Tb	65
bromine	Br	35	holmium	Ho	67	phosphorus	P	15	thallium	Tl	81
cadmium	Cd	48	hydrogen	H	1	platinum	Pt	78	thorium	Th	90
calcium	Ca	20	indium	In	49	plutonium	Pu	94	thulium	Tm	69
californium	Cf	98	iodine	I	53	polonium	Po	84	tin	Sn	50
carbon	C	6	iridium	Ir	77	potassium	K	19	titanium	Ti	22
cerium	Ce	58	iron	Fe	26	praseodymium	Pr	59	tungsten	W	74
cesium	Cs	55	krypton	Kr	36	promethium	Pm	61	uranium	U	92
chlorine	Cl	17	lanthanum	La	57	protactinium	Pa	91	vanadium	V	23
chromium	Cr	24	lawrencium	Lr	103	radium	Ra	88	xenon	Xe	54
cobalt	Co	27	lead	Pb	82	radon	Rn	86	ytterbium	Yb	70
copper	Cu	29	lithium	Li	3	rhenium	Re	75	yttrium	Y	39
curium	Cm	96	lutetium	Lu	71	rhodium	Rh	45	zinc	Zn	30
darmstadtium	Ds	110	magnesium	Mg	12	rubidium	Rb	37	zirconium	Zr	40
dubnium	Db	105	manganese	Mn	25	ruthenium	Ru	44			
dysprosium	Dy	66	meitnerium	Mt	109	rutherfordium	Rf	104	a.n. = atomic number		

Appendix IV: Common Units, Conversion Factors, and Prefixes

Basic Units
Mass (weight)   grams (g)
Length   meters (m)
Volume   liters (L)

Metric Prefixes

Prefix	Symbol	Multiplication factor
mega	M	10^6
kilo	k	10^3
deci	d	10^{-1}
centi	c	10^{-2}
milli	m	10^{-3}
micro	μ	10^{-6}
nano	n	10^{-9}
pico	p	10^{-12}

Apothecary Units
These were once used widely in medicine and are still encountered in nursing, pharmacy, toxicology, and drug analysis.

minim	0.062 mL (milliliter)
dram (fluid)	60 minims or 3.697 mL
ounce (fluid)	8 drams or 480 mimin or 29.57 mL
pint (fluid)	16 ounces or 128 drams or 7,680 minim or 473 mL
gallon (U.S., fluid)	8 pints or 128 ounces or 1,024 drams or 61,440 minim or 3.785 L
grain (weight)	0.0648 g
scruple (weight)	20 grains or 1.296 g
dram (weight)	3 scruples or 60 grains or 3.89 g
ounce (weight)	8 drams or 24 scruples or 480 grains or 31.1 g
pound	12 ounces or 96 drams or 288 scruples or 5,760 grains or 0.373 kg

Websites for Finding Units and Conversions (available as of July 2003)
http://www.sizes.com/indexes.htm
http://www.bartleby.com/61/charts/M0182500.html
http://www.convert-me.com/en/
http://dir.yahoo.com/Science/Measurements_and_Units/Conversion/Online_Converters/
http://www.ex.ac.uk/cimt/dictunit/dictunit.htm
http://www.convertit.com/Go/ConvertIt/
http://www.maribi.com/conversion/
http://www.onlineconversion.com/

Appendix V

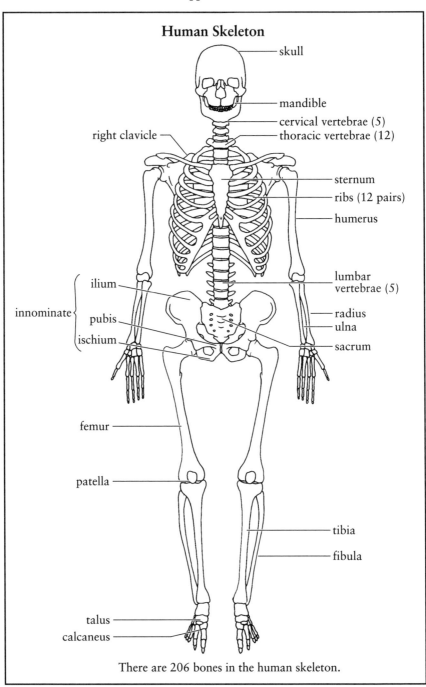

Human Skeleton

- skull
- mandible
- cervical vertebrae (5)
- thoracic vertebrae (12)
- right clavicle
- sternum
- ribs (12 pairs)
- humerus
- ilium
- lumbar vertebrae (5)
- innominate
- pubis
- radius
- ulna
- ischium
- sacrum
- femur
- patella
- tibia
- fibula
- talus
- calcaneus

There are 206 bones in the human skeleton.

Appendix VI

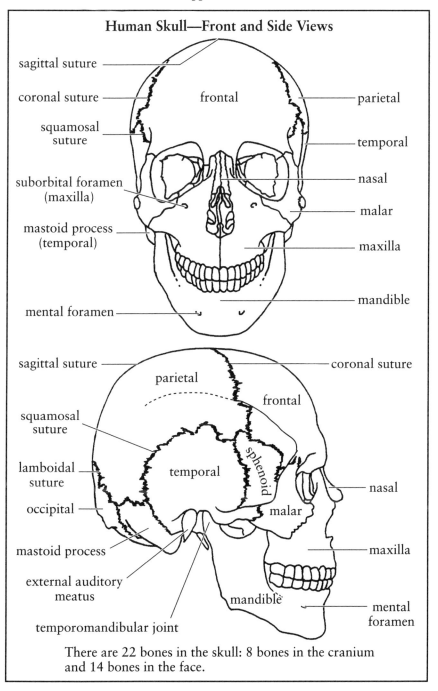

Human Skull—Front and Side Views

sagittal suture

coronal suture

squamosal suture

suborbital foramen (maxilla)

mastoid process (temporal)

mental foramen

frontal

parietal

temporal

nasal

malar

maxilla

mandible

sagittal suture

squamosal suture

lamboidal suture

occipital

mastoid process

external auditory meatus

temporomandibular joint

parietal

frontal

coronal suture

temporal

sphenoid

malar

nasal

maxilla

mandible

mental foramen

There are 22 bones in the skull: 8 bones in the cranium and 14 bones in the face.

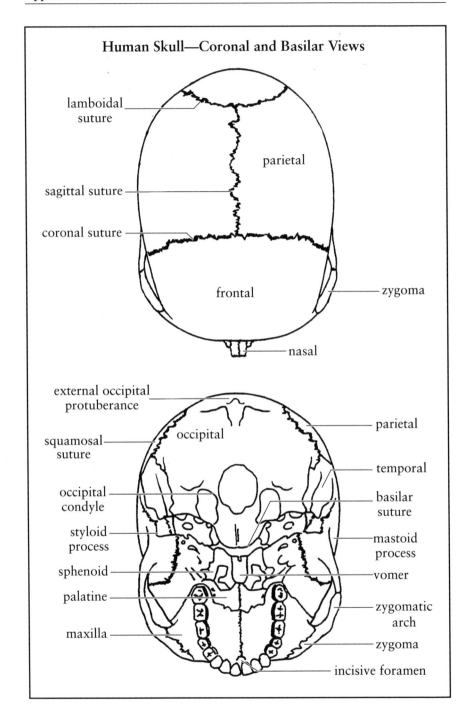

Human Skull—Coronal and Basilar Views

lamboidal suture

parietal

sagittal suture

coronal suture

frontal

zygoma

nasal

external occipital protuberance

parietal

squamosal suture

occipital

temporal

occipital condyle

basilar suture

styloid process

mastoid process

sphenoid

vomer

palatine

zygomatic arch

maxilla

zygoma

incisive foramen